PROTEIN BIOCHEMISTRY, SYNTHESIS, STRUCTURE AND CELLULAR FUNCTIONS

AMYLOIDS

COMPOSITION, FUNCTIONS AND PATHOLOGY

PROTEIN BIOCHEMISTRY, SYNTHESIS, STRUCTURE AND CELLULAR FUNCTIONS

Additional books in this series can be found on Nova's website under the Series tab.

Additional E-books in this series can be found on Nova's website under the E-book tab.

PROTEIN SCIENCE AND ENGINEERING

Additional books in this series can be found on Nova's website under the Series tab.

Additional E-books in this series can be found on Nova's website under the E-book tab.

PROTEIN BIOCHEMISTRY, SYNTHESIS, STRUCTURE
AND CELLULAR FUNCTIONS

AMYLOIDS

COMPOSITION, FUNCTIONS AND PATHOLOGY

IRENE P. HALCHECK
AND
NANCY R. VERNON
EDITORS

Nova Science Publishers, Inc.
New York

For permission to use material from this book please contact us:
Telephone 631-231-7269; Fax 631-231-8175
Web Site: http://www.novapublishers.com

Additional color graphics may be available in the e-book version of this book.

Library of Congress Cataloging-in-Publication Data

Amyloids : composition, functions, and pathology / editors, Irene P. Halcheck and Nancy R. Vernon.
p. ; cm.
Includes bibliographical references and index.
ISBN 978-1-62100-538-4 (hardcover)
I. Halcheck, Irene P. II. Vernon, Nancy R.
[DNLM: 1. Amyloid. 2. Plaque, Amyloid--physiopathology. QU 55]

571.6'4--dc23

2011035664

Published by Nova Science Publishers, Inc. † New York

CONTENTS

PREFACE

Assembly and deposition of physiologically generated, amyloidogenic proteins occur in various human diseases, referred to as amyloidoses, through misfolding that is either partial unfolding or partial folding of the proteins. A growing body of evidence suggests that misfolding of a given protein is driven by point mutations or by the presence of environmental factors favorable for amyloidogenesis. In this book, the authors present topical research in the study of the composition, function and pathology of amyloids, including the effect of yeast red pigment on amyloid fibrils in vivo and in vitro; amyloid precursor protein and its relation to Alzheimer's disease; amyloid precursor protein expression after traumatic brain injury; protein aggregation as a modulatory factor in EPMI and the atomic-resolution structures of prion AGAAAAGA amyloid fibrils.

Chapter 1 - Yeast strains carrying mutations in genes *ADE1* or *ADE2* accumulate red pigment, a result of polymerization of aminoimidazoleribotide (an intermediate of adenine biosynthesis). The authors studied the effect of the red pigment on the formation of amyloid fibrils both in vivo and in vitro. In the experiments in vivo they registered Thioflavine T fluorescence in crude lysates of yeast red pigment accumulating strains carrying *ade1* or *ade2* mutations and in white isogenic strains—either adenine prototrophs or carrying mutations at the first stages of purine biosynthesis. Red pigment presence in cells was shown to lead to a drop of amyloid content as demonstrated by both Thioflavine T fluorescence and protein separation in agarose gel and in PAGE followed by blot-hybridization with Sup35p. The elevated Thioflavine T fluorescence level characteristic to white strains (when compared to the red ones) was transmitted by cytoduction. In vitro experiments demonstrated that the red pigment and its low molecular weight derivative, a result of acid hydrolysis of the original pigment, dramatically diminish the intensity of amyloid bound Thioflavine T fluorescence of both insulin and lysozyme fibrils. A correlation between the decrease of the intensity of Thioflavine T fluorescence and the concentration of both forms of the red pigment was demonstrated. The red pigments were able to outcompete the Thioflavine T molecules already bound to the fibrils. The affinity of the hydrolyzed red pigment to amyloid fibrils turned out to be higher than that of the original pigment. Electron microscopy of growing amyloids permitted to visualize a drop of fibril content in samples due to both forms of the pigment and a decrease in size of the insulin fibrils caused by the native pigment.

Chapter 2 - Although amyloid precursor protein (APP) plays central roles in Alzheimer's disease (AD), the physiological functions of this protein have not been fully elucidated. γ-Secretase was first identified as a protease that cleaves APP within the transmembrane domain and produces Aβ peptides, which are thought to play a role in the pathogenesis of AD and their presence is used as a hallmark of the AD brain. However, the physiological functions of this protease also remain to be clarified. In the canonical Notch signaling pathway, ligands bind to the extracellular domain of Notch on neighboring cells and trigger sequential proteolytic cleavage. Finally, the intracellular domain (ICD) of Notch is released from the cell membrane by γ-secretase and translocates to the nucleus where it modulates gene expression through binding to transcription factors. Therefore, γ-secretase regulates the Notch signaling pathway. Recently, it was demonstrated that many type 1 transmembrane proteins, including APP, are substrates for γ-secretase, and the ICDs of these substrates are released from the cell membrane by γ-secretase. These processes are very similar to those occurring in Notch signaling. Therefore, the common enzyme, γ-secretase, modulates proteolysis and the turnover of possible signaling molecules, suggesting that mechanisms similar to Notch signaling may contribute widely to γ-secretase-regulated signaling pathways. Thus, it is likely that APP also has a signaling mechanism. Indeed, they have shown that the ICD of APP (AICD) markedly alters gene expression and induces neuron-specific apoptosis. These results suggest that APP signaling may be responsible for the onset of AD. This chapter presents an outline of APP and γ-secretase. The authors also focus on the hypothesis thatγ-secretase-regulated mechanisms similar to Notch signaling may play a role in signaling events involving type 1 transmembrane proteins, including APP. Here, they discuss the possibility that APP signaling induces dynamic changes in gene expression, which may be closely correlated with AICD-induced neuron-specific apoptosis, leading to the onset of AD.

Chapter 3 - The β-amyloid protein derived from amyloid precursor protein (APP) is a major component of senile plaque, and β-amyloid protein deposits in Alzheimer disease (AD) are thought to be produced by alternative processing of APP. Therefore, the overexpression of APP is an important factor in the neuronal degeneration of AD. The expression of APP has also been shown to increase in swollen axons and neuronal perikarya after neuronal injury, including ischemia and stab injury. As the pathological findings in dementia pugilistica mimic those in AD, it has been posited that traumatic brain injury (TBI) is an epidemiological risk factor for AD, although other genetic factors, which include the apoE genotype, might act additively. These findings suggest that the overexpression of APP after TBI may potentiate Alzheimer disease pathology. However, the chronological changes in APP expression and cerebral function of APP have not been evaluated after TBI. In this study, the authors immunohistochemically investigated chronogical changes in cellular sources and levels of APP production compared to that of mRNA for APPas assessed by RT-PCR in the rat brain following TBI. Traumatic brain injury increased the expression of APP in the neuronal perikarya and in damaged dystrophic neurites from 1 day to 90 days after injury. Moreover, at 7 days after injury, some macrophages/microglia also were co-localized with APP, which was overproduced by the neuronal perikarya and APP-positive dystrophic neurites after injury and then APP were phagocytosed by macrophages/microglia during this phase. However, astroglia did not express APP immunopositivity after brain injury. These results suggested that long-term overexpression of APP was confirmed by immunohistochemical and biological technique after TBI. Subsequently, the authors investigated the effects of direct infusion of an anti-APP antibody into the damaged brain region on cerebral function and morphologically changes following TBI in the rat. At three days after TBI, there were many TUNEL-positive neurons and astrocytes around damaged area and a significant increase in numbers of TUNEL-positive cells in the PBS group versus the anti-APP group. At seven days after TBI, there was a significant increase in the numbers of large GFAP-positive cells, long elongated projections, and MAP-2-positive cells around damaged area in the anti-APP group versus the PBS group. At seven days after TBI, both the area of the damaged brain region and the arrival time to platform were significantly decreased in the anti-APP group versus the PBS group. Further, after TBI the arrival time to platform in the anti-APP group recovered to that observed in uninjured sham operation group animals. These data suggest that overproduction of APP after TBI inhibits astrocyte activity and reduces neural cell survival around the damaged brain region. On the basis of these findings, the authors speculate that overexpression of APP after TBI is related to Alzheimer type dementia, and is an important risk factor for this disease.

Chapter 4 - The plasma membrane-bound amyloid precursor protein is cleaved by sequential actions of two proteases β- and γ-secretases into a short amyloid-beta peptide (Aβ) fragment. The fragment passes through a dynamic process to precipitate into insoluble fibrous aggregates. These aggregates interact with brain neurons and cause neurotoxicity, leading to neurodegeneration and cognitive and behavioral decline, characteristics of Alzheimer's disease (AD). In this review, the authors focus on how Aβ behaves during self-association and causes neurotoxicity while emphasizing the protective effect of docosahexaenoic acid (DHA; C22:6*n*-3), a highly polyunsaturated fatty acid required for incessant maintenance of neuronal functions, against amyloidogenesis. DHA inhibits in vitro fibrillation of Aβ with concomitant inhibition of Aβ-induced cytotoxic effects in neuronal cells and learning-related memory impairments in AD model rats infused with the peptide in the cerebral ventricle.

Chapter 5 - Conformational disorders, due to pathological changes in protein folding and oligomerization, comprise many neurodegenerative diseases, such as Parkinson's, Alzheimer's, various dementias as well as prion diseases. A common characteristic of these diseases are aberrantly folded proteins, rich in β secondary structure that have a high tendency to aggregate and form amyloid fibrils. The modified amyloid cascade hypothesis of neurodegeneration states that the soluble protein oligomers and prefibrillar aggregates are the species exerting cytotoxicity rather than full-length fibrils or extracellular amyloid plaques. In this chapter, the authors present a brief overview of the protein aggregation status in connection to neurodegenerative disease. Moreover, they provide an insight on how increased protein aggregation may modulate neurodegeneration in the progressive myoclonus epilepsy of type 1 (EPM1). EPM1, also known as Unverricht-Lundborg disease, is a rare type of hereditary epilepsy with features of cerebellar atrophy and increased oxidative stress. The authors' hypothesis from 2005 states that some of the mutants of human stefin B found in EPM1 patients may undergo changed aggregation in cells, thus exerting additional gain in toxic function (apart from loss of normal function as an endogenous cysteine proteases inhibitor). This in turn may influence the patients' phenotype. Their *in vitro* experiments thus far have shown that the fragment R68X forms amyloid-fibrils very fast even at neutral pH, whereas the G4R mutant exhibits a prolonged lag phase, during which the toxic prefibrillar aggregates accumulate. The experiments using cell culture have demonstrated that the fragment R68X forms cytotoxic inclusions of the aggresome type, whereas the G4R mutant forms such aggregates only in conditions of proteasome inhibition. Results on two more missense EPM1 mutants, G50E and Q71P are being collected.

Chapter 6 - Stereoinversion of aspartic acid (Asp) residues in amyloid β (Aβ) peptide is observed in the brain of Alzheimer's dementia patients. The amyloid β 1–42 ($A\beta_{1-42}$) peptide, which has 42 amino acid residues, is considered to be related to Alzheimer's dementia. This peptide includes three Asp residues—Asp1, Asp7, and Asp23. Recently, $A\beta_{1-42}$ peptides containing D-Asp residues were artificially synthesized, and their secondary structures were experimentally observed. Although the experiments indicate that stereoinversion of Asp23 causes β-sheet formation of the $A\beta_{1-42}$ peptide, the atomic-level structure of the $A\beta_{1-42}$ peptide could not be obtained experimentally. Because $A\beta_{1-42}$ containing D-Asp23 tend to aggregate, the atomic-level structural biological experiments are difficult to conduct; thus, structural bioinformatics approaches to the peptides have an important role. In this study, the authors performed molecular dynamics simulation, which is one of the most useful methods in structural bioinformatics, for $A\beta_{1-42}$ peptides containing D-Asp residues. In addition, the abilities of several settings of molecular dynamics methods to evaluate the structural features of Aβ peptides containing D-Asp were compared. The results indicate that stereoinversion of Asp residues induces changes in the secondary structures of the C-terminal regions of $A\beta_{1-42}$ peptides. The usefulness of some implicit solvent models for molecular dynamics simulations of Aβ peptides is also discussed.

Chapter 7 - Assembly and deposition of physiologically generated, amyloidogenic proteins occur in various human diseases, referred to as amyloidoses, through misfolding that is either partial unfolding or partial folding of the proteins. A growing body of evidence suggests that misfolding of a given protein is driven by point mutation(s) or by the presence of environmental factors favorable for amyloidogenesis. One of the most extensively studied partial-folding-type amyloidogenic proteins is amyloid ß-protein (Aß), a proteinaceous component of senile plaques in Alzheimer disease. Although the soluble Aß monomer does not show any fixed secondary structure under physiological conditions, Aß adopts an α-helix-rich structure in the presence of gangliosides in particular membrane microdomains such as lipid rafts. Previous studies suggest that the ganglioside-induced α-helix-rich conformer acts as an endogenous seed for the consecutive assembly of Aß, resulting in the formation of ß-sheet-rich amyloid fibrils in the brain. Notably, it has also been suggested that the alterations of the lipid composition of membrane microdomains likely occur prior to the induction of the partially folded endogenous seeds and the formation of amyloid fibrils. Regarding the emergence of toxicities of protein aggregates, it is widely accepted that amyloid fibrils as well as smaller oligomers, which may be intermediates generated on the pathway to amyloid fibril formation or derivative fragments of amyloid fibrils, attack membranes targeting or not targeting specific proteins expressed on the surface of the membranes. In this review, the impact of the presence of biological membranes in terms of the initiation of amyloid fibril formation and causing amyloid toxicities will be discussed.

Chapter 8 - The aggregation of α-synuclein to form insoluble fibrillar inclusions is the probable key event in the onset of many neurodegenerative diseases including Parkinson's disease and the Lewy body variant of Alzheimer's disease. Previous studies and aggregation prediction programmes have identified key regions within the α-synuclein sequence that are implicated in pathogenic aggregation of the protein. Here, I have examined the self-association properties of short synthetic peptides corresponding to regions of the full-length protein. The peptides examined were selected from within the fibril core based on secondary structure previously identified by NMR. A range of biophysical techniques have been used to try and elucidate the properties of the amyloid formation pathway. One of the major unknown properties of amyloid proteins is the nature of the toxic species. It is commonly thought to be the early oligomeric or protofibrillar species that are the culprit for toxic effects seen in neuronal death rather than insoluble fibrils. The data presented here provides a useful insight into amyloid formation and will be used to guide the synthesis of peptide inhibitors of α-synuclein aggregation and provide potential new therapies for neurodegenerative diseases.

Chapter 9 - The formation of unusual modifications on the sulfhydryl residue of 10th cysteine of transthyretin (TTR) caused to induce the amyloidogenicity. Based on the mass spectrometric analysis of patients' plasma with TTR related amyloidosis and other similar diseases and the profiles of TTR itself and TTR-related octapeptide under the alkaline condition at 37 degrees, the authors illustrated the schema of reaction steps for the formation of TTR-related amyloid fibril but there are still unknown events to well-understand the acquisition of the amyloidogenicity. The authors investigated intermediate reactants *via* an unusual radical reaction following *S*-sulfonation to clarify precursors of proto-amyloidogenic TTR by ultra-high resolution time-of-flight MS (UHR-TOFMS). These reactants were not only mixed disulfide TTRs but also three abnormal structural products, such as dehydroalanine and glycine degradated from 10th cysteine residue of TTR with the reaction of β-elimination and threonine/ homoserine was also degraded on 13th methionine using the TTR modified with cysteine-*S*-sulfonate *in vitro*. Especially the dehydroalanine, α-acryl amino acid, having a high-reactivity, could be randomly polymerized between TTR itself and co-existence other molecules with a covalent bridge. In conclusion, the authors revealed the induction of the dehydroalanine moiety from 10th cysteine residue of TTR irreversibly and progressively must be an essential molecule to form TTR-related amyloid fibril.

Chapter 10 - To the best of the authors knowledge, there is little structural data available on the AGAAAAGA palindrome in the hydrophobic region (113-120) of prion proteins due to the unstable, noncrystalline and insoluble nature of the amyloid fibril, although many experimental studies have shown that this region has amyloid fibril forming properties and plays an important role in prion diseases. In view of this, the present study is devoted to address this problem from computational approaches such as local optimization steepest descent, conjugate gradient, discrete gradient and Newton methods, global optimization simulated annealing and genetic algorithms, canonical dual optimization theory, and structural bioinformatics. The optimal atomic-resolution structures of prion AGAAAAGA amyloid fibils reported in this Chapter have a value to the scientific community in its drive to find treatments for prion diseases or at least be useful for the goals of medicinal chemistry.

Chapter 11 - Amyloidoses are a heterogeneous group of diseases characterized by the deposition of fibrillar protein in extracellular tissue space. Amyloid deposits consist of a stable, insoluble aggregate of precursor proteins that have adopted an antiparallel beta pleated sheet configuration, having an amorphous and eosinophilic appearance. Although the reason for the accumulation of fibrils is poorly understood, several factors may be involved seeing as there are certain characteristics in precursor molecules that make them amyloidogenic, thus predisposing them to amyloid fibril formation. Different types of amyloidosis exist depending on the fibrillar protein precursor. The most common are: AL, which is associated with plasma cell dyscrasia; AA, associated with chronic inflammation; and Aβ, which is found in patients on long-term haemodialysis for renal failure. Amyloid deposits may affect many organs (systemic amyloidoses), thus decreasing the patient's survival if vital organs are affected, or they may occur in a single tissue (localized amyloidoses) with a better prognosis. Biopsy is necessary to confirm the amyloid deposit, which acquires an apple green birefringence under polarized light after staining with Congo red. Abdominal fat aspirate is a non-invasive technique that presents high sensibility, 70-90%. Prompt diagnosis and evaluation for systemic disease is essential prior to treatment.

Chapter 12 - This article discusses approaches for testing monoclonal proteins in clinical laboratories with special emphasis on amylodosis AL and free light chain disease. Where reflex testing is possible good communication between the laboratory and the clinician leads to the most effective work-up. The article briefly reviews the major monoclonal gammopathies, including multiple myeloma, lymphocytoplasmic lymphoma, heavy chain disease, plasmacytoma, plasma cell dyscrasia associated with neuropathy, amyloidosis AL, free light chain disease, and monoclonal gammopathy of unknown significance. There is emphasis on the pathophysiology of amyloidosis AL and free light chain disease and a testing algorithm for identifying amyloidosis AL. Standard clinical laboratory techniques for identifying monoclonal proteins are discussed and examples of serum and urine protein electrophoresis and immunofixation electrophoresis are shown. A new serum free light chain assay that adds additional armament to traditional testing is discussed. The article discusses the strengths and weaknesses of this assay as compared to traditional urine testing for monoclonal free light chains. It is concluded with an algorithm that addresses approaches for monoclonal protein screening.

Chapter 13 - Recent biochemical and molecular analyses suggest that familial amyloid polyneuropathy (FAP) is not as rare as previously thought. Transthyretin Val30Met-associated FAP (FAP ATTR Val30Met) is the most common form of FAP. Although patients with FAP ATTR Val30Met are geographically concentrated to Japan, Portugal, and Sweden, a late-onset form of this type of FAP was discovered in non-endemic areas and is widely distributed throughout the world. In this chapter, the authors describe the clinical, electrophysiologic, and histopathologic features of FAP ATTR Val30Met in Japanese patients by comparing conventional early-onset cases from endemic foci and late-onset cases from non-endemic areas. When compared with the classical phenotype of FAP, the clinicopathologic features of FAP patients from non-endemic areas tend to be nonspecific. As a result, a diagnosis of FAP may be missed unless amyloid deposition is seen on sural nerve biopsy. Therefore, an in depth knowledge of the characteristics of FAP ATTR Val30Met as well as a high degree of clinical suspicion is required to achieve early diagnosis.

Chapter 14 - Biomarkers for early and accurate diagnosis of Alzheimer's disease (AD) have been paid great attention over the past several years. Alzheimer's Disease NeuroImaging Initiative (ADNI), which follows up precise and longitudinal changes in biomarkers, has begun in the USA, Europe, Australia and Japan, and there have been several important findings [1]. Quite a few studies and projects including ADNI revealed that amyloid β (Aβ) should be one of the most important biomarkers of AD. Accumulation and distribution of Aβ in the brain and the measurement of its concentration in the cerebrospinal fluid can be analyzed by imaging techniques using positron emission tomography (PET) and ELISA, respectively. An amyloid imaging modality using a novel PET tracer was developed by Klunk et al. in 2004 [2]. Since then some of the investigators including us have released new amyloid tracers. The authors' group recently developed a novel PET tracer, ^{11}C-labeled 2-(2-[2-dimethylaminothiazol- 5-yl] ethenyl) -6- (2- [fluoro]ethoxy) benzoxazole ([^{11}C]BF-227), and succeeded in *in vivo* detection of amyloid plaques in the AD brains [3,4]. Needless to say, biomarkers in biofluid now also play important roles in the diagnosis of AD. Their group for the first time in the world reported that tau protein in the cerebrospinal fluid (CSF) is elevated in AD patients compared to normal controls and other neurological diseases [5]. In addition to tau, Aβ (especially Aβ1-42) in CSF is a sensitive and specific biomarker of AD. Both the elevated total-tau (t-tau) and phosphorylated-tau (p-tau) and the decreased Aβ1-42 in CSF can distinguish not only AD from normal controls but also can discriminate between the subjects with mild cognitive impairment who convert to AD later (MCI-converters) and the ones with MCI who do not convert to AD (MCI-non-converters) while they are in the MCI stage [6]. It has been reported that Aβ1-42 in CSF and amyloid PET show a change in the pathological process of AD earliest among the biomarkers, therefore these are considered to be the most sensitive ones for the early diagnosis of AD [6].

In: Amyloids
Editors: I. P. Halcheck et al.

ISBN: 978-1-62100-538-4

Chapter 1

THE EFFECT OF YEAST RED PIGMENT ON AMYLOID FIBRILS *IN VIVO* AND *IN VITRO*

O. V. Nevzglyadova[1*], E. V. Mikhailova[1], A. V. Artemov[1], A. G. Mittenberg[1], I. M. Kuznetsova[1], K. K. Turoverov[1], M. A. Khodorkovskii[2] and T. R. Soidla[1]

[1]Institute of Cytology, Russian Academy of Sciences,
St. Petersburg, Russia
[2]St. Petersburg State Polytechnical University,
St. Petersburg, Russia

ABSTRACT

Yeast strains carrying mutations in genes *ADE1* or *ADE2* accumulate red pigment, a result of polymerization of aminoimidazoleribotide (an intermediate of adenine biosynthesis). We studied the effect of the red pigment on the formation of amyloid fibrils both in vivo and in vitro. In the experiments in vivo we registered Thioflavine T fluorescence in crude lysates of yeast red pigment accumulating strains carrying *ade1* or *ade2* mutations and in white isogenic strains—either adenine prototrophs or carrying mutations at the first stages of purine biosynthesis. Red pigment presence in cells was shown to lead to a drop of amyloid content as demonstrated by both Thioflavine T fluorescence and protein separation in agarose gel and in PAGE followed by blot-hybridization with Sup35p. The elevated Thioflavine T fluorescence level characteristic to white strains (when compared to the red ones) was transmitted by cytoduction. In vitro experiments demonstrated that the red pigment and its low molecular weight derivative, a result of acid hydrolysis of the original pigment, dramatically diminish the intensity of amyloid bound Thioflavine T fluorescence of both insulin and lysozyme fibrils. A correlation between the decrease of the intensity of Thioflavine T fluorescence and the concentration of both forms of the red pigment was demonstrated. The red pigments were able to outcompete the Thioflavine T molecules already bound to the fibrils. The affinity of the hydrolyzed red pigment to amyloid fibrils turned out to be higher than that of the

[*]Corresponding author: Institute of Cytology, Tikhoretsky avenue 4, St. Petersburg 194017, Russia; Phone +7 812 2974439, Fax +7 812 2970341, e-mail: oneva43@yahoo.com

original pigment. Electron microscopy of growing amyloids permitted to visualize a drop of fibril content in samples due to both forms of the pigment and a decrease in size of the insulin fibrils caused by the native pigment.

Keywords: amyloid, prion, Thioflavine T, red pigment, *Saccharomyces cerevisiae.*

INTRODUCTION

Animal and human "proteinopathies" (Prusiner, 2001) or "conformational diseases" (Herczenik, Gebbink, 2008; Uversky, 2008) are caused by changes of protein conformation leading to abnormal aggregation and, as a result, to generation of amyloid fibrils. If such an amyloid state can be transmitted from a cell to another cell and from organism to organism, and especially if protein only induction of a disease is proven, such "infectious" proteins are called prions. Prion diseases are practically incurable and constitute a serious challenge for contemporary medicine and veterinary science.

Discovery of prions in the eukaryotic microorganism yeast (cf. Chernoff et al., 2002; Chernoff, 2004, 2007; Wickner, 1994; Wickner et al., 2007, 2010, 2011) opened new vistas for studying conformational diseases by applying powerful techniques of contemporary molecular genetics in vivo available for these organisms. These in vivo studies are usually blended with ex vivo and in vitro studies of the prion material of higher organisms.

All these approaches are often used to search for therapeutic agents able to cure prion diseases (Trevitt, Collinge, 2006). Among the first agents showing encouraging results were derivatives of the dye Congo Red (Demaimay et al., 2000; Webb et al, 2007), chlorpromazine (Korth et al., 2001) and other tricyclic derivatives of acridine and phenothiazine compounds like quinacrine and 9-aminoacridine (Korth et al., 2001; Phuan et al., 2007; Teruya et al., 2009).

During the following years many other potential anti-prionic compounds were discovered, such as squalestatin (Bate et al., 2004), dianilino-binaphthyl-sulfonate (Cordeiro et al., 2004), suramin (Nunziante et al., 2005), two compounds of the N'-benzylidene-benzohydrazide class (Schiffer et al., 2007), 4,5-dianilinophthalamide (Wang H. et al., 2008), phthalocyanine tetrasulfonate (Lamberto et al. 2009), and various 2-aminothiazoles (Ghaemmaghami et al., 2010). Attention was drawn to various polyanions such as glycosaminoglycans (Wong et al., 2001; Adjou et al., 2003; Lawson et al., 2010), participating in prion life cycle and permitting to modify the cycle. It was demonstrated that prion replication is affected by various cationic phosphorus-containing dendrimers (Solassol et al., 2004) and aminosteroids (Charveriat et al., 2009). Amyloid accumulation in cells was shown to be affected by polyphenols (including tannic acid and tea extracts), phenothiazines, antihistamines, statins and antimalarial compounds (Kocisco et al., 2003). Amyloidophilic compounds, notably styrylbenzoazole and phenylhydrazine derivatives were found to modify prion infection (Teruya et al., 2009). A combination of computer modeling and ex vivo experiments permitted to detect several small molecular weight antiprionic compounds (Hosokawa-Muto et al, 2009).

Interesting is a protocol of express-screening for antiprionic compounds using a test system for visualizing modification of effects of yeast prions [*PSI*$^+$] and [*URE3*] proposed by a group of biologists working in France. Yeast prions based express-tests are followed by an analysis of prion action in ex vivo mammalian system. This battery permitted to detect anti-prion activity of various kastellopaolitines, phenanthridine derivatives and analogues of an antihypertensive drug Guanabenz (Saupe, 2003; Tribouillard et al., 2006; Tribouillard-Tanvier et al., 2008). Chernoff (2004) noted that curing yeast cells of prions via manipulating chaperones can lead us to development of new potential anti-prion treatments.

Many efforts were devoted these years to the search for chemical compounds able to block amyloid propagation. For example, it was demonstrated (Webb et al., 2007) that several Congo Red derivatives influenced amyloid formation by addressing it to an alternative pathway. Use of these compounds leads to formation of non-toxic intermediates that are not harmful to the cell. Unfortunately, many of the compounds used in such experiments are toxic themselves. A most interesting approach was demonstrated by (Ehrnhoefer et al., 2008; Bieschke et al., 2010) who found that polyphenol (-)-epigallocatechine gallate from green tea converts large mature fibrils into small amorphous protein aggregates preventing their conversion into toxic intermediates.

Many other attempts at searching for anti-amyloid and anti-prionic compounds can be listed here (see Trevitt, Collinge, 2006; Riemer et al., 2008).

As not a single one among the above mentioned compounds have been introduced into clinical practice (Trevitt, Collinge, 2006; Charveriat et al., 2009), it seems to be obvious that further search of antiprionic compounds is most actual.

We have investigated potential antiprionic properties of yeast red pigment. The red pigment is a polymer of aminoimidazoleribotide (AIR) molecules containing a few aminoacid residues either in the backbone of the polymer or as side chains (Smirnov et al., 1967). We used a modified Smirnov's method (Smirnov et al., 1967) to purify the pigment that turned out to be a polymer with molecular weight not exceeding 10 kDa. Hydrolysis by HCl according to Smirnov leading to removal of ribosyl groups does not inhibit polymerization of the residues or discolor the pigment. This process permitted us to obtain a derivative of the red pigment with M.W. 2 kDa (This implies a mild decrease of polymer size together with ribosyl removal). We studied the effect of red pigment on prion life cycle in vivo in yeast cells and the effect of both the native red pigment and its derivative on insulin and lysozyme fibril growth and Thioflavine T binding to the fibrils in vitro.

This paper is an autoreview of results reported in a series of our recent publications, augmented with new experiments of the same type and illustrated by figures and tables that have not been published earlier in this form.

Materials and Methods

The yeast strains used in this work are presented in Table 1 that contains genotypes and plasmotypes. The strain AH/B was kindly provided by Dr. A. Borchsenius, the strains K-5 and K-13—by Dr. B.F. Yarovoi, 168t by Dr. T.S. Karpova. The strain 74-D694 was kindly donated by Dr. L. Mironova, the strain GT26 by Prof. Y. Chernoff.

The cultural media and basic methods, including preparation and analysis of cell lysate pellets, electrophoresis and western blot, two-dimensional polyacrylamide gel electrophoresis, mass spectrometry (MALDI TOF) and protein identification, measuring fluorescence intensity, red pigment purification, model experiments with insulin and lysozyme fibrils, model experiments with protein aggregates of [PSI+] strains in vitro, TEM, and statistical methods used were reported in our publications (Nevzglyadova et al., 2009, 2010a, b, c, 2011, Mikhailova et al., 2011)

Table 1. *Saccharomyces cerevisiae* strains used in this paper

	Strain	Genotype and plasmatype
1	74-D694/Gu1	*MATa ade1-14 ura3 his3 trp1 leu 2* [*psi⁻ pin⁻*]
2	AH/B	*MATα ade1-14 ura3 his3 lys9 kar1* [*psi⁻ pin⁻*]
3	GT26	*MATa aro7-1 cys1-72 his4-166 leu2-1 lys2-187 met8-1 trp5-48 ura3-1* [*PSI⁺ PIN⁺*]
4	168tester	*MATa his5* [*psi⁻*]
5	K-13	*MATa his5 kar1-1* [*PSI⁺*]
6	K- 5	*MATα his4 ade2 kar1-1* [*psi⁻*]

CONCLUSION

A Starting Point of the Study

Earlier (Nevzglyadova et al., 2008) we demonstrated that the determination of fluorescence intensity of amyloid bound Thioflavine T can be used for estimating amyloid content in lysates of yeast cells. As there can be other compounds (besides amyloid) that can be stained by Thioflavine T, we compared strains with identical genetic background. In these experiments we used strains containing or not containing a well-known yeast prion, the [PSI^+] factor. Like most other investigators working with [PSI^+], we used strains containing a nonsense mutation in *ADE1*, coding for SAICAR synthase. This mutation leads to the accumulation of red pigment, the product of polymerization of AIR—aminoimidazole ribotide (Smirnov et al., 1967).

Nonsense mutations in *ADE1* or *ADE2* (leading to red colony color) are considered to be especially suitable for the prionization studies involving Sup35p (a.k.a. eukaryotic translation termination factor eRF3). Prionization of Sup35p inactivates the majority of soluble Sup35p molecules, which results in a partial loss of translation termination activity and as a result in low level suppression: reading through the nonsense codons, leading to emergence of sufficient amount of full-sized and functional SAICAR-synthase or AIR-carboxylase molecules. This lifts the block of adenine biosynthesis and halts accumulation of the red pigment. In this way the color of yeast colonies (together with related ability to grow in minimal medium without adenine) permits to make half-quantitative express conclusions about the level of nonsense suppression and hence about Sup35p prionization.

We demonstrated (Nevzglyadova et al., 2010a) that the intensity of amyloid bound Thioflavine T fluorescence (IF) for the lysate pellet proteins of [PSI^+] cells is higher than for the corresponding proteins of pink [psi^-] cells. We explained the strain difference by different amount of prion amyloid in these strains, supposing that [PSI^+] contains more amyloid than [psi^-]. This was confirmed by in vitro experiments with proteins extracted from pellets of crude lysates that were stained with Thioflavine T or separated in 1.5 % agarose gels. At the same time other experiments with pairs of isogenic strains differing only by their color led us to realization that not only the prion composition, but also the red pigment presence determines the amyloid amount in yeast cells. Red pigment does not eliminate [*PSI*$^+$] amyloids. We learned that the non-suppressible adenine mutations carrying red pigment producing strains retain the [*PSI*$^+$] factor. Nevertheless, our results also demonstrated that the red pigment inhibits amyloid interaction with Thioflavine T and putatively depresses amyloid level in yeast cells.

Amyloid Content in the Pellet Fraction of Crude Lysates Appears to Depend on Red Pigment Accumulation in Yeast Cells

To study the red pigment role in protein amyloidization, several white/red isogenic pairs of strains, sharing the same genetic background and presumably differing only by mutations in some genes of adenine biosynthesis pathway, was used. Mutations blocking red pigment formation in Ade$^-$ strains or leading to pigment synthesis in the Ade$^+$ ones were induced using UV-mutagenesis. Cell survival rate during mutagenetic treatment was higher than 1 %. For allelism tests with the induced mutations we used strains carrying previously characterized mutations in various genes involved in adenine synthesis. Standard procedures were used for UV-mutagenesis and for the allelism test (Sherman et al., 1986). We learned that the white mutations induced in Ade$^-$ strains affected genes controlling the first steps of adenine biosynthesis and red mutations in Ade$^+$ strains affected *ADE1* or *ADE2*. These experiments are described in more details in (Nevzglyadova et al., 2010b). We compared lysate pellet proteins in the white/red pairs of isogenic strains and also in pairs of strains whose red pigment content was modified physiologically: red strains were grown either on YEPD with no additional adenine ("normal adenine") or on YEPD with extra adenine 100 mg/l (YEPD-ADE) that repressed the formation of the red pigment.

The effects of both ways effects of changing colony color, either by additional mutations or by blocking *de novo* purine synthesis, are demonstrated in Figure 1. It is easy to see that if a red strain is converted to a white one (either by additional adenine mutation or by growth on extra adenine) the IF estimates are augmented. When *ade1* or *ade2* mutations are induced in an originally white strain (leading to accumulation of the red pigment) IF drops considerably. This hints at red pigment being able to decrease amyloid content in yeast cells. A full set of data of all the pairwise combinations tested can be found in (Nevzglyadova et al., 2009). We would like to notice that amyloid content depends on red pigment even in [*psi*$^-$] strains, so the effect must involve not only Sup35p amyloid, but possibly also some anonymous amyloids or prions present in the strains studied here.

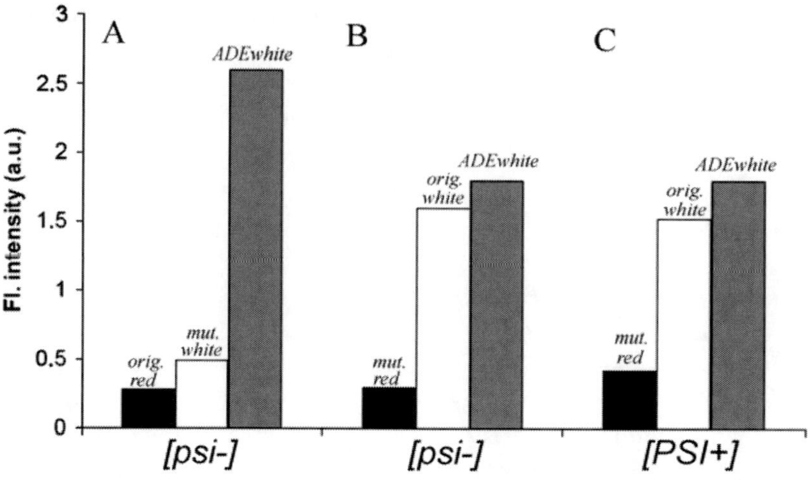

Figure 1. Distribution of fluorescence intensity (IF) of Thioflavine T bound to pellet proteins of isogenic strains differing in prion content and red pigment accumulation. A: [psi$^-$] strains: original K-5 (red), its white mutant K-5/w1 and K-5/ADE—the original K-5 strain displaying white colony color due to growth on a medium with high adenine content. B: [psi$^-$] strains: red mutant 168t/r1, the original white 168t and 168t/r1/ADE—the mutant 168t/r1displaying white colony color due to growth on a medium with high adenine content. C: [*PSI*$^+$] strains: red mutant GT26/r1, the original white GT26 and GT26/r1/ADE—the mutant GT26/r1 displaying white colony color due to growth on a medium with high adenine content.

Some interesting features of the red pigment effect can be seen when comparing the white strains that lack red pigment due to their genotype and the strains carrying one adenine mutation either in *ADE1* or *ADE2* but turned white by extra adenine. One can see that when the original strain was red (A), the IF data for two different types of white cells—those, belonging to the white mutant and to the original strain "whitened" due to growth on an extra adenine are drastically different. If the original strain was white, the genotypically and phenotypically white strains differ very slightly.

The originally white [*psi*$^-$] and [*PSI*$^+$] strains (B and C) are characterized by mitotically stable high amyloid level, permitting to suppose that at least the [*psi*$^-$] strain (but most likely both of them) contain some anonymous prions. The white mutant strains obtained from the original red ones (A) have much lower amyloid level than white strains of B and C. Most likely the amount of prions in the originally red strains has stably decreased and some (but not all) anonymous prions have been lost. Extra adenine in this case leads to much higher IF than in the white mutant. We suppose that the growth on extra adenine can lead to emergence of nonheritable amyloids corresponding to some anonymous prions lost in the original red strains. Generation of nonheritable amyloids was described in the work by (Salnikova et al., 2005, Kushnirov et al., 2007). This hypothesis seems to deserve further consideration.

We compared the pellet proteins of white/red strain pairs in 1.8 % agarose gel. The results indicated that in all pairwise combinations the amount of polymeric material of high molecular weight is clearly reduced in red strains (Nevzglyadova et al., 2010b). The data on agarose electrophoresis closely match the data on intensity of fluorescence of Thioflavine T bound to pellet proteins of red and white strains. The increase of IF in white strains correlates with enhancement of the fraction of polymeric proteins of high molecular weight (see Figure 2).

This result suggested an experiment on identification of the proteins whose presence in pellets depended on the red pigment using 2D electrophoresis and MALDI.

2D Electrophoresis and MALDI Analysis of Pigment-Dependent Proteins

We were looking forward to confirming a connection between the red pigment effect and amyloid "life cycle" in yeast cells by MALDI analysis.

Earlier we developed an approach to deal with such topics: comparison of 2D electrophoresis data for pellet proteins of [PSI^{+}] and [psi^{-}] strains—of a cytoductant and its parental recipient strain. More than 50 proteins that were present among pellet proteins of [PSI^{+}] and absent in [psi^{-}] strains were identified by 2D electrophoresis and subsequent MALDI. We supposed them to be physically prion-associated and mobilized to pellets by prions. These proteins belonged mainly to the following groups: chaperones, proteins involved in oxidative stress, glucose metabolism and in translation (Nevzglyadova et al., 2009, 2010a, c).

To learn if there exists also a related group of red pigment-dependent proteins we compared 2D electrophoregrams of pellet proteins of a red strain AH/B [psi^{-}] and its white mutant AH/B/w, and white [PSI^{+}] strain K-13 and of its red mutant K-13/r, and [psi^{-}] strain AH/B grown on YEPD and YEPD-ADE media. The data showed that the white strains contain several lysate pellet proteins lacking in the pellet (and so presumably not aggregated) in red strains. This experiment was followed by MALDI analysis that permitted us to identify 23 pellet proteins in strains AH/B/w, K5-13, and GT26 that were classified as pigment-dependent proteins, for they were missing in the corresponding red strains (Nevzglyadova et al., 2010b).

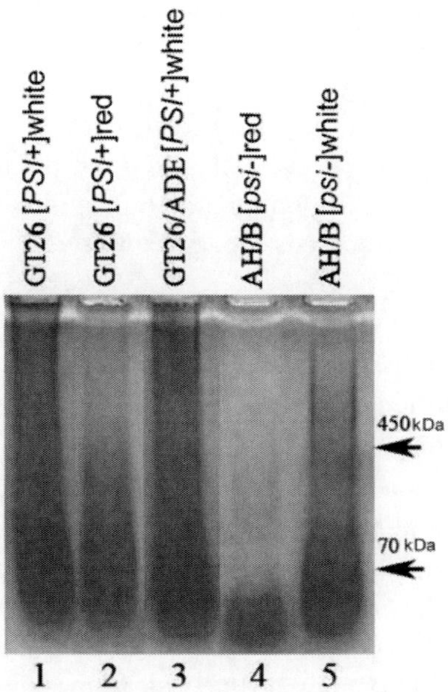

Figure 2. Protein electrophoregrams of lysate pellets proteins of the red and white yeast strains in 1.8 % agarose gel. ADE denotes strains grown on the high adenine (YEPD-ADE) medium.

In further studies we extended the approach to more strains. Table 2 lists 48 pigment-dependent proteins identified in our studies. We did not find any established prions. Present in the list are numerous chaperones, glucose metabolism enzymes, and proteins involved in cellular response to stress. Notable are also proteins involved in translation and in proteolysis. As documented in Table 2 the data on pigment-dependent proteins in yeast crude lysate pellets are rather similar to our previous results on prion-dependent proteins—some key proteins coincide, others belong to the same functional groups.

As we have demonstrated (Nevzglyadova et al., 2010c) the list of identified proteins is closely related to the lists of proteins identified in yeast amyloid aggregates by other authors (Wang Y. et al., 2007; Bagriantsev et al., 2008; Erjavec et al., 2008). Especially important seems to be the correspondence with the list of proteins mobilized to yeast aggresomes (Wang Y. et al., 2009). Aggresomes seem to be involved in cellular defense against misfolded proteins. This correspondence seems not to be accidental, as one of the key proteins of aggresomes, Cdc48p, was also found among the pigment-dependent proteins. Chaperon action, stress response and proteolysis are also closely related to the aggresome functions.

As shown in Table 2 in our experiments we have identified 19 different chaperones, 21 stress response proteins and 6 proteins involved in proteolysis, that are either prion- or pigment-dependent. Among the red pigment-dependent proteins that have not been identified as prion-dependent ones, we would like to single out a glucose metabolism proteins Glk1p, stress response-related proteins Sod1p, Gvp36p, Mcr1p, Ola1p and chaperone Hsc82p.

We have identified (Nevzglyadova et al., 2009, 2010a, 2010b, 2010 c, 2011) numerous prion- and red-pigment dependent proteins, but have no sufficient evidence yet about these proteins (except for Sup35p and Rnq1p) being true amyloid prions. At the same time our data demonstrate that not only Sup35p but also some other anonymous amyloids are seemingly transmitted in the cytoduction experiments. One of the most likely candidates of such amyloids is superoxide dismutase. An age-related degenerative disease—amyotropic multiple sclerosis—involves the accumulation of aggregated superoxide dismutase of *SOD1* with amyloid-like structure and the aggregates were shown to propagate in a prion-like manner in neuronal cells (Banci et al., 2008; Elam et al., 2003; Munch et al., 2011).

Superoxide dismutase (Sod1p) was identified as a red-pigment dependent protein and its isoform Sod2p—as a prion-dependent protein in our studies (Nevzglyadova et al., 2009, 2010c). In these papers we discussed the possibility of yeast superoxide dismutase prionization in more details.

Higher IF Values Characteristic of White Strains Compared with the Red Ones Are (Partly) Transmitted by Cytoduction

The data of measuring of amyloid bound Thioflavine T fluorescence together with results of lysate pellet proteins separation in agarose and polyacrylamide gels (Nevzglyadova et al., 2010b; 2011) suggest that several proteins of white strains in contrast with the red ones have undergone excessive amyloidization. We made an attempt to learn whether white strains are able to transmit their higher IF via cytoplasm using cytoduction, that is, if the anonymous amyloids involved in the white strain high IF level are inheritable? To answer this question, we obtained cytoductants with a nucleus derived from one (recipient) strain and extra cytoplasm derived from another (donor) strain exhibiting either white or red colony color.

Table 2. List of pigment- and prion-dependent proteins identified by 2D-electrophoresis followed by MALDI

Pigment-dependent protein name	NCBI Acc. No	Class		Prion-dependent protein name	NCBI Acc. No	Class
Established prions						
				Sup35p		Prion [PSI$^+$] protein
				Rnq1p	NP_009902	Prion [PIN$^+$] protein
Chaperones						
Hsp78p	NP_010544	Chaperone (Mito) Oxidative stress response, prevents aggregation of misfolded proteins, resolubilizes protein aggregates		Hsp78p	NP_010544	Chaperone (Mito) Oxidative stress response, prevents aggregation of misfolded proteins, resolubilizes protein aggregates
Hsp26p	NP_009628	Chaperone, suppresses unfolded proteins aggregation Cyt+nucl		Hsp26p	NP_009628	Chaperone, suppresses unfolded proteins aggregation Cyt+nucl
				Hsp42p	NP_010456	Chaperone
Ssa1p (HSP70 Family)	NP_009396	Chaperone Cyt+nucl, cell wall		Ssa1p (HSP70 Family)	NP_009396	Chaperone Cyt+nucl, cell wall
Ssa2p (HSP70 Family)	EDV09289	Chaperone Cell wall, vacuole Cyt+mito				
				Ssb1p (HSP70 family)	NP_010052	Chaperone
				Ssc1p (HSP70 family)	NP_012579	Chaperone
				Sse1p (HSP70 family)	NP_015219	Chaperone
Hsc82p (HSP90 family)	NP_013911	Chaperone Proteasome assembly, response to stress, refolding, Cyt				
				Hsp104p	NP_013074	Chaperone

Table 2. (Continued)

Pigment-dependent protein name	NCBI Acc. No	Class		Prion-dependent protein name	NCBI Acc. No	Class
Cpr3p	NP_013633	Chaperone, Apoptosis Mito, cyclophilin				
				Sec53p	NP_116609	Mannose metabolism Folding and glycosylation of secretory proteins
Eno2p	NP_012044	Co-chaperone Glucose metabolism		Eno2p	NP_012044	Co-chaperone Glucose metabolism
Glucose metabolism enzymes						
Eno2p	NP_012044	Co-chaperone Glucose metabolism		Eno2p	NP_012044	Co-chaperone Glucose metabolism
Eno1p	NP_011770	Glucose metabolism		Eno1p	NP_011770	Glucose metabolism
Adh1p	NP_014555	Glucose metabolism		Adh1p	NP_014555	Glucose metabolism
Cdc19p	NP_009362	Glucose metabolism (TCA) Cyt		Cdc19p	NP_009362	Glucose metabolism (TCA) Cyt
Glk1p	NP_009890	Glucose metabolism, Cyt				
Gpm1p	NP_012770	Glucose metabolism Cyt+mito				
Fba1p	NP_012863	Glucose metabolism		Fba1p	NP_012863	Glucose metabolism
Pgk1p (=Pgk72)	NP_009938	Glucose metabolism		Pgk1p (=Pgk72)	NP_009938	Glucose metabolism
Tpi1p	NP_010335	Glucose metabolism		Tpi1p	NP_010335	Glucose metabolism
				Pdc1p	NP_013145	Glucose metabolism
YGL253Wp-like or Hxk2p	EDZ72350 NP_011261	Glucose metabolism				

Pigment-dependent protein name	NCBI Acc. No	Class		Prion-dependent protein name	NCBI Acc. No	Class
Ald4p	NP_015019	Glucose metabolism Ethanol metabolism, Mito		Ald4p	NP_015019	Glucose metabolism Ethanol metabolism, Mito
Aco1p	NP_013407	TCA cycle Mito+cyt Protein catabolism		Aco1p	NP_013407	TCA cycle Mito+cyt Protein catabolism
Aco2p	NP_012335	TCA cycle, Mito				
Cit1p	NP_014398	TCA cycle, Mito		Cit1p	NP_014398	TCA cycle, Mito
Kgd1p	NP_012141	TCA cycle, Mito				
Tdh1p	NP_012483	Glucose metabolism				
				Tdh2p	NP_012542	Glucose metabolism ROS metabolism
Tdh3p	NP_011708	Glucose metabolism ROS metabolism		Tdh3p	NP_011708	Glucose metabolism ROS metabolism
Proteins involved in stress response						
				Tdh2p	NP_012542	Glucose metabolism ROS metabolism
Tdh3p	NP_011708	Glucose metabolism ROS metabolism		Tdh3p	NP_011708	Glucose metabolism ROS metabolism
				Act1p	NP_116614	Cytoskeleton, Oxidative stress responseActin
				Yhb1p	NP_011750	Oxidative stress response
Sod1p	NP_012638	Oxidative stress response, ROS metabolism				
Aco1p	NP_013407	TCA cycle Mito+cyt Protein catabolism		Aco1p	NP_013407	TCA cycle Mito+cyt Protein catabolism
				Sod2p	NP_011872	Oxidative stress response ROS metabolism

Table 2. (Continued)

Pigment-dependent protein name	NCBI Acc. No	Class		Prion-dependent protein name	NCBI Acc. No	Class
Mcr1p	NP_012772	Oxidative stresss response, Mito respiration				
Ola1p	NP_009581	Proteasome ATPase, induced by hydrogen peroxide, Cyt				
Tma19p	NP_012867	Oxidative stress response, Translation, Cyt+mito		Tma19p	NP_012867	Oxidative stress response, Translation, Cyt+mito
Zeo1p	NP_014532	Response to stress, Cell wall integrity, Mito+cyt		Zeo1p	NP_014532	Response to stress, Cell wall integrity, Mito+cyt
Hsp78p	NP_010544	Chaperone (Mito) Oxidative stress response, prevents aggregation of misfolded proteins, resolubilizes protein aggregates		Hsp78p	NP_010544	Chaperone (Mito) Oxidative stress response, prevents aggregation of misfolded proteins, resolubilizes protein aggregates
Hsc82p (HSP90 family)	NP_013911	Chaperone Proteasome assembly, response to stress, refolding, Cyt				
Gvp36p	NP_012223	Stress resistance Golgi, Vacuole biogenesis				

Pigment-dependent protein name	NCBI Acc. No	Class		Prion-dependent protein name	NCBI Acc. No	Class
Psa1p	NP_010228	Stress resistance Protein glycosylation Mannose, cell wall, Cyt				
				Por1p	NP_014343	Respiration Mito organization, influences oxidative stress response
Translation proteins						
Tma19p	NP_012867	Oxidative stress response, Translation, Cyt+mito		Tma19p	NP_012867	Oxidative stress response, Translation, Cyt+mito
Efb1p	NP_009398	Translation (elongation)				
Tef2p	NP_009676	— " —				
Tuf1p	NP_014830	— " — (Mito)				
Asc1p	NP_013834	Translation regulation Signal transducing G-protein, Cyt				
				Pab1p	NP_011092	Translation initiation Poly(A) binding protein
				Rpl2Ap or Rpl2Bp	NP_116688 NP_012246	Translation Ribosomal protein
Pigment-dependent protein name	NCBI Acc. No	Class		Prion-dependent protein name	NCBI Acc. No	Class
				Rpl4A	NP_009587	— " —
				Rpl8bp	NP_013055	— " —
				Rps6bp	NP_009740	— " —
				Rpl10p	NP_013176	— " —
				Rpl9Ap	NP_014332	— " —
				Grs1p	NP_009679	Translation Glycyl-tRNA ligase Transcription termination

Table 2. (Continued)

Pigment-dependent protein name	NCBI Acc. No	Class		Prion-dependent protein name	NCBI Acc. No	Class
Ura1p	NP_012706	Ribosome biogenesis Pyrimidine biosynthesis Cyt				
Proteins involved in proteolysis						
Aco1p	NP_013407	TCA cycle Mito+cyt Protein catabolism		Aco1p	NP_013407	TCA cycle Mito+cyt Protein catabolism
Cdc48p	NP_010157	Chaperone, ATPase Aggresome, Apoptosis Protein degradation				
Hsc82p (HSP90 family)	NP_013911	Chaperone Proteasome assembly, response to stress, refolding, Cyt				
Ola1p	NP_009581	Proteasome ATPase, induced by hydrogen peroxide, Cyt				
Tfs1p	NP_013279	Proteolysis regulation Signaling, kinase A pathway				
				Pep4p	NP_015171	Proteinase
Other proteins						
Ach1p	NP_009538	Acetate metabolism, Mito				
Shm1p	NP_009822	One-carbon compound metabolism Purine, pyrimidine, amino acid, and lipid biosynthesis Mito				
Shm2p	NP_013159	— " — (Cyt)				

Pigment-dependent protein name	NCBI Acc. No	Class		Prion-dependent protein name	NCBI Acc. No	Class
Ipp1p	NP_009565	Inorganic pyro-phosphatase Prot homodimerization				
Atp1p	NP_009453	ATP synthase sub-unit precursor Mito				
Atp2p	NP_012655	— " — (beta subunit)				
Sam2p	NP_010790	Methionine metabolism				
Fsh1p	NP_011915	Serine hydrolase Cyt+nucl				
Gdh1p	NP_015020	Nitrogen metabolism, glutamate synthesis, Cyt+nucl				
Rgi2p/Yil057cp	NP_012207	Unknown, induced by carbon limitation				
Pigment-dependent protein name	NCBI Acc. No	Class		Prion-dependent protein name	NCBI Acc. No	Class
				Ilv5p	NP_013459	Leucine, isoleucine, and valine biosynthesis
				Leu4p	NP_014295	Leucine, isoleucine, and valine biosynthesis
				Lpd1p	NP_116635	Leucine, isoleucine, and valine catabolism
				Cus1p	NP_013967	Spliceosome assembly
				Msc1p	NP_013578	Meiotic recombination
				Qcr2p	NP_015517	Respiration
				Qcr7p	NP_010818	Respiration

In some experiments that showed prion- and pigment-dependent proteins but did not allow to discriminate between them we identified several other proteins belonging to chaperones—Ssb2p, Hsp60p, Cpr1p, Fpr1p, Tsa1p; glucose metabolism proteins—Adh4p, Fum1p, Lat1p; proteins involved in stress responses—Ahp1p, Ctt1p, Yhb1p, Pst2p, Tsa1p; a translation protein—Tif35p and several unclassified proteins. These additional proteins are documented in Tables 1 and 4 in (Nevzgladova et al, 2010c).

As a result in our experiments (Nevzglyadova et al, 2009, 2010a, b, c, 2011, Mikhailova et al., 2011) aimed at finding prion- and pigment-dependent proteins we have identified 19 different chaperones, 22 glucose metabolism enzymes, 21stress proteins, 15 translation proteins, 6 proteins involved in proteolysis, 2 prions, and 26 unclassified proteins.

Table 3 shows the results for 8 different crosses. The table compares the fluorescence intensity of Thioflavine T bound to pellet proteins of cytoductants that obtain the cytoplasm from the red or white parental strain. In Part 1 the cytoductants were selected by transfer of the [*PSI*$^{+}$]-factor. In two independent crosses (A and B) the role of recipient strain (i.e., the strain transferring its nucleus to cytoductant) is played by red strain AH/B [*psi*$^{-}$]. This strain carries a suppressible nonsense mutation in the *ADE1* gene. At the same time, [*PSI*$^{+}$]-donors of red color contain in their genotype mutations in genes *ADE1* (GT26red) and *ADE2* (K5-13/red) that are not suppressible and therefore lead to accumulation of red pigment.

Table 3. Comparison of fluorescence intensity of Thioflavine T bound to pellet proteins of cytoductants that obtain the cytoplasm from the red or white parental strain

Part 1. Crosses involving [*PSI*$^{+}$] donor strains

A: (1) GT26/r1[*PSI*$^{+}$] **red** x AH/B[psi^{-}] **red**		**A**: (2) GT26[*PSI*$^{+}$] **white** x AH/B[*psi*$^{-}$] **red**	
	IF		**IF**
[*PSI*$^{+}$] **pink cytoductants**	**100 %**	[*PSI*$^{+}$] **pink cytoductants**	**158 %**
Donor parent GT26/r1 **red**	100 %	Donor parent GT26 **white**	617 %
B: (1) K-13/r3[*PSI*$^{+}$] **red** x AH/B[*psi*$^{-}$] **red**		**B**: (2) K-13[*PSI*$^{+}$] **white** x AH/B[*psi*$^{-}$] **red**	
	IF		**IF**
[*PSI*$^{+}$] **pink cytoductants**	**100 %**	[*PSI*$^{+}$] **pink cytoductants**	**132 %**
Donor parent K-13/r3 **red**	**100 %**	Donor parent K-13 **white**	**749 %**

A: The difference between IF of cytoductants obtained in two types of cytoduction is statistically significant at $p < 0.01$. B: The difference between IF of cytoductants obtained in two types of cytoduction is statistically significant at $p < 0.05$.

Part 2. Crosses involving [*psi*$^{-}$ *rho*$^{+}$] donor strains

A: (1) K-5[*psi*$^{-}$*rho*$^{+}$] **red** x 74-D[*psi*$^{-}$*rho*^{0}] **white**		**A**: (2) K-5/w1[*psi*$^{-}$*rho*$^{+}$] **white** x 74-D[*psi*$^{-}$*rho*^{0}] **white**	
	IF		**IF**
[*psi*$^{-}$*rho*$^{+}$] **white cytoductants**	**100 %**	[*psi*$^{-}$*rho*$^{+}$] **white cytoductants**	**156 %**
Donor parent K-5 **red**	**100 %**	Donor parent K-5/w1 **white**	**187 %**
B: (1) 168t/r1 [*psi*$^{-}$*rho*$^{+}$] **red** x K-5/w1 [*psi*$^{-}$*rho*^{0}] **white**		**B**: (2) 168t [*psi*$^{-}$*rho*$^{+}$] **white** x K-5/w1 [*psi*$^{-}$*rho*^{0}] **white**	
	IF		**IF**
[*psi*$^{-}$*rho*$^{+}$] **white cytoductants**	**100 %**	[*psi*$^{-}$*rho*$^{+}$] **white cytoductants**	**137 %**
Donor parent 168t/r1 **red**	**100 %**	Donor parent 168t **white**	**449 %**

A: The difference between the IF of cytoductants obtained in two types of cytoduction is statistically significant at $p < 0.01$. B: The difference between the IF of cytoductants obtained in two types of cytoduction is statistically significant at $p < 0.05$.

Cytoductants were selected as [PSI^+] haploid heterokaryotic clones with recipient strain nuclear markers exhibiting suppression of a nonsense mutation in *ADE1*. The mean IF values in six cytoductants that have obtained cytoplasm from white donor parental strain exceeds the IF of cytoductants with cytoplasm of the red isogenic donor strain taken as 100%.

The color of the cytoductants was slightly pink, showing no differences due to the recipient strain marker ade1-14 suppression by transmitted [PSI^+] factors. The lack of difference in color between the [PSI^+] cytoductants suggests that the excess of Thioflavine T staining material transmitted by white strains is possibly due to the presence of inheritable prions other than [PSI^+] in these strains.

To examine the inheritance of amyloid amount in white [psi^-] strains, we used classical protocol when the donor and recipient strains differed by their [*rho*] factor (Table 3, Part 2). In such experiments, cytoductants were selected by acquisition of [rho^+], undamaged mitochondrial DNA providing respiratory competence.

The data obtained in two independent crosses (A and B) are presented in Table 3, part 2. In both types of crosses, the recipient strains 74-D/Gu1 (A) and K-5 (B) have no mitochondrial DNA and do not accumulate red pigment.

One can see that the mean IF values for eight cytoductants that received the [rho^+] cytoplasm of red donor strain are unambiguously below the mean IF values for cytoductants which obtained the cytoplasm from isogenic white strain. This demonstrates that the excess of Thioflavine T staining material can be transmitted by white strains in absence of the [PSI^+] factor and, quite likely, is due to the contribution of anonymous prions.

Sup35p Prionization Differs in Red and White Strains According to Blot Hybridization

In experiments with independent [PSI^+] red/white strain pairs, we used Sup35p antibodies and compared the amount of Sup35p polymers in their lysate pellets (Nevzglyadova et al., 2011).

Pellet proteins were separated in 1.5% agarose gel and transferred to nitrocellulose membranes according to (Bagriantsev et al., 2006). Figure 3A demonstrates data obtained with adenine prototroph K-13w (lane 3) and with *ade2* mutant of the same strain, K-13/r3 (lane 1). One can notice the difference in response of the original white strain and its red derivatives to the antibodies. We also compared red strains growing on standard medium YEPD and on YEPD-ADE blocking the red pigment accumulation. Comparing lanes 1/2, one can notice the more pronounced Sup35p polymer signal for white cells.

We made an attempt to find out if the differences observed here are due to the different expression of Sup35p in red and white cells. Figure 3B shows the results of blot-hybridization of Sup35p antibodies with PAGE-separated proteins of lysates normalized for their protein content and monomerized by boiling (lanes 1-2) and with pellet fractions obtained by centrifugation of the lysates (lanes 3-4). Lanes 1-2 demonstrate that the synthesis of Sup35p in red strains and in strains grown on high concentration of adenine (and, as a consequence, lacking red pigment) displays no differences. At the same time the amount of Sup35p aggregated and found in sediment fraction is clearly higher in white strains than in red ones (compare lanes 4/3).

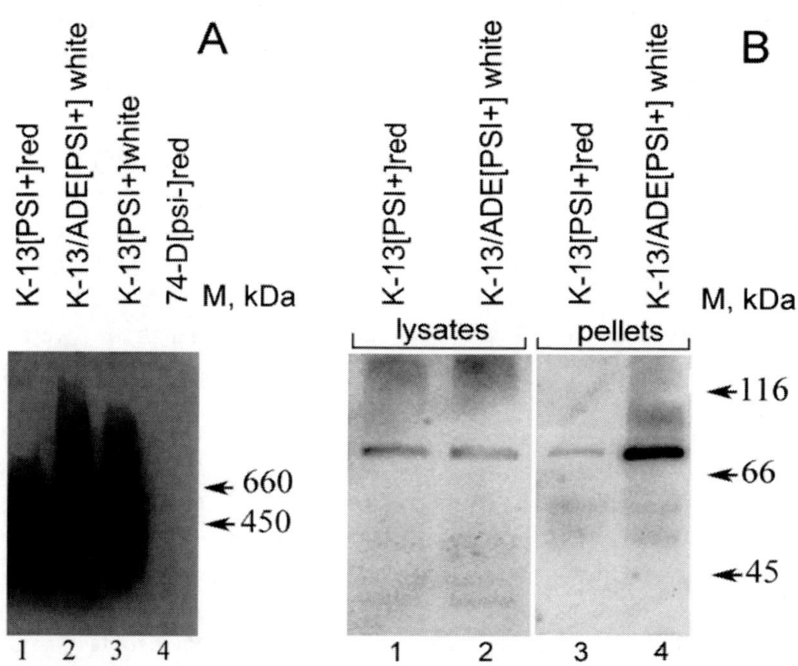

Figure 3 A. Blot hybridization of pellet proteins of [PSI^+] adenine prototrophic strain K13 and its red mutant separated in 1.5 % agarose gel and challenged with Sup35p antibodies. Lane 4 serves as a negative control. B. Blot hybridization of lysate and pellet proteins of [PSI^+] adenine prototrophs and their red mutants separated in 10 % PAGE and challenged with Sup35p antibodies. ADE denotes strains grown on the high adenine (YEPD-ADE) medium.

So, the data shown in Figure 3B provide evidence favoring different amounts of prionized Sup35p in red and white derivatives of [PSI^+] strains. At the same time, the agarose gel data in Figure 3A suggest that red pigment presence leads to shorter Sup35 polymers.

Characteristics of the Red Pigment and Its Two Forms

The red pigment is a polymer of AIR (aminoimidazole) with a few amino acids attached (Smirnov et al., 1967). The polymerization pathway is supposed to be as follows. AIR is conjugated with glutathione and moved to the vacuole. Such transfer to the vacuole seems to be necessary for AIR polymerization. The pigment in the vacuole functions as an endogenous fluorophore and is often used in studies of vacuolar morphogenesis (Weisman et al., 1987; Wada et al., 1992; Chauduri et al., 1996).

The size of the pigment polymer was estimated to be less than 10 kDa. The chemical nature of amino acid attachment was not determined. Nevertheless, it was shown that even after several rounds of deproteinization the pigment still contains several amino acids – glycine, glutamine (or glutamic acid) and asparagine (aspartic acid) being the major ones among them (Smirnov et al., 1967).

We (Nevzglyadova et al., 2011) determined the molecular weight and amino acid composition of our pigment preparation as well as its fluorescence spectrum. Yeast cells carrying mutations in the ADE1 gene were used as donors of the red pigment. We determined the amount of amino acids in three independent samples of red pigment, using HPLC. In all cases studied, the concentration of asparagines (or aspartic acid), glutamine (or glutamic acid) and glycin was higher than for other ones: 12.2, 27.5 and 16.4 % respectively, of the sum total of all amino acids. These numbers are in close agreement with the data of (Smirnov et al., 1967): 15.6, 33.9 and 21.9. We also got the fluorescence spectrum of a sample of red pigment. The maximum of fluorescence at ca. 520 nm agrees well with the existing data on the fluorofore accumulated in ade2 mutants (Weisman et al., 1987).

We obtained two variants of red pigment differing for molecular weight by a factor of 5-10. The native version of the pigment (P1) was purified from yeast cells by a modified method of Smirnov. TOF-MALDI analysis of the obtained compound demonstrates that it is composed of polymers with M.W. 1-10 kDa. The acid hydrolysis of P1 permitted to get its derivative P2 composed of polymers in range from 0.3-2 kDa according to the TOF-MALDI data (Nevzglyadova et al., 2011, Mikhailova et al., 2011).

In our experiments we used different concentrations of pigment, estimated by OD_{490}. OD_{490}=4.5 corresponds to 20 mg/ml.

Red Pigment Action on Yeast Crude Lysate Pellet Protein Aggregates

We studied the action of purified pigment on ethanol- precipitated aggregates of lysate pellet proteins of two different Ade^+ [*PSI^+*] strains (Nevzglyadova et al., 2011).

We demonstrated that red pigment addition decreases intensity of Thioflavine T fluorescence bound to aggregated proteins.

During the process of removing the pigment by washing, a large amount of amyloid material is solubilized. The same is true for the red pigment, as washed pellets are discolored. Results with two independent cultures (Nevzglyadova et al., 2011) demonstrate that fluorescence curves of Thioflavine T bound to the red pigment treated and untreated pellet proteins grow rather similarly after washing. The effectiveness of washing in these experiments is obviously high.

Studying red pigment action in model experiments (illustrated by Figure 4), demands additional control to ensure that the observed decrease in Thioflavine T fluorescence is not due to light absorbance by the red pigment. The absorption spectrum of Thioflavine T in presence of amyloid fibrils treated and untreated with the red pigment is shown in the insert.

One can see that the peaks of absorption curves are localized in the region of 350-419 nm but no significant absorption due to the red pigment is registered in the regions of Thioflavine T exciting (450 nm) or emission (486 nm). This result shows that the red pigment effect on amyloid bound Thioflavine T fluorescence is most likely not related to absorption properties of the pigment.

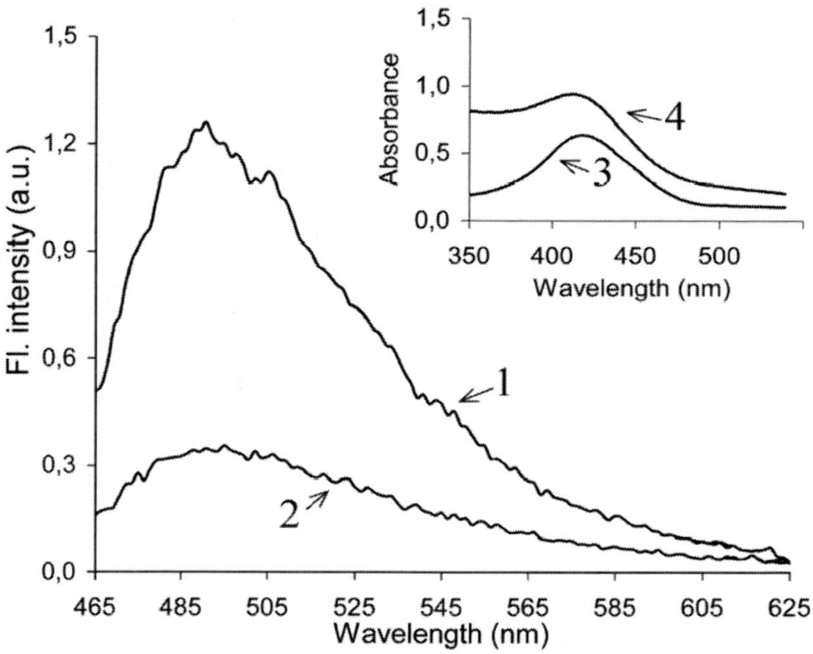

Figure 4. Effect of the red pigment on spectral properties of Thioflavine T in the presence of lysate pellet proteins of [*PSI*$^+$] strain GT26. Fluorescence spectra of Thioflavine T in the presence of untreated pellet proteins (1), pellet proteins treated with red pigment (2). Insert. Absorption spectra of Thioflavine T in the presence of lysate pellet proteins (3) and the same treated with red pigment (4).

Model Experiments with Insulin and Lysozyme Fibrils

Insulin was selected as a low molecular weight protein capable of the amyloid formation. Amyloid deposits of insulin have been observed in vivo in patients with diabetes after repeated subcutaneous insulin infusion and in normal aging (Brange et al., 1997).

In solution at physiological pH the protein has α–helical structure and consists of hexamers, dimers, and monomers. At high concentration of insulin monomers and low pH insulin is able to form fibrils in vitro. Pathways of the insulin fibril formation are described by Ahmad et al. (2005) and Ivanova et al. (2009). Low pH at high concentration of insulin monomers according to these authors was used for amyloid fibril formation in our experiments.

Lysozyme is also able to give rise to amyloid fibrils (Canet et al., 1999). Fibril formation is stimulated by certain mutant variants of the protein giving rise to familial amyloidosis (Morozova-Roche et al. 2000). To form lysozyme fibrils we used a method of Vernaglia et al. (2004) applying a buffer with high concentration of guanidine hydrochloride at 50^0 C.

Red pigment action was studied by using the fibrillar insulin and lysozyme (Nevzglyadova et al., 2011; Mikhailova et al., 2011). Figure 5 demonstrates clear differences between fluorescence spectra of Thioflavine T bound to untreated insulin (Figure 5A) and lysozyme (Figure 5B) fibrils and to fibrils treated with red pigment. The treatment with both pigment variants—P1 and P2—leads to a considerable drop of fluorescence intensity. The inserts demonstrate that at the wavelength of maximal effect of red pigment on amyloid bound Thioflavine T fluorescence the absorption curves of Thioflavine T+amyloid and

Thioflavine T+amyloid+red pigment show practically no difference. There is obviously no difference between P1 and P2 action on IF.

At the same time, peaks of fluorescence spectra of Thioflavine T mixed with either the red pigment, insulin and lysozyme monomers or monomers incubated with pigment alone do not exceed 0.1 used arbitrary units and exhibit no significant difference (data not shown). The effect of red pigment is observed practically immediately after its addition and seems to reflect competition between the red pigment and Thioflavine T for some sites on amyloid fibrils.

The conclusion seems to be supported by concentration dependence of red pigment application. The degree of inhibition of amyloid-bound Thioflavine T fluorescence depends on red pigment concentration. The correlation is demonstrated on Figure 6 for both forms of the pigment.

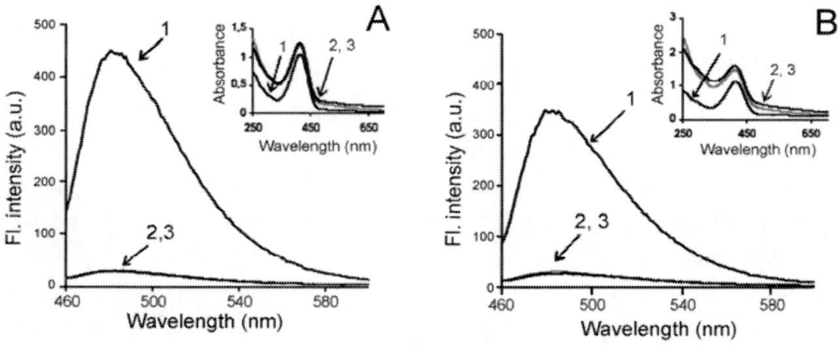

Figure 5. Effect of the red pigment on spectral properties of Thioflavine T in the presence of insulin and lysozyme fibrils. A. Fluorescence spectra of Thioflavine T bound to insulin fibrils (1), insulin fibrils treated with the red pigments P1 (2) and P2 (3). B. Fluorescence spectra of Thioflavine T bound to lysozyme fibrils (1), lysozyme fibrils treated with the red pigments P1 (2) and P2 (3). Inserts: Absorbance spectra of Thioflavine T bound to native insulin (A) and lysozyme (B) fibrils (1), fibrils treated with the red pigments P1 (2) and P2 (3).

Considerable drop of Thioflavine T fluorescence is apparent at pigment concentration 0.015 mg/ml for both P1 and P2. P1 seems to be more active than P2, but the tendency is not significant according to the Student's criterion. The data on Figure 6 apply to insulin, the corresponding data for lysozyme are quite similar (data not shown).

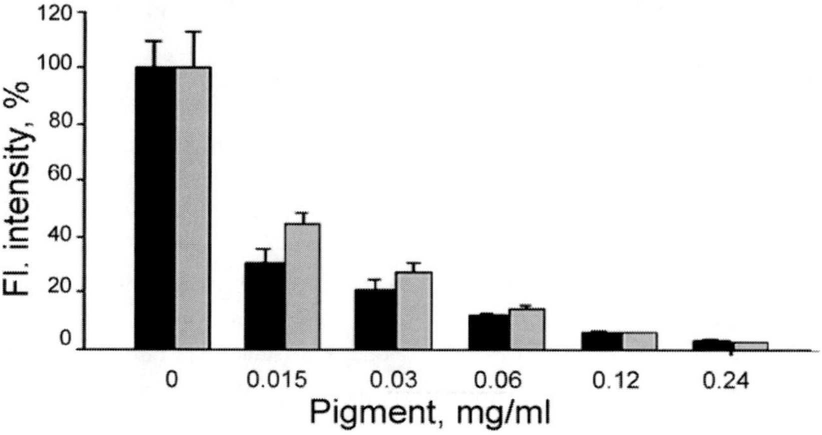

Figure 6. Influence of various concentrations of red pigment forms P1 (black) and P2 (grey) on fluorescence of Thioflavine T bound to insulin fibrils.

We also changed the order of addition of Thioflavin T and red pigment to the fibrils. Even in the case when P1 or P2 were added after the dye, the drop of amyloid bound Thioflavin T fluorescence was pronounced (usually more than tenfold). This seems to mean that red pigment is able to outcompete Thioflavin T even when the dye is already bound to the fibrils (Mikhailova et al., 2011).

To study the relative affinities of P1 and P2 further we performed experiments on washing the fibrils to remove the red pigment before Thioflavine T addition. Fibrils were harvested by centrifugation and supernatant containing red pigment and also polypeptide oligomers was discarded. Figures 7A and B demonstrate that the IF curve obtained after washing insulin and lysozyme fibrils containing P1 is higher than that for the P2 containing sample. This seems to be related to different affinity of the two red pigment forms with fibrils. P1 is likely to be more easily washed out and replaced by Thioflavine T than P2, resulting in the difference of the curves.

The next question we tried to answer was if the pigment derivatives effect the growth of fibrils. The increase of optical density (light scattering) characterizes fibril growth. We determined the difference of optical density at the beginning and end of growth: OD_{280} at zero (0) time and after 16 hours. The growth of fibrils was estimated as (OD16 – OD0)/OD16. The values obtained are shown in Figure 8.

The results demonstrate that both variants of the red pigment inhibit growth of insulin and lysozyme fibrils. We found no evidence for gross differences between influence of P1 and P2 on fibril growth.

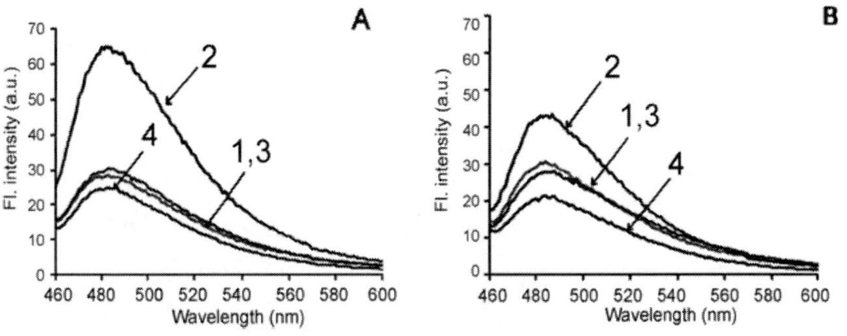

Figure 7. Effect of the red pigment forms P1 and P2 on spectral properties of the fluorescence of Thioflavine T bound to insulin (A) and lysozyme (B) fibrils. Fibrils treated with P1 (1), fibrils treated with P1 followed by washing (2), fibrils treated with P2 (3), fibrils treated with P2 followed by washing (4).

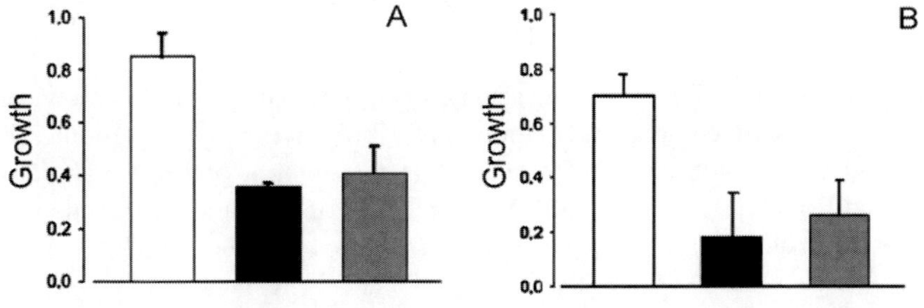

Figure 8. Fibril growth of insulin (A) and lysozyme (B) with or without red pigments P1 and P2. Growth of amyloid fibrils in control (white), in presence of P1 (black) and P2 (grey).

The other approach to study fibril growth is to register the IF of Thioflavine T bound to fibrils. Using this kind of data for characterization of fibril growth inhibition by red pigment leads to some problems. Red pigment addition leads to an immediate drop of amyloid bound Thioflavine T fluorescence.

To tackle the problem we measured not only the fluorescence of fibrils grown in the presence of the pigment, but also the fluorescence of control fibrils after the post-incubation addition of the same quantity of pigment. We expect that if the growth in the presence of pigment does not affect the length or amount of fibrils the fluorescence data in both cases will not differ. They will depend only on competition between Thioflavine T and pigment molecules for some unspecified sites on amyloid molecules. But this is clearly not the case.

In more than 10 independent experiments the growth in presence of various amounts of P1 and P2 (OD_{490} 2.5-5.0) leads to a more pronounced decrease in IF than in the case of post-incubation addition of the pigment. This is illustrated on Figure 9, presenting the data of one typical experiment with lysozyme fibrils. The data with insulin fibrils (Mikhailova et al., 2011) are quite similar. The results suggest that both forms of pigments inhibit fibril formation.

The data obtained in a study of lysozyme fibrils mobility in agarose gels seem to support our conclusions on red pigment action. The lysozyme fibrils were grown in a buffer in the absence or in the presence of red pigment variants P1 and P2 (OD_{450} 5) as well as in two times diluted pigment solutions. The aliquots were centrifuged, washed, dissolved in TAE buffer with 3 % SDS and 0.1 M DTT and loaded to 0.9 % agarose gel.

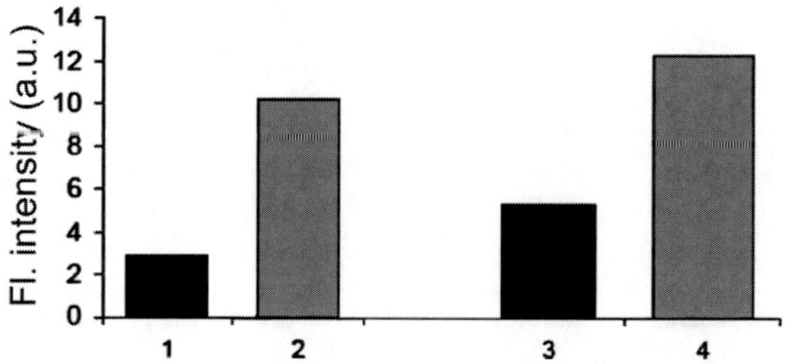

Figure 9. Intensity of fluorescence of Thioflavine T bound to lysozyme fibrils treated with red pigments according to different protocols. Fibrils are grown in presence of P1 (1), fibrils are grown without pigment and then are treated with P1 (2), fibrils are grown in presence of P2 (3), fibrils are grown without pigment and then are treated with P2 (4).

Figure 10 shows that fibrils were dissociated to state of oligomers of M.W. close to 70 kDa. This allows us to compare the amount of fibril material after Coomassie staining (without a need to use blotting). One can see that the amount of fibril material in control exceeds those after P1 (cf. lanes 1 and 2) and P2 (cf. lanes 1 and 4) treatments. At lower pigment level the differences are less pronounced (cf. lanes 1and 3, 1 and 5).

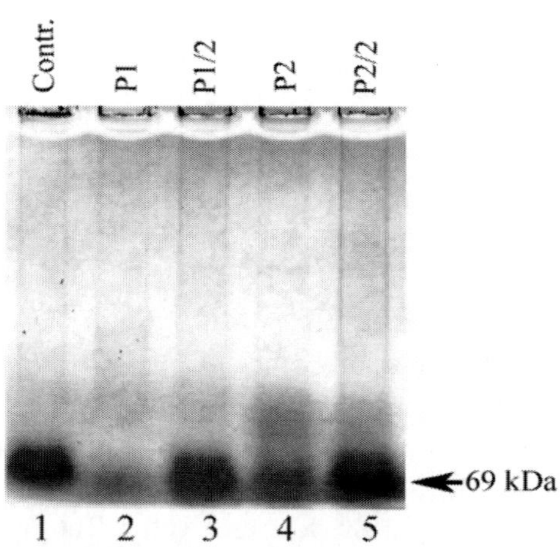

Figure 10. The electrophoregram of lysozyme fibrils in 0.9 % agarose gel. Control fibrils grown in absence of the red pigment (1), fibrils grown in the presence of P1 (2), fibrils grown in the presence of P1 diluted twofold (3), fibrils grown in the presence of P2 (4), fibrils grown in the presence of P2 diluted twofold (5).

Electron Microscopy

In these experiments (Mikhailova et al., 2011) we visualized the inhibition of amyloid formation by growing insulin and lysozyme fibrils in the presence of red pigment. The normalized material of the grown fibrils at magnification x 8000 demonstrates the red pigment effect on fibril growth: a pronounced decrease in amount of the mature fibrils of insulin (Figure 11, compare B and C with A) and the same of lysosyme Figure 12.

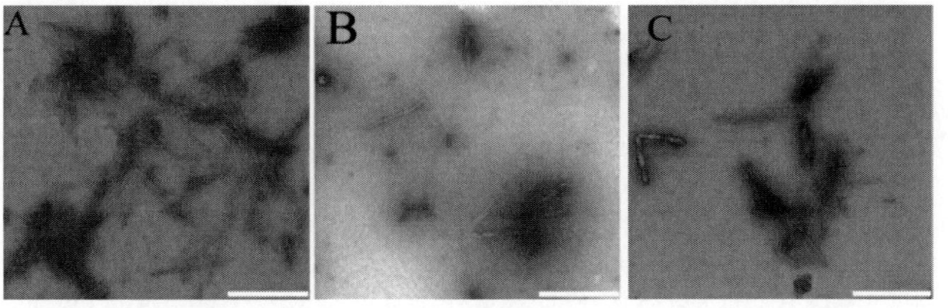

Figure 11. Negative-stain EM images of insulin fibrils grown without red pigment (A) and in the presence of P1 (B) and P2 (C). Scale bar in all images represents 1 mcm. x 8000.

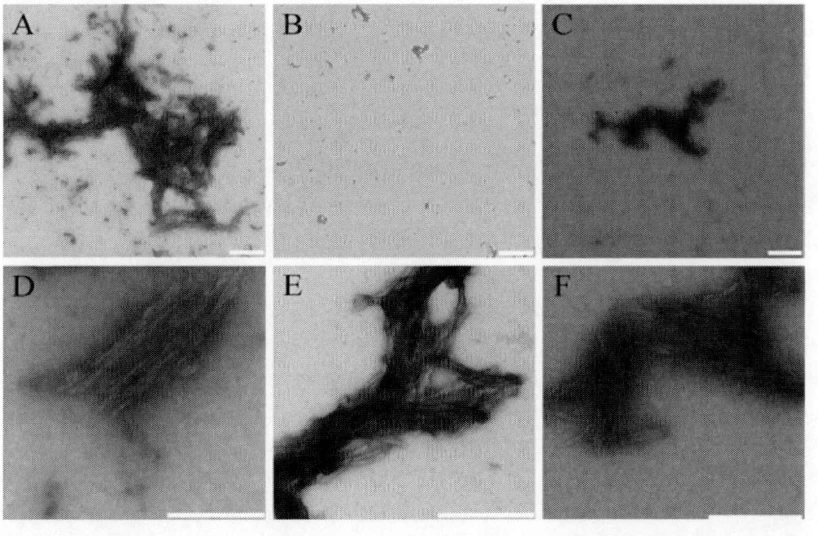

Figure 12. Negative-stain EM images of lysozyme fibrils grown without red pigment (A,D) and in the presence of P1 (B,E) and P2 (C,F) at different magnifications. A,B,C x 8000; D,E,F x 20000. Scale bar in all images represents 500 nm.

Comparison of insulin fibrils between control growth without pigment (A) and growth in presence of different forms of the pigment (B and C) permitted us to demonstrate that when P1 leads to clearcut shortening of the fibrils, P2 has smaller effect.

To look for the difference in length between lysozyme fibrils grown with and without pigment at magnification of x 20000 we used the material of pellet after centrifugation.

In the case of lysozyme P1 and P2 influence only the amount of the fibrils formed. P1 is more effective than P2.

CONCLUSION

Our study of white and red yeast isogenic strains permits us to conclude that the presence of the red pigment leads to mitotically stable and transmittable decreased amyloid level. We suppose that in living cells the mechanism of the decrease is mostly indirect. This is confirmed by comparison of prion-dependent and pigment-dependent proteins.

The 2D electrophoretic comparison of pellet proteins of red and white isogenic strains, followed by MALDI, allowed identification of 48 pigment-dependent proteins. These proteins mostly belong to functional classes of chaperones, proteins involved in glucose metabolism, and proteins involved in stress response. They correspond rather closely to prion-dependent proteins characterized in our previous work. Also present were proteins involved in translation, and in proteolysis.

The overlap can be best explained if seemingly pigment-dependent proteins are amyloid-dependent. We suppose that the affinity to amyloid (or to other amyloid-associated proteins) mobilizes them to pellet fraction of yeast crude lysates. The red pigment decreases amyloid accumulation and/or blocks some sites on amyloids, impairing association between the identified proteins and amyloids. As a result, the presence of these proteins in the pellet fraction depends on accumulation of the red pigment. Their amount is lower in the red than in the white cells.

The level of amyloid in cytoductants carrying additional cytoplasm derived from white donor cells is increased when compared with the case of a red-colored donor. This means that the pigment partly cures red donor cells of transmissible amyloids. The result makes sense in the light of our hypothesis on red pigment action via binding amyloid fibrils. As the red pigment block some sites on the polymeric prionizable protein molecule, this can prevent its breaking to seeds, which are needed for efficient intracellular propagation of prions and for prion transmission in cytoduction. According to this hypothesis the red pigment binding to amyloid fibrils can cause a decrease of amyloid levels indirectly, via impeding prion polymer contacts with chaperones which lead to generation of seeds (propagons) needed for prion spreading (Chernoff et al., 1995; Jung et al., 2000; Jones and Masison, 2003; Kryndushkin et al., 2003; Song et al., 2005; Rikhvanov et al., 2007). Indeed, Table 2 demonstrates that the presence of 8 chaperones and co-chaperones are red pigment dependent. This seems to support our hypothesis.

At the same time, we have demonstrated by hybridization on agarose gels (Figure 3) that red pigment leads to clearly smaller aggregates of Sup35p. This effect is possibly caused by direct interaction of the red pigment with amyloid fibers.

This raises an important question how the red pigment, which is localized in the vacuole, modifies amyloid molecules localized in the cytoplasm. Recently, an interesting possibility to explain this "anomaly" has emerged. Tyedmers et al. (2010a) demonstrated that different forms (dots and rings) of the aggregates of Sup35p prion domain fused to GFP are localized at the perivacuolar insoluble protein deposit (IPOD). This site was involved in de novo prion induction. These data suggest how the red pigment localized in the vacuole can possibly influence prion "life cycle". Since many, if not all, prion amyloids are connected with this site (Kaganovich et al., 2008, Tyedmers et al., 2010a, 2010b) in yeast, and with a related site in mammals, the effects akin to the red pigment described here may turn out to be quite universal. We would like to add that in spite of being apparently confined to vacuoles

(Chauduri et al., 1996; Wada et al., 1992; Weisman et al., 1987) the red pigment was demonstrated to have another potential physiological effect not limited to vacuole—modification of DNA cleavage (Meskauskas et al., 1985).

Some direct effects of red pigment were shown in vitro using insulin and lysozyme fibrils. Both forms of the red pigment (native P1) and hydrolyzed P2 can bind amyloid efficiently competing with Thioflavine T. The red pigment is able to outcompete Thioflavin T even when the dye is already bound to the fibrils.

To study the relative affinities of P1 and P2 further we performed experiments on washing the fibrils to remove the red pigment before Thioflavine T addition. P1 is likely to be more easily washed out and replaced by Thioflavine T than P2, possibly due to lower affinity to amyloid fibrils.

Most important is the effect of red pigment on fibril formation. Both forms of the red pigment lead to significant decrease of amyloid fibril growth.

We suppose that the red pigment of baker's yeast, an organism utilized in human foods for several thousand years, can possibly pretend for a position of one of future anti-amyloidic therapeutic compounds. This possibility will be studied further.

ACKNOWLEDGMENTS

We thank Dr. L. Mironova, Dr. Y. Chernoff, Dr. A.Zekhnov and Dr. G. Zhouravleva for consultations and for donating plasmids and strains. This work was supported by a grant of the Russian Foundation for Basic Research (project 09-04-00750-a) and Program "Molecular and Cell Biology" by the Russian Academy of Sciences (2008-2012). The work was carried out in compliance with the current Russian Federation laws regulating genetic experimentation.

REFERENCES

Adjou, K.T., Simoneau, S., Sales, N., Lamoury, F., Dormont, D., Papy-Garcia, D., Barritault, D., Deslys, J.-P., Lasmezas, C.I. (2003). A novel generation of heparan sulfate mimetics for the treatment of prion diseases. *J. Gen. Virology 8,* 2595-2603.

Ahmad, A., Uversky, V.N., Hong, D., Fink, A.L. (2005). Early events in the fibrillation of monomeric insulin. *J. Biol. Chem. 280,* 42669–42675.

Bagriantsev, S.N., Kushnirov, V.V., Liebman, S,W. (2006). Analysis of amyloid aggregates using agarose gel electrophoresis. *Methods Enzymol. 412*, 33-48.

Bagriantsev, S. N., Gracheva, E. O., Richmond, J. E. and Liebman, S. W. (2008). Variant-specific [PSI^+] infection is transmitted by Sup35 polymers within [PSI^+] aggregates with heterogeneous protein composition. *Mol. Biol. Cell*, *19*, 2433-2443.

Banci, L., Bertini, I., Boca, M., Girotto, S., Martinelli, M., Valentine, J.S., Vieru, M. (2008). SOD1 and amyotrophic lateral sclerosis: mutations and oligomerization. *PLoS ONE 3*, e1677.

Bate, C., Salmona, M., Diomede, L., Williams, A. (2004). Squalestatin Cures Prion-infected Neurons and Protects Against Prion Neurotoxicity. *J. Biol. Chem. 279,* 14983–14990.

Bieschke, J., Russ, J., Friedrich, R.P., Ehrnhoefer, D.E., Wobst, H., Neugebauer, K., Wanker, E.E.. (2010). EGCG remodels mature α-synuclein and amyloid-β fibrils and reduces cellular toxicity. *Proc. Natl. Acad. Sci. USA 107:* 7710–7715.

Brange, J., Andersen, L., Laursen E.D., Meyn G., Rasmussen E. (1997) Toward understanding insulin fibrillation. *J. Pharm. Sci.* 86:517–525.

Canet, D., Sunde, M., Last, A.M., Miranker, A., Spencer, A., Robinson, C.V., Dobson, C.M. (1999) Mechanistic studies of the folding of human lysozyme and the origin of amyloidogenic behavior in its disease-related variants. *Biochemistry 38,* 6419–6427

Charveriat, M., Reboul, M., Wang, Q., le Picoli, C., Lenuzza, N., Montagnac, A., Nhiri, N., Jacquet, E., Gueritte, F., Lallemand, J-Y., Deslys, J.-P., Mouthon, F. 2009. New inhibitors of prion replication that target the amyloid precursor. *J. Gen. Virology 90,* 1294–1301.

Chauduri, B., Ingavale, S. and Bachhawat, A. K. (1996). apd1$^+$, a gene required for red pigment formation in ade6 mutants of Schizosaccharomyces pombe, encodes an enzyme required for glutathione biosynthesis: A role for glutathione and a glutathione-conjugate pump. *Genetics*, *145*, 75-83.

Chernoff, Y. O. (2004). Cellular control of prion formation and propagation in yeast. In: *Prions and prion diseases: current perspectives,* Telling, G. C. (ed.), 257-303. Horizon Bioscience: Wymondham.

Chernoff Y.O., Ed. (2007). *Protein-based inheritance.* Austin, Texas: Landes Bioscience. 137 p.

Chernoff, Y.O., Lindquist, S.L., Ono, B., Inge-Vechtomov, S.G., Liebman, S.W. (1995). Role of the chaperone protein hsp104 in propagation of the yeast prion-like factor [*PSI*$^+$]. *Science 268***,** 880-884.

Cordeiro,Y., Lima, L.M.T.R., Gomes, M.P.B., Foguel, D., Silva, J.L. (2004). Modulation of prion protein oligomerization, aggregation, and β-sheet conversion by 4,4'-Dianilino-1,1'-binaphthyl-5,5'-sulfonate (bis-ANS). *J. Biol. Chem. 279*, 5346–5352.

Demaimay, R., Chesebro, B., Caughey, B. (2000). Inhibition of formation of proteaseresistant prion protein by Trypan Blue, Sirius Red, other Congo Red analogs. *Arch. Virol. Suppl. 16*, 277–83.

Elam, J.S., Taylor, A.B., Strange, R., Antonyuk, S., Doucette, P.A., Rodriguez, J.A., Hasnain, S.S., Hayward, L.J., Valentine, J.S., Yeates, T.O., Hart, P.J. (2003). Amyloid-like filaments and water-filled nanotubes formed by SOD1 mutant proteins linked to familial ALS. *Nature Struct. Biol. 10*, 461-467.

Erjavec, N., Larsson, L., Grantham, J., Nystrom, T. (2007). Accelerated aging and failure to segregate damaged proteins in Sir2 mutants can be suppressed by overproducing the protein aggregation-remodeling factor Hsp104p. *Genes. Dev. 21***,** 2410-2421.

Ernhoefer, D.E., Biesche, J., Boeddrich, A., Herbst, M., Masino, L., Lurz, R., Engemann, S., Pastore, A., Wanker, E.E. (2008). EGCG redirects amyloidogenic polypeptides into unstructured, off-pathway oligomers. *Nature Struct. Mol. Biol. 15*, 558–566.

Ghaemmaghami, S., May, B.C.H., Renslo, A.R., Prusiner, S.B. (2010). Discovery of 2-Aminothiazoles as potent antiprion compounds. *J. Virology 84,* 3408–3412.

Herczenik, E., Gebbink, M.F.B.G. (2008). Molecular and cellular aspects of protein misfolding and disease. *FASEB J. 22*, 2115–2133.

Hosokawa-Muto, J., Kamatari, Y.O., Nakamura, H.K., Kuwata, K. (2009). Variety of antiprion compounds discovered through an in silico screen based on cellular-form prion protein structure: correlation between antiprion activity and binding affinity. *Antimicrobial agents and chemotherapy 53*, 765–771.

Ivanova, M.I., Sievers, S.A., Sawaya, M.R., Wall, J.S., Eisenberg, D. (2009). Molecular basis for insulin fibril assembly. *Proc. Natl. Acad. Sci. USA 106*, 18990-18995.

Jones, G.W. and Masison, D.C. (2003). Saccharomyces cerevisiae Hsp70 mutations affect [PSI+] prion propagation and cell growth differently and implicate Hsp40 and tetratricopeptide repeat cochaperones in impairment of [PSI+]. *Genetics*, *163*, 495-506.

Jung, G., Jones, G., Wegrzyn, R. D. and Masison, D.C. (2000). A role for cytosolic Hsp70 in yeast [PSI^+] prion propagation and [PSI^+] as a cellular stress. *Genetics*, *156*, 559-570.

Kaganovich, D., Kopito, R., Frydman, J. (2008). Misfolded proteins partition between two distinct quality control compartments. *Nature 454*, 1088-1095.

Kocisko, D.A., Baron, G.S., Rubenstein, R., Chen, J., Kuizon, S., Caughey, B. (2003). New inhibitors of scrapie-associated prion protein formation in a library of 2,000 drugs and natural products. *J. Virology 77*, 10288–10294.

Korth, C., May, B.C., Cohen, F.E., Prusiner, S.B. (2001). Acridine and phenothiazine derivatives as pharmacotherapeutics for prion disease. *Proc. Natl. Acad. Sci. USA 98*, 9836–41.

Kryndushkin, D. S., Alexandrov, I. M., Ter-Avanesyan, M. D. and Kushnirov, V. V. (2003). Yeast [PSI^+] protein aggregates are formed by small Sup35 polymers fragmented by Hsp104. *J. Biol. Chem.*, *278*, 49636-49643.

Kushnirov, V. V., Vishnevskaya, A. B., Alexandrov, A. M. and Ter-Avanesyan, M. D. (2007). Prion and nonprion amyloids: A comparison inspired by the yeast Sup35 protein. In *Protein-based inheritance*, Chernoff, Y. O. (ed), 73-82. Landes Bioscience: Austin, Texas.

Lamberto, G.R., Binolfi, A., Orcellet, M.L., Bertoncini, C.W., Zweckstetter, M., Griesinger, C., Fernandez, C.O. (2009). Structural and mechanistic basis behind the inhibitory interaction of PcTS on a-synuclein amyloid fibril formation. *Proc. Natl. Acad. Sci. USA 106,* 21057–21062.

Lawson, V.A., Lumicisi, B., Welton, J., Machalek, D., Gouramanis, K., Klemm, H.M., Stewart, J.D., Masters, C.L., Hoke, D.E., Collins, S.J., Hill, A.F. (2010). Glycosaminoglycan Sulphation Affects the Seeded Misfolding of a Mutant Prion Protein. *PLoS One 5 (8)*, e12351.

Meskauskas, A., Ksenzenko, V., Shlyapnikov, M., Kryukov, V., Citavicius, D. (1985). 'Red pigment' from ADE-2 mutants of S. cerevisiae prevents DNA cleavage by restriction endonucleases. *FEBS Letters 182*, 413-414.

Mikhailova, E.V., Artemov, A.V., Snigirevskaya, E.S., Artamonova, T.O., Khodorkovskii, M. A., Soidla, T.R., Nevzglyadova, O.V. (2011) The effect of red pigment on insulin fibril formation in vitro. *Tsitologiya 53 (10)*, 52–58.Morozova-Roche, L.A., Zurdo, J., Spencer, A., Noppe, W., Receveur, V., Archer, D.B., Joniau, M., Dobson, C.M. (2000). Amyloid fibril formation and seeding by wild-type human lysozyme and its disease-related mutational variants. *J. Struct. Biol. 130*, 339-51.

Munch, C., O'Brien, J., Bertolotti, A. (2011). Prion-like propagation of mutant superoxide dismutase-1 misfolding in neuronal cells. *Proc. Natl. Acad. Sci. USA 108*, 3548–3553.

Nevzglyadova, O.V., Kuznetsova, I.M., Artemov, A.V., Mikhailova, E.V., Turoverov, K.K., Soidla, T.R. (2008). Comparative assay of amyloid and prion contents in yeast cells. *Cell. Tissue Biology 2*, 71-80. (Translated from *Tsitologiya 50*, 40-48.).

Nevzglyadova, O.V., Artemov, A.V., Mittenberg, A.G., Solovyov, K.V., Kostyleva, E.I., Mikhailova, E.V., Kuznetsova, I.M., Turoverov, K.K., Soidla, T.R. (2009). Prion-associated proteins in yeast: comparative analysis of isogenic [PSI$^+$] and [psi$^-$] strains. *Yeast 26*, 611–631.

Nevzglyadova, O.V., Artemov, A.V., Mittenberg, A.G., Kostyleva, E.I., Mikhailova, E.V., Solovev, K.V., Kuznetsova, I.M., Turoverov, K.K., Soidla, T.R. (2010a). Comparison of crude lysate pellets of isogenic strains of yeast with different prion composition: Identification of prion-associated proteins. *Cell. Tissue Biology 4*, 36 53. (Translated from *Tsitologiya 52,* 63–79.)

Nevzglyadova, O.V., Artemov, A.V., Mittenberg, A.G., Mikhailova, E.V., Kuznetsova, I.M., Turoverov, K.K., Soidla, T.R. (2010b). The effect of red pigment on amyloidization of yeast proteins. *Cell. Tissue Biology 4,* 152-166. (Translated from *Tsitologiya 52*, 80–93.)

Nevzglyadova, O.V., Artemov, A.V., Mittenberg, A.G., Mikhailova, E.V., Kuznetsova, I.M., Turoverov, K.K., Soidla, T.R. (2010c). *Chapter 6. Yeast Protein Aggregates, Containing Chaperones and Glucose Metabolism Enzymes. In Handbook of Molecular Chaperones: Roles, Structures and Mechanisms, Durante P and Colucci L. (eds).* Nova Science Publishers: New York, USA; 241-270.

Nevzglyadova, O.V., Kuznetsova, I.M., Mikhailova, E.V., Artamonova, T.O., Artemov, A.V., Mittenberg, A.G., Kostyleva, E.I., Turoverov, K.K., Khodorkovskii, M.A., Soidla, T.R. (2011). The effect of red pigment on amyloidization of yeast proteins. *Yeast 28*, 505-526.Nunziante, M., Kehler, C., Maas, E., Kassack, M.U., Groschup, M., Schatz, H.M. (2005). Charged bipolar suramin derivatives induce aggregation of the prion protein at the cell surface and inhibit PrPSc replication . *J. Cell. Sci. 118*, 4959-4973.

Phuan, P.-W., Zorn, J.A, Safar, J., Giles, K., Prusiner, S.B., Cohen, F.E., May, B.C.H. (2007). Discriminating between cellular and misfolded prion protein by using affinity to 9-aminoacridine compounds. *J. Gen. Virol. 88*, 1392–1401.

Prusiner, S.B. (2001). Shattuck lecture - neurodegenerative diseases and prions. *N. Engl. J. Med. 344*, 1516–26.

Riemer, C., Burwinkel, M., Schwarz, A., Gultner, S., Mok, S.W.F., Heise, I., Holtkamp, N., Baier, M. (2008). Evaluation of drugs for treatment of prion infections of the central nervous system *J. Gen. Virol. 89*, 594–597.

Rikhvanov, E. G., Romanova, N. V. and Chernoff, Y. O. (2007). Chaperone effects on prion and nonprion aggregates. In *Protein-based inheritance*, Chernoff, Y. O. (ed), 83-92. Landes Bioscience: Austin, Texas.

Salnikova, A.B., Kryndushkin, D.S., Smirnov, V.N., Kushnirov, V.V., Ter-Avanesyan, M.D. (2005). Nonsense suppression in yeast cells overproducing sup35 (eRF3) is caused by its non-heritable amyloids. *J. Biol. Chem., 280,* 8808 - 8812.

Saupe, S.J. (2003). New anti-prion drugs make yeast blush. *Trends in Biotechnology 21 (12)*, 516-519.

Schiffer, N.W., Broadley, S.A., Hirschberger, T., Tavan, P., Kretzschmar, H.A., Giese, A., Haass, C., Hart, F.U., Schmid, B. (2007). Identification of anti-prion compounds as efficient inhibitors of polyglutamine protein aggregation in a zebrafish model. *J. Biol. Chem. 282*, 9195–9203.

Sherman, F., Fink, G. R. and Hicks, J. B. (1986). *Laboratory course manual for methods in yeast genetics*. Cold Spring Harbor Laboratory Press: *Cold Spring Harbor*, NY.

Smirnov, M.N., Smirnov, V.N., Budowsky, E.I., Inge-Vechtomov, S.G., Serebrjakov, N.G. (1967). Red pigment of adenine-deficient yeast *Saccharomyces cerevisiae*. *Biochem. Biophys. Res. Com. 27*, 299-304.

Solassol, J., Crozet, C.,Perrier, V., Leclaire, J., Beranger, F., Caminade, A-M., Meunier, B., Dormont, D., Majoral, J-P., Lehmann, S. (2004). Cationic phosphorus-containing dendrimers reduce prion replication both in cell culture and in mice infected with scrapie. *J. Gen. Virology 85*, 1791–1799.

Song, Y., Wu, Y. X., Jung, G., Tutar, Y., Eisenberg, E., Greene, L. E. and Masison, D. C. (2005). Role for Hsp70 chaperone in *Saccharomyces cerevisiae* prion seed replication. *Eukarotic Cell*, *4*, 289-297.

Teruya, K., Kawagoe, K., Kimura, T., Chen, C.-J., Sakasegawa, Y., Doh-ura, K. (2009). Amyloidophilic compounds for prion diseases. *Infectious Disorders — Drug Targets* 9, 15–22.

Trevitt C. R., Collinge J. (2006). A systematic review of prion therapeutics in experimental models. *Brain 129*, 2241–2265.

Tribouillard, D., Bach, S., Gug, F., Desban, N., Beringue, V., Andrieu, T., Dormont, D., Galons, H., Laude, H., Vilette, D., Blondel, M. (2006). Using budding yeast to screen for anti-prion drugs. *Biotechnol. J. 1*, 58-67.

Tribouillard-Tanvier, D., Beringue, V., Desban, N., Gug, F., Bach, S., Voisset, C., Galons, H., Laude, H., Vilette, D., Blondel, M. (2008). Antihypertensive drug guanabenz is active in vivo against both yeast and mammalian prions. *PLoS One 3 (4)*, e1981.

Tyedmers, J., Treusch, S., Dong, J., McCaffery, J.M., Bevis, B., Lindquist, S. (2010a). Prion induction involves an ancient system for the sequestration of aggregated proteins and heritable changes in prion fragmentation. *Proc. Natl. Acad. Sci. USA 107*, 8633-8538.

Tyedmers, J., Mogk, A., Bukau, B. (2010b). Cellular strategies for controlling protein aggregation. *Nature Rev. Mol. Cell. Biol. 11*, 777-788.

Uversky, V.N. (2008). Amyloidogenesis of natively unfolded proteins. *Curr. Alzheimer Res. 5*, 260–287.

Vernaglia, B.A., Huang, J., Clark, E.D. (2004). Guanidine hydrochloride can induce amyloid fibril formation from hen egg-white lysozyme. *Biomacromolecules 5*, 1362-70.

Wada, Y., Ohsumi, Y., Anraku, Y. (1992). Genes for directing vacuolar morphogenesis in *Saccharomyces cerevisiae*. I. Isolation and characterization of two classes of *vam* mutants. *J. Biol. Chem. 267*, 18665-18670.

Wang, H,. Duennwald, M.L., Roberts, B.E., Rozeboom, L.M., Zhang, Y.L., Steele, A.D., Krishnan, R., Su, L.J., Griffin, D., Mukhopadhyay, S., Hennessy, E.J., Weigele, P., Blanchard, B.J., King, J., Deniz, A.A., Buchwald, S.L., Ingram, V.M., Lindquist, S., Shorter, J. (2008). Direct and selective elimination of specific prions and amyloids by 4,5-dianilinophthalimide and analogs. *Proc. Natl. Acad. Sci. USA 105*, 7159–7164.

Wang, Y., Meriin, A.B., Costello, C.E., Sherman, M.Y. (2007). Characterization of proteins associated with polyglutamine aggregates: A novel approach towards isolation of aggregates from protein conformation disorders. *Prion 1*, 128–135.

Wang, Y., Meriin, A.B., Zaarur, N., Romanova, N.V., Chernoff, Y.O., Costello, C.E., Sherman, M.Y. (2009). Abnormal proteins can form aggresome in yeast: aggresome-targeting signals and components of the machinery. *FASEB J. 23*, 451– 463.

Webb, S., Lekishvili, T., Loeschner, C., Sellarajah, S., Prelli, F., Wisniewski, T., Gilbert, I.H., Brown, T.R. (2007). Mechanistic Insights into the Cure of Prion Disease by Novel Antiprion Compounds, *J. Virol. 81*, 10729–10741.

Weisman, L. S., Bacallao, R. and Wickner, W. (1987). Multiple methods of visualizing the yeast vacuole permit evaluation of its morphology and inheritance during the cell cycle. *The Journal of Cell Biology*, *105*, 1539-1547.

Wickner, R. B. (1994). [URE3] as an altered URE2 protein: Evidence for an prion analog in S.cerevisiae. *Science*, *264*, 566-569.

Wickner, R.B., Edskes, H.K., Shewmaker, F., Nakayashiki, T., Engel, A., McCann, L., Kryndushkin, D. (2007). Yeast prions: Evolution of the prion concept. *Prion 1,* 94-100.

Wickner, R.B., Edskes, H.K., Kryndushkin, D., McGlinchey, R., Bateman, D., Kelly, A. (2011). Prion diseases of yeast: Amyloid structure and biology. *Semin. Cell. Dev. Biol. 22, 469-475.* Wickner R.B., Shewmaker F., Edskes H., Kryndushkin D., Nemecek J., McGlinchey R., Bateman D., Winchester C-L. (2010). Prion amyloid structure explains templating: how proteins can be genes. *FEMS Yeast Res. 10,* 980-991.

Wong, C., Xiong, L.W., Horiuchi, M., Raymond, L., Wehrly, K., Chesebro, B., Caughey, B. (2001). Sulfated glycans and elevated temperature stimulate PrPSc-dependent cell free formation of protease-resistant prion protein. *EMBO J. 20*, 377–86.

Zakharov, I. A. and Yarovoy, B. (1977). Cytoduction as a new tool in studying the cytoplasmic heredity in yeast. *Mol. Cell. Biochem*, *14*, 15-18.

In: Amyloids
Editors: I. P. Halcheck et al.
ISBN: 978-1-62100-538-4

Chapter 2

AMYLOID PRECURSOR PROTEIN: ITS SIGNALING ASPECT AND ALZHEIMER'S DISEASE

Kohzo Nakayama*[*a], *Hisashi Nagase*[b], *Chang-Sung Koh*[c], and *Takeshi Ohkawara*[a]

[a]Department of Anatomy
[b]Department of Immunology and Infectious Diseases,
Shinshu University, School of Medicine
[c]Department of Biomedical Sciences, Shinshu University,
School of Health Sciences, Matsumoto, Nagano 390-8621, Japan

ABSTRACT

Although amyloid precursor protein (APP) plays central roles in Alzheimer's disease (AD), the physiological functions of this protein have not been fully elucidated. γ-Secretase was first identified as a protease that cleaves APP within the transmembrane domain and produces Aβ peptides, which are thought to play a role in the pathogenesis of AD and their presence is used as a hallmark of the AD brain. However, the physiological functions of this protease also remain to be clarified. In the canonical Notch signaling pathway, ligands bind to the extracellular domain of Notch on neighboring cells and trigger sequential proteolytic cleavage. Finally, the intracellular domain (ICD) of Notch is released from the cell membrane by γ-secretase and translocates to the nucleus where it modulates gene expression through binding to transcription factors. Therefore, γ-secretase regulates the Notch signaling pathway. Recently, it was demonstrated that many type 1 transmembrane proteins, including APP, are substrates for γ-secretase, and the ICDs of these substrates are released from the cell membrane by γ-secretase. These processes are very similar to those occurring in Notch signaling. Therefore, the common enzyme, γ-secretase, modulates proteolysis and the turnover of possible signaling molecules, suggesting that mechanisms similar to Notch signaling may contribute widely

* Corresponding author: Kohzo Nakayama, Department of Anatomy, Shinshu University, School of Medicine, Matsumoto, Nagano 390-8621, Japan, Telephone/Fax: +81-263-37-2594, E-mail: kohzona@shinshu-u.ac.jp. Present address: Department of Developmental and Regenerative Medicine, Mie University, Graduate School of Medicine, Tsu, Mie 514-8507, Japan.

to γ-secretase-regulated signaling pathways. Thus, it is likely that APP also has a signaling mechanism. Indeed, we have shown that the ICD of APP (AICD) markedly alters gene expression and induces neuron-specific apoptosis. These results suggest that APP signaling may be responsible for the onset of AD.

This chapter presents an outline of APP and γ-secretase. We also focus on the hypothesis thatγ-secretase-regulated mechanisms similar to Notch signaling may play a role in signaling events involving type 1 transmembrane proteins, including APP. Here, we discuss the possibility that APP signaling induces dynamic changes in gene expression, which may be closely correlated with AICD-induced neuron-specific apoptosis, leading to the onset of AD.

ABBREVIATIONS

ACE, angiotensin-converting enzyme;
AD, Alzheimer's disease;
APP, amyloid precursor protein;
AICD, the intracellular domain of APP;
APLP, APP like protein;
Aph-1, anterior pharynx defective-1;
bHLH, basic helix-loop-helix;
CAA, cerebral amyloid angiopathy;
C. elegans, Caenorhabditiselegans;
Dll, Delta-like protein;
Dll1IC, the intracellular domain of Dll1;
ECE, endothelin-converting enzyme;
FAD, familial AD;
GO, gene ontology,
Hes, hairy/enhancer of split;
ICD, intracellular domain;
IDE, insulin-degrading enzyme;
JM, juxtamembrane;
NCT, necastrin;
NICD, the intracellular domain of Notch;
RIP, the regulated intramembrane proteolysis;
NEP, Neprilysin;
PS, presenilin;
Pen-2, PS enhancer-2;
RA, all-*trans*-retinoic acid;
TM, transmembrane;

1. INTRODUCTION

Although amyloid precursor protein (APP) plays central roles in Alzheimer's disease (AD) [1,2], the physiological functions of this protein have not been fully elucidated [3].

Indeed, probably due to redundancy, APP knockout mice have relatively minor phenotypic effects [4], for example they show impaired long-term potentiation and memory loss without general neuron loss. On the other hand, γ-secretase was first identified as a protease that cleaves APP within the transmembrane domain (TM) and produces Aβ peptides [5], which are the main hallmarks of AD and are thought to be involved in the pathogenesis in the AD brains. However, the physiological functions of this protease also remain to be clarified [6,7].

The signaling hypothesis suggesting that the primary function of γ-secretase is to regulate the signaling of type 1 membrane proteins was proposed by the analysis of Notch signaling[8]. Notch is a family of evolutionarily conserved single transmembrane spanning proteins (type 1 transmembrane proteins; the amino terminus is extracellular and the carboxy terminus is cytoplasmic) that mediate the fates of numerous cells in both invertebrates and vertebrates [9,10]. Notch signaling mediates somitogenesis, lymphoid cell differentiation, and neural development [11]. For example, cells expressing the major ligand Delta inhibit the neural differentiation of neighboring Notch-expressing cells during neurogenesis [9,10]. Notch signaling has attracted a great deal of interest due to its relation to AD. *Caenorhabditiselegans*(*C. elegans*) sel-12, which appears to facilitate the reception of signals mediated by lin-12 (*C. elegans* Notch), was identified by screening for a suppressor of lin-12 gain-of-function mutation (12). Sel-12 is a counterpart of human presenilin (PS), which was identified by genetic linkage analyses as the gene responsible for several forms of early onset familial AD (FAD) [13-15], and is a catalytic component of γ-secretase [16]. Therefore, it was suggested that the Notch signaling pathway may be related to AD. Indeed, as described below, the Notch signaling pathway was shown to be regulated by γ-secretase-mediated proteolysis.

The Notch signaling pathway has a unique mechanism in that it is controlled by proteolytic cleavage reactions [17,18]. In the canonical Notch signaling pathway, ligands bind to the extracellular domain of Notch expressed on neighboring cells, and trigger sequential proteolytic cleavage. Finally, the intracellular domain (ICD) of Notch (NICD) is released from the cell membrane by γ-secretase and translocates into the nucleus where it modulates gene expression through binding to transcription factors. Thus, γ-secretase plays a central regulatory role in Notch signaling.

As mentioned above, the physiological functions of γ-secretase have yet to be determined. More than 60 type 1 transmembrane proteins have been reported as substrates for γ-secretase [19] and their ICDs are also released from the cell membrane [8,20]. These processes are very similar to those involved in Notch signaling. Therefore, these observations indicating that the common enzyme γ-secretase modulates the proteolysis and turnover of putative signaling molecules suggested that mechanisms similar to the Notch signaling pathway may contribute toγ-secretase-regulated signaling pathways [8,21,22].

Interestingly, ICD of APP (AICD), which is released from the cell membrane by γ-secretase, has also been reported to undergo translocation to the nucleus [23-25] and may function as a transcriptional regulator [26,27]. Several AICD-interacting proteins, which are thought to regulate AICD stability and cellular localization, have been identified to date [3,28], and a number of models of APP signaling have also been proposed. For example, AICD recruits Fe65 protein and translocates into the nucleus where the AICD-Fe65-Tip60 ternary complex may control transcription of target genes [29]. As AICD has been shown to have apoptotic potential [3,28], it is likely that APP signaling induces cell death, which leads to AD.

To explore APP signaling, we established embryonic carcinoma P19 cell lines overexpressing AICD [30]. Although neurons differentiated from these cell lines, AICD expression induced neuron-specific apoptosis. Furthermore, we showed that AICD induces dynamic changes in the gene expression profile through this process of neuron-specific cell death [31]. Thus, it is likely that APP also has a signaling mechanism and that AICD may play a role in APP signaling,leading to AD.

This chapter presents a summary of current knowledge regarding APP and γ-secretase. We also focus on the possibility thatγ-secretase-regulated mechanisms similar to Notch signaling may play roles in signaling events involving type 1 transmembrane proteins, including APP. Our results indicating that AICD induces dynamic changes in gene expression and neuron-specific apoptosis are also discussed. Based on these observations, we discuss the possibility that APP signaling is responsible for the onset of AD.

2. APP

(a) Overview of APP

APP was first identified as a cDNA cloned using a partial amino acid sequence of the Aβ fragment from the amyloid plaque of AD brains [32]. This cDNA encodes a type 1 membrane protein that is expressed in many tissues, which is accumulated in the synapses of neurons. The human APP gene has a total length of 240 kb containing at least 18 exons [33] and is localized to the long arm of chromosome 21 [32]. Several alternative splicing isoforms of APP ranging in length from 365 to 770 amino acids have been reported, which differ mainly in the absence (APP-695, predominantly expressed in neurons) or presence (APP-751 and APP-770) of the Kunitz protease inhibitor (KPI) domain located toward the N-terminus of the protein [34].

As discussed in detail below, APP undergoes sequential proteolytic cleavage reactions by the regulated intramembrane proteolysis (RIP) mechanism to yield the extracellular fragment, intracellular fragment (AICD), and Aβ fragment located in the membrane-spanning region, the latter of which is thought to be the main component responsible for the onset of AD.

Counterparts of this protein have been identified in other organisms, including *Drosophila* [35,36], *C.elegans* [37], and all mammals [38]. Mammals also possess two other APP family proteins called APP-like protein 1 (APLP1) and 2 (APLP2) [38]. APLP1 expression is restricted to neurons but that of APLP2, although highly enriched in the brain, is also detected in most other tissues. These APP family proteins share several conserved motifs, such as the E1 and E2 domains in the extracellular region [34,39,40]. The E1 domain contains several subdomains including a growth factor-like domain, a metal-binding motif, and a serine protease inhibitor domain that is absent from the isoform differentially expressed in the brain [39]. The E2 domain possesses a coiled coil dimerization motif and may bind proteoglycans in the extracellular matrix [40]. Interestingly, although the highest degree of sequence conservation within the APP family is found in the ICD, the sequence of the Aβ fragment is not highly conserved and is unique to APP [38].

(b) Putative Physiological Functions of APP

Although APP plays a central role in the pathogenesis of AD [1,2], the physiological functions of this protein have yet to be fully elucidated [3]. However, three putative functions of APP are widely accepted. First, APP has been suggested to act as a cell adhesion molecule, and thus to play a role in cell–cell interaction. Indeed, the E1 and E2 domains have been shown to interact with extracellular matrix proteins andheparan sulfate proteoglycans [39,40]. In addition, the extracellular domains of APP family proteins can potentially interact *in trans* suggesting that these molecules can bind to each other in a homo- or heterophilic manner to promote cell–cell adhesion [41].

Second, there is good evidence for roles of APP in synapse formation and repair, and these putative functions may be related to its adhesive properties as described above [42-44]. APP expression is induced during neuronal differentiation and after neural injury [44]. APP is transported in an anterograde manner to the synaptic sites, where the amount of secreted APP is correlated with synaptogenesis.

Third, APP has been suggested to act as a cell-surface receptor. As described in detail below, it is likely that APP also has a signaling mechanism. The suggestion that APP may act as a cell-surface receptor is of interest from the viewpoint of APP signaling. Indeed, Aβ can bind to APP and may therefore be a candidate ligand for this molecule [45]. F-spondin (46) and Nogo-66 receptor [47] have also been reported to bind to the extracellular domain of APP and regulate the production of Aβ. In addition, APP family members can bind to each other in a homo- or heterophilic manner *in trans* as mentioned above.

Similar to Notch signaling, AICD has been reported to undergo translocation to the nucleus and may function as a transcriptional regulator [28]. Thus, AICD may play an important role in APP signaling. Indeed, as mentioned above, APP homologs show the highest degrees of sequence conservation in the ICDs [30], which may reflect the functional importance of AICD. Moreover, AICD is thought to form complexes with the nuclear adaptor protein Fe65 and the histone acetyltransferase Tip60 [26]. These complexes can bind to the *cis*-acting DNA sequence of the tetraspanin protein KAI1 gene and regulate transcriptional activity [48].

(c) AβAmyloid and Its Degradation Enzymes

Aβ is the main constituent of amyloid plaque, which is thought to be the main hallmark and a major cause of the pathogenesis in the AD brain. As discussed below, Aβ is generated after sequential cleavage of APP by β- and γ-secretases, and ranges in length from 36 to 43 amino acid residues. On the other hand, sequential cleavage by a combination of α- and γ-secretase generates non-amyloidogenic p3 fragments. Although the shorter form of Aβ may typically be produced by cleavage that occurs in the endoplasmic reticulum, the longer forms may be produced in the *trans*-Golgi network [49]. Aβ40 and Aβ42 are the most common isoforms: Aβ40 is predominant over Aβ42, but Aβ42 is more amyloidogenic [50] and is therefore thought to be associated with AD.

Similar amyloid plaques are found in particular variants of Lewy body dementia [51] and in the muscle disease inclusion body myositis [52]. Aβ also forms aggregates coating cerebral blood vessels in cerebral amyloid angiopathy (CAA), present in over 90% of AD patients [53].

Aβ accumulation in the AD brain is thought to occur due to imbalance between the production of Aβ and its removal from the brain through various clearance mechanisms, including enzyme-mediated degradation [54]. Indeed, several candidate enzymes for Aβ degradation are expressed in both neurons and within the cerebral vasculature. Neprilysin (NEP) and insulin-degrading enzyme (IDE), both of which have been studied extensively, are expressed in both neurons and within the vasculature [54]. Generally, IDE is localized to the cytosol and peroxisomes [55], but is thought to be capable of passing through the cell membrane and thus IDE may be released from leaky cells. The levels of both enzymes are reduced in AD. In addition, it has been reported that *APOE* e4, which is a well-established genetic risk factor for the onset of AD and CAA, is associated with reduced levels of both enzymes [56,57].

Many other candidates for Aβ degradation enzymes have been proposed, including endothelin-converting enzymes 1 and 2 (ECE-1 and -2) [58] and angiotensin-converting enzyme (ACE) [59]. The levels of plasmin and plasminogen activators (uPA and tPA) and ECE-2 have also been shown to be reduced in the AD brain [54].

3. γ-SECRETASEMEDIATESNOTCHSIGNALING

Notch is a type 1 transmembrane protein that plays important roles in many aspects of development, including neuronal differentiation and the development of the immune system [9,10]. In the canonical Notch signaling pathway, ligands such as Delta bind to the extracellular domain of Notch on neighboring cells and trigger sequential proteolytic cleavage reactions (Figure 1).

This process is called the RIP mechanism [60], and requires sequential cleavage steps within the juxtamembrane (JM) and TM domains catalyzed by the metalloproteases and γ-secretase, respectively [61]. γ-Secretase releases NICD from the cell membrane into the cytoplasm and NICD is then translocated to the nucleus where its activity is expressed through binding to transcription factors [17,61]. Thus, γ-secretase plays a central regulatory role in Notch signaling.

Members of the family of CSL transcription factors (CBF1/RBP-jκ in mammals, Su(H) in *Drosophila*, and Lag-1 in *C. elegans*) are major downstream targets of Notch signaling [17]. NICD binds to CSL transcription factors, and six tandem ankyrin-like repeats in NICD have been shown to be essential for binding to CSL transcription factors [62].

NICD also binds to Mastermind-like proteins (MAML family in mammals) [63], thus forming the CSL-NICD-MAML complex. The formation of these complexes results in dispersal of co-repressors from CSL and recruitment of co-activators, such as P/CAF and P300 [63,64]. Through this process, the CSL complexes are converted from transcriptional repressors to activators.

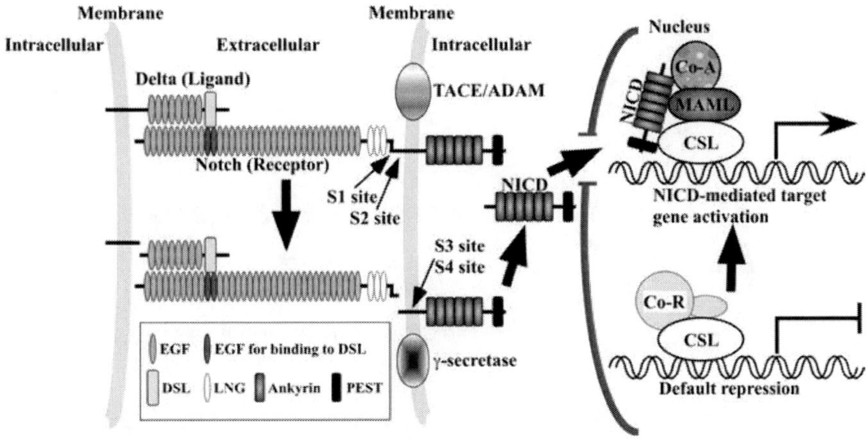

Figure 1. Notch signaling pathway. Notch proteins are expressed on the cell surface as heterodimers after cleavage at the S1 site by furin. The binding of Notch to the ligand triggers sequential proteolytic cleavage of RIP. When Notch binds to the ligand, Notch is cleaved at the S2 site in the juxtamembrane region by TACE or ADAM protease. Next, the remaining protein stub is further cleaved by γ-secretase at the S3 and S4 sites within the transmembrane domain and NICD is released from the membrane. Then, NICD translocates into the nucleus and binds to the CSL together with MAML. The resultant CSL–NICD–MAML complex removes co-repressors from CSL transcription factor and recruits a co-activator, resulting in conversion from repressor to activator. Finally, the complexes of CSL-NICD-MAML-Co-activator promote transcription of the target genes.

Finally, CSL complexes bind to the *cis*-acting DNA sequences of target genes, such as Hes (Hairy/Enhancer of split in *Drosophila*) [65], which encode the basic helix-loop-helix (bHLH) transcription factors, and promote their transcriptional activity.

4. APP Processing Is Very Similar to That of Notch

The detailed steps involved in Notch processing are shown in Figure 2. After translation, Notch is cleaved in the *trans*-Golgi network by furin-like covertase at the S1 site, and the two resulting fragments remain associated to form a functional heterodimer that is expressed on the cell surface [66]. As mentioned above, the RIP mechanism involving sequential proteolytic cleavage is initiated by ligand binding and shedding at the S2 site by TACE or ADAM protease making the truncated Notch [67,68]. Truncated Notch is further cleaved by γ-secretase in at least two sites within the TM domain, *i.e.*, at the S3 site to release NICD and at the S4 site to release the remaining small peptide (Nβ) [69-71].

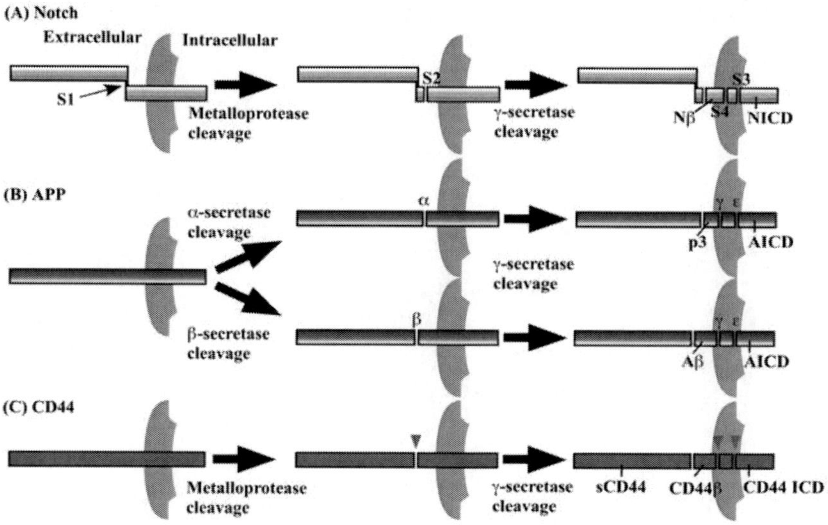

Figure 2. Similarities in the proteolytic processes among Notch, APP, and CD44. (A) In response to ligand binding, Notch undergoes shedding due to metalloprotease cleavage at the S2 site within the JM domain. After shedding the extracellular domain, the remaining Notch stub is further cleaved by γ-secretase at S3 and S4 sites within the TM domain. This sequential proteolysis produces NICD and Nβ fragment. (B) Cleavage of APP by α-secretase or β-secretase at the α-site or β-site, respectively, within the JM domain results in shedding of almost the entire extracellular domain and generates membrane-tethered α- or β-carboxy terminal fragments (CTFs). Several zinc metalloproteinases and BACE2 can cleave APP at the α-site, while BACE1 cleaves APP at the β-site. After shedding the extracellular domain, the remaining stub is further cleaved at least twice within the TM domain at γ- and ε-sites by γ-secretase, producing either p3 peptide (in combination with α-secretase) or Aβ (in combination with BACE1), respectively, and AICD. (C) Several stimuli, such as PKC activation and Ca^{2+} influx, trigger ectodomain cleavage of CD44 by a metalloprotease at the site within the JM domain, resulting in the secretion of soluble CD44 (sCD44). After shedding the extracellular domain, the remaining CD44 stub is further cleaved by γ-secretase at two sites within the TM domain. This sequential proteolysis produces the CD44 ICD and CD44β, an Aβ-like peptide.

As shown in Figure 2, the proteolytic process of APP is very similar to that of Notch and also follows the RIP mechanism. Cleavage of APP by α-secretase [72] or β-secretase [73] at the α-site or β-site, respectively, within the JM region results in shedding of almost the entire extracellular domain and generates membrane-tethered α- or β-carboxy terminal fragments (CTFs). Several zinc metalloproteinases, such as TACE and ADAM protease [74,75], and the aspartyl protease BACE2 [76] can cleave APP at the α-site, while BACE1 (β-site APP cleaving enzyme) cleaves APP at the β-site [73]. Once the extracellular domain has been shed, the remaining stub is further cleaved at least twice by γ-secretase within the TM domain at γ- and ε-sites resulting in production of either non-amyloidogenic p3 peptide (in combination with α-secretase) or amyloidogenic Aβ (in combination with BACE1), respectively, and AICD [6,7]. In addition, AICD was shown to be a substrate of caspase and to be cleaved at the group III caspase consensus sequence 16 aaamino acidsfrom the membrane border within the AICD.

Several γ-secretase substrates also follow the RIP mechanism with release of their ICDs from the cell membrane [8,22]. For example, as shown in Figure 2, the process of sequential proteolytic cleavage of CD44, which is important for immune system function, is very similar to those of Notch and APP [20]. In addition, the ICD of this protein (CD44ICD) is also translocated to the nucleus (Figure 2).

5. γ-SECRETASE

(a) Overview of γ-Secretase

Although γ-secretase was first identified as a protease that cleaves APP within the TM domain and produces Aβ peptides [5], which are thought to have pathogenic roles in AD, the physiological functions of γ-secretase remain to be clarified [6,7]. The γ-secretase is a complex composed of PS, nicastrin (NCT), anterior pharynx defective-1 (Aph-1), and PS enhancer-2 protein (Pen-2) [6,7,16]. PS is recognized as a catalytic component of the γ-secretase complex, and the two PS genes,PS1 gene (*PSEN1*) located on chromosome 14 [13] and PS2 gene (*PSEN2*) located on chromosome 1 [14,15], were identified by genetic linkage analyses as the genes responsible for several forms of early-onset FAD. *PSEN1* and *PSEN2* encode polytopictransmembrane proteins of 467 and 448 amino acids, respectively, which show about 65% sequence identity between the two proteins. While PS1 expression level is higher than that of PS2, both proteins are expressed ubiquitously in the brain and peripheral tissues of adult mammals [77]. Although several structural models for PS with six to ten transmembrane domains have been proposed, the model with eight or nine transmembrane domains is generally accepted [78]. According to this model, PS have a hydrophilic loop domain between the putative 6th and 7th transmembrane domains facing the cytoplasm and are cleaved by an as yet unidentified protease within this loop resulting into two fragments, NTF and CTF, that remain associated as a heterodimer [79]. This proteolytic cleavage is thought to occur when nascent PS assembles with NCT, Aph-1, and Pen-2 as a γ-secretase complex and activates PS as the catalytic component of aspartyl protease [6,7,16].The single-pass membrane protein NCT is thought to be important for recognizing the substrate proteins of γ-secretase [80,81]. The extracellular domain of NCT resembles an aminopeptidase, but lacks catalytic residues, and can interact with the N-terminal stubs of γ-secretase substrates after ectodomain shedding [81]. Thus, shedding of type 1 transmembrane proteins may be essential for the production of free N-termini of these proteins retained in the membrane, which can then be recognized by NCT.Aph-1 is thought to act as a scaffold during the process of γ-secretase complex assembly, and Pen-2 was suggested to act as a trigger for the proteolytic cleavage of PS to regulate PS activity [6,7].

(b) Does γ-Secretase Mediate Signaling Events of Type 1 Transmembrane Proteins?

While the physiological functions of γ-secretase have not been clarified, this protease is known to cleave a diverse array of type 1 transmembrane proteins that have shed their extracellular domains [82]. As reflected by the flexible sequence specificity of γ-secretase activity, more than 60 type 1 transmembrane proteins have been reported as substrates of γ-secretase [19].

As shown in Table 1, these substrates also have a wide range of functions, including roles in cell differentiation (Notch, Delta, and Jagged), cell adhesion (N-cadherin, E-cadherin, and CD44), synaptic adhesion (Nectin-1α), ion conductance regulation (voltage-gated sodium channel β2 subunit), axon guidance and tumor suppression (DCC), neurotrophin receptor (P75NTR), and its homolog (NRADD), lipoprotein receptor (ApoER2), and growth factor-dependent receptor tyrosine kinase (ERBB4).

Several γ-secretase substrates, such as APP and CD44 (Figure 2), were recently reported to follow the RIP mechanism [20]. The ICDs of these substrates are released from the cell membrane by γ-secretase, and these ICDs translocate to the nucleus. As mentioned above, these processes are very similar to those involved in Notch signaling. Therefore, the observations that the common enzyme, γ-secretase, modulates proteolysis and the turnover of possible signaling molecules led to the signaling hypothesis suggesting that mechanisms similar to those occurring in the Notch signaling pathway may contribute widely to γ-secretase-regulated signaling pathways [8,21,22].

Indeed, Delta-like protein 1 (Dll1), which is a major ligand of Notch, is cleaved sequentially by proteases, probably including ADAM and γ-secretase, and ICD of Dll1 (Dll1IC) is released from the cell membrane and undergoes translocation to the nucleus [83,84]. We have further shown that Dll1IC binds to Smad 2 and 3, which are transcription factors for the TGF-β/Activin signaling pathway, and enhances transcription of specific genes that accelerate neuronal differentiation [85]. Therefore, Dll1 also may have a signaling mechanism similar to Notch signaling.

(c) Is γ-Secretase A Proteasome of the Membrane?

As mentioned above, more than 60 type 1 membrane proteins have been reported as substrates of γ-secretase. This raises the simple question of why so many type 1 membrane proteins can transmit signals to the nucleus. In contrast to the signaling hypothesis, Kopan and Ilagan [6] suggested that γ-secretasemay be a proteasome of the membrane. They noted that the ICDs of several of these substrates, such as Syndecan-3, Nectin-1α, p75, and DCC, which are released by γ-secretase, are rapidly degraded. As AICD is also rapidly degraded [23,86], it is usually difficult to detect AICD by western blotting *in vivo*. Furthermore, ectodomain shedding seems to be constitutive for some substrates, and ligand binding has been reported to enhance only intramembrane cleavage of Notch [70], Delta [85], Syndecan-3 [87], and ERBB4 [88]. In addition, Kopan and Ilagan noted that a considerable amount of the evidence supporting the signaling hypothesis was obtained in overexpression assays that differ somewhat from physiological conditions[6]. Based on these observations, they proposed the proteasome hypothesis that the primary function of γ-secretase is to facilitate the selective disposal of type 1 membrane proteins [6].

Table 1. Substrates for γ-secretase

Substrate	Function	PS or ICD function
ApoER2	Lipoprotein receptor, neuronal migration	Activate nuclear reporter
APP	Precursor to Aβ, adhesion, trophic properties, axonal transport?	Aβ generation, release of ICD, complex with Fe65/Tip60, Cell death?
APLP1/2	Cell adhesion?	Form complex with Fe65 and Tip60
E-cadherin	Cell adhesion	Promote disassembly of adhesion complex
N-cadherin	Cell adhesion	Promote CBP degradation
β-catenin	Transduce Wnt signals stabilize adherens junctions	Facilitate phosphorylation
CD43	Signal transduction	Signaling molecule?
CD44	Cell adhesion	Activate TRE-mediated nuclear transcription
CSF1-R	Protein tyrosine kinase	Unknown
CXCL16 & CX3CL1	Membrane chemokine ligands	Unknown
DCC	Axon guidance, tumor suppressor	Activate nuclear reporter
Delta	Notch ligand	Transcriptionalregulation
ERBB4	Receptor tyrosine kinase	Regulate heregulin-induced growth inhibition
HLA-A2	MHC class I molecule	Unknown
IGIF-R	Receptor tyrosine kinase	Unknown
IFN-αR2	Subunit of type I IFN-α receptor	Transcriptional regulation
IL-1RI	Cytokine receptor	Unknown
IL-1RII	Cytokine receptor	Unknown
Jagged	Notch ligand	Modulate AP-1 mediated transcription

Table 1. (Continued)

Substrate	Function	PS or ICD function
LDLR	Lipoprotein receptor	Unkown
LRP	Scavenger and signaling receptor	Activate nuclear reporter
Na channelβ-subunit	Cell adhesion, an auxiliary subunit of voltage-gated Na channel	Alter cell adhesion and migration
Nectin-1α	Adherens junction, synapse receptor	Remodeling of cell junctions?
Notch1-4	Signaling receptor	Transcriptional regulation
NRADD	Apoptosis in neuronal cells	Modulate glycosylation/matutaion of NRADD
P75NTR	Neurotrophin co-receptor, dependence receptor	Modulate p75-TrkA complex? Nuclear singaling?
γ-protocadherin	Cell adhesion, neuronal differentiation	Regulation of gene transcription?
Syndecan-3	Cell surface proteoglycan co-receptor	Regulation of membrane-targeting of CASK
Telencephalin	Cell adhesion	Turnover of telencephalin
Tyrosinase, Tyrosinase-related protein 1/2	Pigment synthesis	Intracellular transport of Post-Golgi Tyr-containing vesicles

PS, presenilin; ICD, intracellular domain; APLP, APP like protein; CBP, CREB (cAMP-responsive element binding protein)-binding protein; TRE, TPA (12-o-tetradecanoylphorbol 13-acetate)-responsive element; AP-1, activator protein-1; CASK, calmodulin-dependent serine kinase; Tyr, Tyrosinase.

While the proteasome hypothesis of γ-secretase is persuasive, the γ-secretase-regulated signaling pathways of some substrates, such as Notch [17,21,61], are also firmly established. Therefore, although further studies are required to resolve these questions, it is possible that γ-secretases are not uniform complexes but that different γ-secretase complexes may exist in different combinations with components such as Aph-1, Pen2, and/or PS isoforms, with different cellular functions, such as roles in signaling or degradation [6]. In addition, it is not clear whether a large proportion of the ICDs of these substrates that are released by γ-secretase are required for the signaling mechanisms. Generally, γ-secretase substrates such as APP are considerably more abundant than transcription factors, which are usually rare molecules. Although a large proportion of ICDs of these substrates are rapidly degraded, it seems that a small amount of the remaining ICDs may fulfill their functions with a small quantity of transcription factors. Thus, the majority of ICDs of these substrates may be degraded and only a small proportion may play a role in signaling.

In relation to this issue, it has been proposed that a specific stimulus may trigger Fe65 binding to AICD to prevent AICD degradation by the proteasome as shown in Figure 3 [89]. Binding to Fe65 is thought to be essential for translocation of AICD to the nucleus. Therefore, as non-phosphorylated AICD binds to Fe65 and forms complexes, alone or with Tip60, these complexes can immediately translocate to the nucleus, where they mediate the expression of certain genes in association with Tip60. On the other hand, as other stimuli induce phosphorylation of AICD, which strongly inhibits binding to Fe65, AICD without Fe65 cannot translocate to the nucleus. Phosphorylated AICD left in the cytosol is rapidly degraded, most likely by the proteasome and/or IDE [86], which is widely expressed with the highest levels of expression in the liver, testes, muscle, and brain. Indeed, it has been reported that when phosphorylated at Thr^{668} in the APP-695 isoform, AICD cannot bind to Fe65 [90].

As mentiond below, since AICD induces neuron-specific apoptosis, it seems to be closely related to the onset of AD. Thus, from the viewpoint of the signaling mechanism, it is likely that AICD has no effect in the normal brain, because almost all AICD is rapidly degraded. However, both AICD and Aβ are overproduced in the AD brain. Therefore, although the majority of AICD is degraded, it is possible that a small amount of the remaining AICD may show its pathological function and cause neuron-specific cell death in the AD brain.

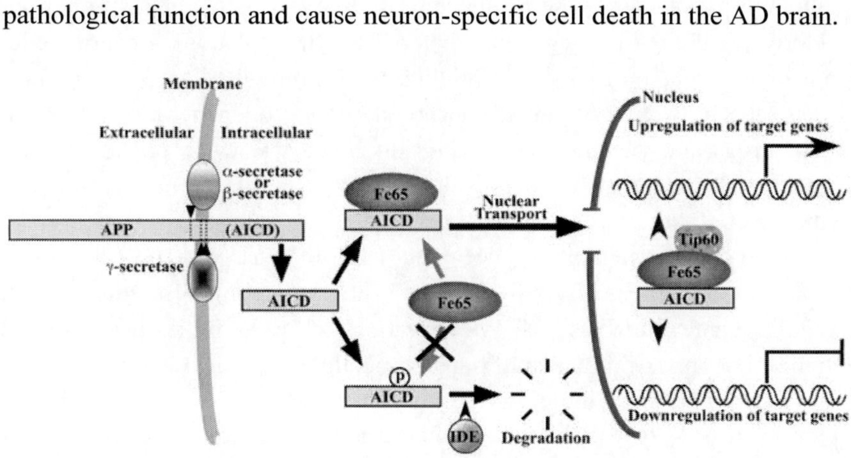

Figure 3. Putative APP signaling pathway. After cleavege within the JM domain by α- or β-secretase, AICD is released from the membrane by γ-secretase. Non-phosphorylated AICD binds to the nuclear adaptor protein Fe65, which is thought to be essential for translocation of AICD to the nucleus.

AICD/Fe65 complexes immediately translocate to the nucleus, where they meidate up- and downregulation of certain target genes in association with histone acetyltransferase Tip60. On the other hand, phosphorylation of AICD strongly inhibits binding to Fe65 and thus also inhibits translocation to the nucleus. Phospholylated AICD left in the cytosol is rapidly degraded, probably by insulin-degrading enzyme (IDE).

6. Is Aβ Sufficient to Clarify All Aspects of the Onset of AD?

The human APP gene is located on the long arm of chromosome 21 [32], which is present in an extra copy in individuals with Down's syndrome (trisomy 21). Subjects with Down's syndrome develop AD by 40 years old, probably due to this gene dosage effect [91]. Autosomal dominant mutations in and around the Aβ region of the APP gene are responsible for hereditary early-onset AD [92], probably due to acceleration of proteolytic processing. In addition, both PS1 and PS2, which are catalytic components of the γ-secretase complex, were identified by genetic linkage analyses as the genes responsible for FAD [13-15]. Therefore, both APP itself and its proteolytic processing appear to be responsible for the onset of AD.

As mentioned above, Aβ is a constituent of amyloid plaque, which is thought to play a major role in the pathogenesis of AD and its presence is a hallmark of the AD brain. Thus, the amyloid hypothesis is generally accepted as the mechanism of the onset of AD. The traditional amyloid hypothesis is that overproduced Aβ forms insoluble amyloid plaques, which are commonly observed in the AD brain and are believed to be the toxic form of APP responsible for neurodegeneration [93]. Although an alternative hypothesis has also been proposed that soluble Aβ oligomers rather than insoluble amyloid plaques are responsible for the onset of AD because the soluble form of Aβ oligomer is toxic for neurons [94,95], Aβ still plays a central role in this hypothesis.

However, several questions have been raised regarding these hypotheses that Aβ is the main component involved in the pathogenesis of AD. One of the most significant arguments against the amyloid hypothesis is the presence of high levels of Aβ deposition in many non-demented elderly people (96), suggesting that Aβ amyloid plaques are not toxic. Indeed, transgenic mice overproducing Aβ show amyloid deposition with a high concentration of Aβ mimicking that seen in the AD brain but do not show neurodegeneration (4). Furthermore, several anti-Aβ drugs and vaccines have failed to show efficacy in phase III clinical trials [97,98]. These results suggest that both soluble and insoluble forms of Aβ may not be involved in the onset of AD.

Based on these observations, it has been suggested that AD may be caused by an APP-derived protein, but not necessarily Aβ [98]. As both extracellular fragment and AICD are generated at the same time as Aβ, acceleration of proteolytic processing leads to overproduction of not only Aβ but also of both extracellular fragment and AICD.

From this viewpoint, very important results havebeen shown that APP is a ligand of Death receptor 6 (DR6) [99], which mediates cell death and is expressed at high levels in the human brain regions most affected by AD. APP is cleaved by β-secretase, releasing the extracellular domain (sAPPβ), which is further cleaved by an as yet unknown mechanism to release a 35-kDa N-terminal fragment (N-APP). This N-APP fragment binds DR6 to trigger degeneration through caspase 6 in axons and caspase 3 in cell bodies. These results suggest that N-APP may be involved in the onset of AD [99].

In this chapter, we focus on the signaling aspects of APP. As mentioned below, AICD induces neuron-specific apoptosis. These results suggest that AICD may also be involved in the onset of AD. Indeed, it has been reported that transgenic mice expressing AICD show pathological features similar to those associated with AD [100]. Taken together, these observations suggest that AD may be caused by multiple mechanisms mediated by multiple APP-derived fragments, not only Aβ but also N-APP and AICD.

7. Roles of AICD

(a) AICD Induces Neuron-Specific Apoptosis

There is accumulating evidence that AICD plays an important role in APP signaling similar to that of Notch and contributes to the onset of AD. Indeed, elevated AICD levels have also been reported in the AD brain [100]. As mentioned above, AICD is thought to form complexes with Fe65 and Tip60 in the nucleus and to control the expression of certain target genes. Indeed, transgenic mice overexpressing AICD and Fe65 show abnormal activity of glycogen synthase kinase 3 beta (*Gsk3b* protein) [101], leading tohyperphosphorylation and aggregation of TAU, resulting in microtubule destabilization and reduction of nuclear β-catenin levels with resultant loss of cell–cell contact mechanisms that may contribute to neurotoxicity in AD. Similar transgenic mice showed abnormal spiking events in their electroencephalograms as well as susceptibility to kainic acid-induced seizures independent of Aβ [102]. Furthermore, c-Abl kinase has also been shown to affect transcriptional regulation of AICD, and c-Abl modulates AICD-dependent cellular responses, transcriptional induction, and apoptotic responses [103]. Therefore, it is likely that AICD-dependent differential gene expression, reflecting the involvement of APP signaling, induces cell death, which may lead to AD pathology.

We established several AICD-overexpressing P19 cell lines in which to examine the APP signaling hypothesis [30]. All of these cell lines could differentiate into neurons after aggregation culture with all-*trans*-retinoic acid (RA) treatment. As shown in Figure 4, undifferentiated AICD-overexpressing P19 cells retained an epithelial cell-like morphology similar to that of control cells carrying the vector alone, while the differentiated cells became round and showed a bipolar morphology with neurite extension. However, AICD expression induced neuron-specific cell death. Indeed, while neurons differentiated from control cells were healthy, almost all neurons derived from AICD-overexpressing P19 cells showed severe degeneration 4 days after induction of differentiation (Figure 4).

Moreover, DNA fragmentation was detected, and all terminal deoxynucleotidyltransferase (TdT)-mediated deoxyuridine triphosphate (dUTP)-biotin nick end-labeling (TUNEL)-positive cells were also Tuj1-positive neurons. These results indicated that AICD can induce neuron-specific apoptosis [30].

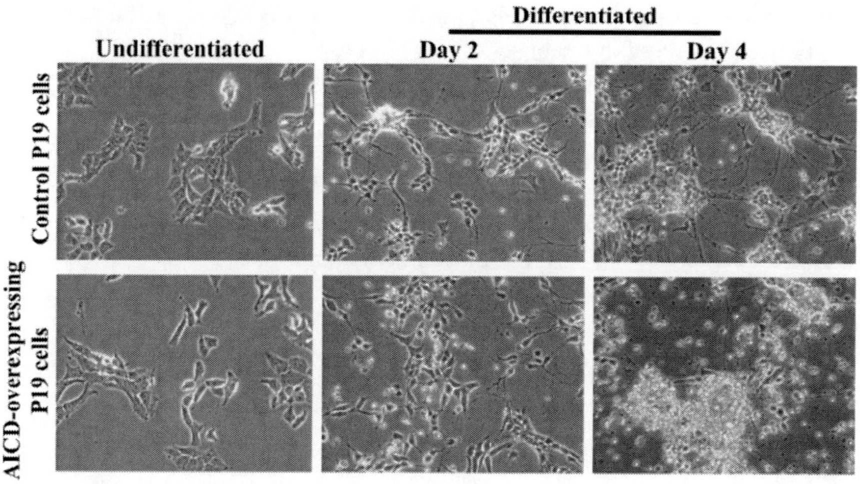

Figure 4. Overexpression of AICD in P19 cells induced neuronal cell death. After aggregation culture with RA, AICD-overexpressing P19 and control P19 cells carrying vector alone were replated and cultured for the indicated periods on dishes and allowed to differentiate. Undifferentiated AICD-overexpressing P19 cells retained epithelial cell-like morphology similar to control cells, while the differentiated cells became round and showed a bipolar morphology with neurite extension. Two days after replating (Day 2), all cell lines grew well and neurons with long neurites appeared. Four days after replating (Day 4), control cells still grew well as clusters and many neurons had differentiated from these cells. However, many AICD-overexpressing P19 cells showed severe degeneration, becoming spherical with numerous vacuoles and detached from the culture dishes.

The effects of AICD were restricted to neurons, with no effects observed on undifferentiated non-neural cells. Therefore, these results suggest that AICD plays a central role in APP signaling, which leads to AD.

(b) AICD Induces Dynamic Changes in the Gene Expression Profile

If APP signaling does occur, AICD should alter the expression of certain genes. To test this possibility and identify the genes involved in this process of neuron-specific apoptosis, we again employed both AICD-overexpressing P19 cells and control P19 cells to evaluate AICD-induced changes in gene expression at 3 time points during culture: the undifferentiated state, after 4 days of aggregation with RA (aggregated state), and 2 days after replating(differentiated state) [31]. The levels of expression of 41256 transcripts and transcript variants of more than 20000 independent genes were monitored by DNA microarray analysis and estimated as relative numerical values from the intensity of hybridization signals [31]. As shown in Figure 5, the expression levels of 277 genes were upregulatedby more than 10-fold in the presence of AICD, while 341 genes showed downregulation of expression to less than 10% of the original level.

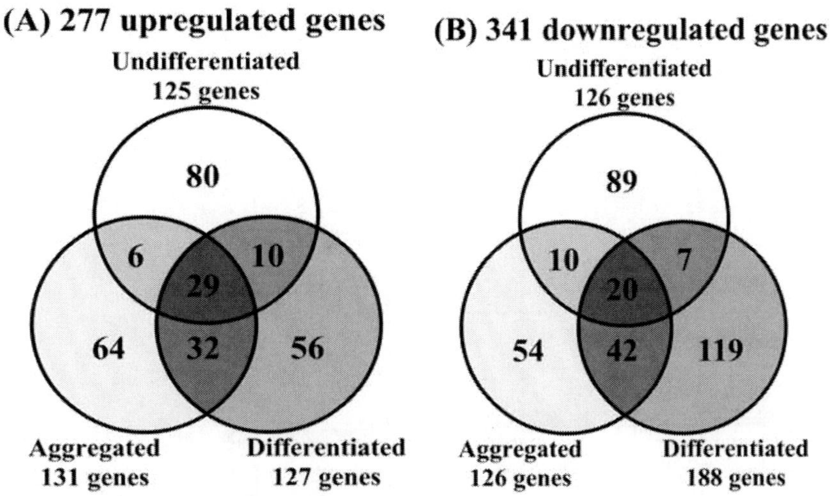

Figure 5. Venn diagrams showing the total numbers of genes upregulated by more than 10-fold in the presence of AICD (A) and genes downregulated to less than 10% of their original level (B) at three states of neural differentiation in P19 cells: undifferentiated, aggregated, and differentiated.

Several genes showed extremely strong upregulation in the presence of AICD (Table 2) [31]. For example, protein tyrosine phosphatase receptor T (*Ptprt*) expression was below the limit of detection at all sampling points in control P19 cells, but AICD-overexpressing P19 cells showed strong *Ptprt* gene expression at all sampling points: *i.e.*, 906-fold, 204-fold, and 116-fold upregulation in the undifferentiated, aggregated, and differentiated states, respectively.In contrast to these upregulated genes, AICD also strongly inhibited the expression of certain genes (Table 2) [31]. Although hairy and enhancer of split 5 (*Hes5*) expression was undetectable in undifferentiated control P19 cells, their levels of expression were markedly increased through neural differentiation: *i.e.*, increases in expression level of almost 300-fold were observed in the aggregated and differentiated states.

However, this remarkable induction through differentiation in control P19 cells was not detected in AICD-overexpressing P19 cells (Table 2), indicating that AICD strongly inhibits this induction. As these results indicated that AICD mediates both upregulation and downregulation of many genes, it is likely that AICD plays an important role in APP signaling.

These up- and downregulated genes were classified according to Gene Ontology (GO) terms at several hierarchical levels [31]. A few of the up- and downregulated genes were classified into GO terms related to cell death, but many of these genes had no relation to cell death. In addition, we also evaluated AICD-induced changes in expression of genes involved in cell death in AD, but found no significant changes in expression of these genes. Therefore, it seems like possible that AICD does not directly induce the expression of genes involved in cell death, but the extreme dynamic changes in gene expression described here disrupt the homeostasis of certain neurons and thus give rise to neuron-specific cell death.

Table 2. Expression levels of 7 upregulated and 7 downregulated genes, as well as 3 housekeeping genes

Gene Symbol	Gene Name	Function	Relative Expression Levels (fold)		
			Undifferentiated	Aggregated	Differentiated
Non-regulated genes (housekeeping genes)					
Actb	β-actin	cytoskeleton protein	-1.2	1.2	1
Sdha	succinate dehydrogenase subunit A	electron transporter in the TCA cycle and respiratory chain	-1.1	-1.6	-1.2
Eef1a1	eukaryotic translation elongation factor-1 alpha 1	essential component for the elongation phase during protein translation	1	-1.1	1.2
Upregulated genes					
Ptprt	protein tyrosine phosphatase receptor T	protein tyrosine phosphatase that regulates STAT3 activity	906	204	116
Cpb1	carboxypeptidase B1	hydrolysis of C-terminal end of basic amino acid peptide bond	16	296	222
Nr2e1	tailless homolog	transcription factor that is essential for neural stem cell proliferation and self-renewal	5.8	244	54
Myh1	myosin heavy chain 1	one of the component of motor protein myosin	-4.2	259	-1.1
Dnahc7c	axonemal dynein heavy chain	essential for motility of cilia and flagella	133	41	43
Alkbh3	alkylation repair homolog 3	AlkB enzyme that repairs methylation damage in DNA and RNA	69	80	43

Table 2. Continued

Gene Symbol	Gene Name	Function	Relative Expression Levels (fold)		
			Undifferentiated	Aggregated	Differentiated
Ctgf	connective tissue growth factor	skeletogenesis/vasculogenesis by modulating BMP, Wnt, and IGF-I signals	90	54	40
Downregulated genes					
Hes5	hairy and enhancer of split 5	transcription factor that inhibits neurogenesis	-8.7	-3039	-2515
Slc10a6	sodium-dependent organic anion transporter	transport of sulfoconjugated steroid hormones and bile acids	-145	-785	-1212
Nid1	nidogen-1	extracellular matrix linker protein	-304	-165	-507
LOC213332	analog of Na+ dependent glucose transporter 1	putative glucose transporter	-232	-325	-306
Dtx1	Deltex1	regulator of Notch signaling pathway	-30	-85	-691
Rbp4	retinol-binding protein 4	retinol transporter from the liver to extrahepatic tissues	-525	-100	-24
Col3a1	collagen type III alpha 1	extracellular matrix protein	4.1	-29	-234

Relative expression levels (-fold) were estimated from the intensity of hybridization signals. Housekeeping gene expression was unaltered in AICD-overexpressing P19 and control P19 cells in all states, suggesting that these genes may not be affected by AICD. These results also indicated that the observed differences in expression were not due to technical problems, such as uneven hybridization or poor RNA quality.

(c) Comparison of Expression Data from AICD-Overexpressing Cells and the AD Brain

If APP signaling is related to the onset of AD, our expression results from AICD-overexpressing p19 cells are expected to be correlated to that of AD. Therefore, we compared our results [31] with several previous reports of AD-specific DNA microarray data. Blalock *et al.* divided AD patients into 3 classes based on cognition level: incipient, moderate, and severe. They compared the expression profiles in the hippocampus of these classes with those of normal controls and extracted genes showing differential expression [104]. However, there were no significant correlations between our data and their results. Liang *et al.* also reported the expression profiles of 6 regions of the brain (entorhinal cortex, hippocampus, medial temporal gyrus, posterior cingulate, superior frontal gyrus, and primary visual cortex) and compared the profiles of AD brains with those of normal controls (105). We also compared our data with their results, and again found no significant correlations.

As there were no significant correlations between our results and those of previous studies of the AD brain, we compared the expression data reported by Blalock *et al.*[104] with those reported by Liang *et al.* [105]; surprisingly, we also found no correlations between these two data sets.

Although the reasons for these conflicting results are not clear, they are likely to be at least partially due to the variable quality of RNA isolated from the AD brain because it is technically difficult to isolate high-quality RNA from dying cells such as AD neurons. Therefore, the AICD-expressing P19 cells described in this chapter may be a useful model not only for the study of APP signaling but also for cellular and molecular studies of AD.

CONCLUSION

Although APP plays central roles in AD, the physiological functions of this protein have yet to be fully elucidated. γ-Secretase was first identified as a protease that cleaves APP within the TM domain and produces Aβ peptides, the main constituents of amyloid plaque, which are thought to play a role in the pathogenesis of AD and the presence of which is a hallmark of the AD brain. However, the physiological functions of this enzyme also remain to be determined.

As summarized here, the Notch signaling pathway is regulated by γ-secretase; intramembrane cleavage of Notch by γ-secretase serves to release ICD, which shows its activity in the nucleus through binding to transcription factors. Many type 1 transmembrane proteins, including APP, were recently shown to be substrates for γ-secretase resulting in the release of ICDs of these proteins from the cell membranes. Some of these released ICDs translocate to the nucleus. These processes are very similar to those involved in Notch signaling. Therefore, the observation that the common enzyme, γ-secretase, modulates proteolysis and the turnover of possible signaling molecules suggested that mechanismssimilar to the Notch signaling pathwaymay contribute widely toγ-secretase-regulated signaling.

This chapter focused on the signaling aspects of APP and its pathological roles. Indeed, as described here, we have shown that AICD alters gene expression and induces neuron-specific apoptosis. Therefore, it is likely that APP plays a role in signaling events and that γ-secretase-regulated APP signaling is at least partially responsible for the onset of AD.

REFERENCES

[1] Hardy, J. (1997) Amyloid, the presenilins and Alzheimer's disease. *Trends Neurosci*, 20, 154-159.

[2] Selkoe, D.J. (2001) Alzheimer's disease: genes, proteins, and therapy. *Physiol. Rev*, 81, 741-766.

[3] Zheng, H. and Koo, E.H. (2006) The amyloid precursor protein: beyond amyloid. *Mol.Neurodegener*, 1, 5.

[4] McGowan, E. (ed.) (2003) *Alzheimer animal models: models of Abeta deposition in transgenic mice*. ISN Neuropath Press, Basel.

[5] Haass, C. and Selkoe, D.J. (1993) Cellular processing of beta-amyloid precursor protein and the genesis of amyloid beta-peptide. *Cell*, 75, 1039-1042.

[6] Kopan, R. and Ilagan, M.X. (2004) Gamma-secretase: proteasome of the membrane? *Nat Rev Mol Cell Biol*, 5, 499-504.

[7] Selkoe, D.J. and Wolfe, M.S. (2007) Presenilin: running with scissors in the membrane. *Cell*, 131, 215-221.

[8] Nakayama, K., Nagase, H., Hiratochi, M., Koh, C.S. and Ohkawara, T. (2008) Similar mechanisms regulated by gamma-secretase are involved in both directions of the bi-directional Notch-Delta signaling pathway as well as play a potential role in signaling events involving type 1 transmembrane proteins. *Curr. Stem Cell Res.Ther*, 3, 288-302.

[9] Artavanis-Tsakonas, S., Matsuno, K. and Fortini, M.E. (1995) Notch signaling. *Science*, 268, 225-232.

[10] Lewis, J. (1998) Notch signalling and the control of cell fate choices in vertebrates. *Semin. Cell Dev.Biol*, 9, 583-589.

[11] Bolos, V., Grego-Bessa, J. and de la Pompa, J.L. (2007) Notch signaling in development and cancer. *Endocr Rev*, 28, 339-363.

[12] Levitan, D. and Greenwald, I. (1995) Facilitation of lin-12-mediated signalling by sel-12, a Caenorhabditiselegans S182 Alzheimer's disease gene. *Nature*, 377, 351-354.

[13] Sherrington, R., Rogaev, E.I., Liang, Y., Rogaeva, E.A., Levesque, G., Ikeda, M., Chi, H., Lin, C., Li, G., Holman, K. *et al.* (1995) Cloning of a gene bearing missense mutations in early-onset familial Alzheimer's disease. *Nature*, 375, 754-760.

[14] Levy-Lahad, E., Wasco, W., Poorkaj, P., Romano, D.M., Oshima, J., Pettingell, W.H., Yu, C.E., Jondro, P.D., Schmidt, S.D., Wang, K. *et al.* (1995) Candidate gene for the chromosome 1 familial Alzheimer's disease locus. *Science*, 269, 973-977.

[15] Rogaev, E.I., Sherrington, R., Rogaeva, E.A., Levesque, G., Ikeda, M., Liang, Y., Chi, H., Lin, C., Holman, K., Tsuda, T. *et al.* (1995) Familial Alzheimer's disease in kindreds with missense mutations in a gene on chromosome 1 related to the Alzheimer's disease type 3 gene. *Nature*, 376, 775-778.

[16] Iwatsubo, T. (2004) The gamma-secretase complex: machinery for intramembrane proteolysis. *Curr.Opin.Neurobiol*, 14, 379-383.

[17] Artavanis-Tsakonas, S., Rand, M.D. and Lake, R.J. (1999) Notch signaling: cell fate control and signal integration in development. *Science*, 284, 770-776.

[18] Justice, N.J. and Jan, Y.N. (2002) Variations on the Notch pathway in neural development. *Curr.Opin.Neurobiol*, 12, 64-70.

[19] McCarthy, J.V., Twomey, C. and Wujek, P. (2009) Presenilin-dependent regulated intramembrane proteolysis and gamma-secretase activity. *Cell Mol Life Sci*, 66, 1534-1555.

[20] Nagase, H., Koh, C.S. and Nakayama, K. (2011) gamma-Secretase-regulated signaling pathways, such as notch signaling, mediate the differentiation of hematopoietic stem cells, development of the immune system, and peripheral immune responses. *Curr. Stem Cell Res.Ther*, 6, 131-141.

[21] Koo, E.H. and Kopan, R. (2004) Potential role of presenilin-regulated signaling pathways in sporadic neurodegeneration. *Nat. Med*, 10 Suppl, S26-33.

[22] Nakayama, K., Nagase, H., Koh, C.S. and Ohkawara, T. (2011) gamma-Secretase-Regulated Mechanisms Similar to Notch Signaling May Play a Role in Signaling Events, Including APP Signaling, Which Leads to Alzheimer's Disease. *Cell Mol.Neurobiol.*

[23] Cupers, P., Orlans, I., Craessaerts, K., Annaert, W. and De Strooper, B. (2001) The amyloid precursor protein (APP)-cytoplasmic fragment generated by gamma-secretase is rapidly degraded but distributes partially in a nuclear fraction of neurones in culture. *J.Neurochem*, 78, 1168-1178.

[24] Gao, Y. and Pimplikar, S.W. (2001) The gamma -secretase-cleaved C-terminal fragment of amyloid precursor protein mediates signaling to the nucleus. *Proc. Natl. Acad.Sci.USA*, 98, 14979-14984.

[25] Kimberly, W.T., Zheng, J.B., Guenette, S.Y. and Selkoe, D.J. (2001) The intracellular domain of the beta-amyloid precursor protein is stabilized by Fe65 and translocates to the nucleus in a notch-like manner. *J. Biol.Chem*, 276, 40288-40292.

[26] Cao, X. and Sudhof, T.C. (2001) A transcriptionally active complex of APP with Fe65 and histone acetyltransferase Tip60. *Science*, 293, 115-120.

[27] Guenette, S.Y. (2002) A role for APP in motility and transcription? *Trends PharmacolSci*, 23, 203-205; discussion 205-206.

[28] Muller, T., Meyer, H.E., Egensperger, R. and Marcus, K. (2008) The amyloid precursor protein intracellular domain (AICD) as modulator of gene expression, apoptosis, and cytoskeletal dynamics-relevance for Alzheimer's disease. *Prog.Neurobiol*, 85, 393-406.

[29] Slomnicki, L.P. and Lesniak, W. (2008) A putative role of the Amyloid Precursor Protein Intracellular Domain (AICD) in transcription. *ActaNeurobiol. Exp. (Wars)*, 68, 219-228.

[30] Nakayama, K., Ohkawara, T., Hiratochi, M., Koh, C.S. and Nagase, H. (2008) The intracellular domain of amyloid precursor protein induces neuron-specific apoptosis. *NeurosciLett*, 444, 127-131.

[31] Ohkawara, T., Nagase, H., Koh, C.S. and Nakayama, K. (2011) The amyloid precursor protein intracellular domain alters gene expression and induces neuron-specific apoptosis. *Gene*, 475, 1-9.

[32] Kang, J., Lemaire, H.G., Unterbeck, A., Salbaum, J.M., Masters, C.L., Grzeschik, K.H., Multhaup, G., Beyreuther, K. and Muller-Hill, B. (1987) The precursor of Alzheimer's disease amyloid A4 protein resembles a cell-surface receptor. *Nature*, 325, 733-736.

[33] Yoshikai, S., Sasaki, H., Doh-ura, K., Furuya, H. and Sakaki, Y. (1990) Genomic organization of the human amyloid beta-protein precursor gene. *Gene*, 87, 257-263.

[34] Sisodia, S.S., Koo, E.H., Hoffman, P.N., Perry, G. and Price, D.L. (1993) Identification and transport of full-length amyloid precursor proteins in rat peripheral nervous system. *J Neurosci*, 13, 3136-3142.

[35] Rosen, D.R., Martin-Morris, L., Luo, L.Q. and White, K. (1989) A Drosophila gene encoding a protein resembling the human beta-amyloid protein precursor. *Proc. Natl. Acad. Sci. USA*, 86, 2478-2482.

[36] Luo, L.Q., Martin-Morris, L.E. and White, K. (1990) Identification, secretion, and neural expression of APPL, a Drosophila protein similar to human amyloid protein precursor. *J.Neurosci*, 10, 3849-3861.

[37] Daigle, I. and Li, C. (1993) apl-1, a Caenorhabditiselegans gene encoding a protein related to the human beta-amyloid protein precursor. *Proc. Natl. Acad. Sci. USA*, 90, 12045-12049.

[38] Coulson, E.J., Paliga, K., Beyreuther, K. and Masters, C.L. (2000) What the evolution of the amyloid protein precursor supergene family tells us about its function. *Neurochem.Int*, 36, 175-184.

[39] Dahms, S.O., Hoefgen, S., Roeser, D., Schlott, B., Guhrs, K.H. and Than, M.E. (2010) Structure and biochemical analysis of the heparin-induced E1 dimer of the amyloid precursor protein. *Proc. Natl. Acad Sci. USA*, 107, 5381-5386.

[40] Wang, Y. and Ha, Y. (2004) The X-ray structure of an antiparallel dimer of the human amyloid precursor protein E2 domain. *Mol. Cell*, 15, 343-353.

[41] Soba, P., Eggert, S., Wagner, K., Zentgraf, H., Siehl, K., Kreger, S., Lower, A., Langer, A., Merdes, G., Paro, R. *et al.* (2005) Homo- and heterodimerization of APP family members promotes intercellular adhesion. *EMBO J*, 24, 3624-3634.

[42] Bibel, M., Richter, J., Schrenk, K., Tucker, K.L., Staiger, V., Korte, M., Goetz, M. and Barde, Y.A. (2004) Differentiation of mouse embryonic stem cells into a defined neuronal lineage. *Nat.Neurosci*, 7, 1003-1009.

[43] Leyssen, M., Ayaz, D., Hebert, S.S., Reeve, S., De Strooper, B. and Hassan, B.A. (2005) Amyloid precursor protein promotes post-developmental neuritearborization in the Drosophila brain. *EMBO J.*, 24, 2944-2955.

[44] Hung, A.Y., Koo, E.H., Haass, C. and Selkoe, D.J. (1992) Increased expression of beta-amyloid precursor protein during neuronal differentiation is not accompanied by secretory cleavage. *Proc. Natl. Acad. Sci. USA*, 89, 9439-9443.

[45] Lorenzo, A., Yuan, M., Zhang, Z., Paganetti, P.A., Sturchler-Pierrat, C., Staufenbiel, M., Mautino, J., Vigo, F.S., Sommer, B. and Yankner, B.A. (2000) Amyloid beta interacts with the amyloid precursor protein: a potential toxic mechanism in Alzheimer's disease. *Nat.Neurosci*, 3, 460-464.

[46] Ho, A. and Sudhof, T.C. (2004) Binding of F-spondin to amyloid-beta precursor protein: a candidate amyloid-beta precursor protein ligand that modulates amyloid-beta precursor protein cleavage. *Proc. Natl. Acad. Sci. USA*, 101, 2548-2553.

[47] Park, J.H., Gimbel, D.A., GrandPre, T., Lee, J.K., Kim, J.E., Li, W., Lee, D.H. and Strittmatter, S.M. (2006) Alzheimer precursor protein interaction with the Nogo-66 receptor reduces amyloid-beta plaque deposition. *J.Neurosci.*, 26, 1386-1395.

[48] Baek, S.H., Ohgi, K.A., Rose, D.W., Koo, E.H., Glass, C.K. and Rosenfeld, M.G. (2002) Exchange of N-CoRcorepressor and Tip60 coactivator complexes links gene expression by NF-kappaB and beta-amyloid precursor protein. *Cell*, 110, 55-67.

[49] Hartmann, T., Bieger, S.C., Bruhl, B., Tienari, P.J., Ida, N., Allsop, D., Roberts, G.W., Masters, C.L., Dotti, C.G., Unsicker, K. *et al.* (1997) Distinct sites of intracellular production for Alzheimer's disease A beta40/42 amyloid peptides. *Nat. Med.*, 3, 1016-1020.

[50] Jarrett, J.T., Berger, E.P. and Lansbury, P.T., Jr. (1993) Thecarboxy terminus of the beta amyloid protein is critical for the seeding of amyloid formation: implications for the pathogenesis of Alzheimer's disease. *Biochemistry*, 32, 4693-4697.

[51] Lippa, C.F., Smith, T.W. and Swearer, J.M. (1994) Alzheimer's disease and Lewy body disease: a comparative clinicopathological study. *Ann.Neurol*, 35, 81-88.

[52] Askanas, V., Engel, W.K. and Alvarez, R.B. (1992) Light and electron microscopic localization of beta-amyloid protein in muscle biopsies of patients with inclusion-body myositis. *Am. J.Pathol*, 141, 31-36.

[53] Haan, J. and Roos, R.A. (1990) Amyloid in central nervous system disease. *Clin. Neurol.Neurosurg*, 92, 305-310.

[54] Miners, J.S., Baig, S., Palmer, J., Palmer, L.E., Kehoe, P.G. and Love, S. (2008) Abeta-degrading enzymes in Alzheimer's disease. *Brain Pathol*, 18, 240-252.

[55] Authier, F., Bergeron, J.J., Ou, W.J., Rachubinski, R.A., Posner, B.I. and Walton, P.A. (1995) Degradation of the cleaved leader peptide of thiolase by a peroxisomal proteinase. *Proc. Natl. Acad. Sci. USA*, 92, 3859-3863.

[56] Miners, J.S., Van Helmond, Z., Chalmers, K., Wilcock, G., Love, S. and Kehoe, P.G. (2006) Decreased expression and activity of neprilysin in Alzheimer disease are associated with cerebral amyloid angiopathy. *J.Neuropathol. Exp.Neurol*, 65, 1012-1021.

[57] Cook, D.G., Leverenz, J.B., McMillan, P.J., Kulstad, J.J., Ericksen, S., Roth, R.A., Schellenberg, G.D., Jin, L.W., Kovacina, K.S. and Craft, S. (2003) Reduced hippocampal insulin-degrading enzyme in late-onset Alzheimer's disease is associated with the apolipoprotein E-epsilon4 allele. *Am. J.Pathol*, 162, 313-319.

[58] Turner, A.J. and Murphy, L.J. (1996) Molecular pharmacology of endothelin converting enzymes. *BiochemPharmacol*, 51, 91-102.

[59] Erdos, E.G. and Skidgel, R.A. (1987) The angiotensin I-converting enzyme. *Lab. Invest*, 56, 345-348.

[60] Brown, M.S., Ye, J., Rawson, R.B. and Goldstein, J.L. (2000) Regulated intramembrane proteolysis: a control mechanism conserved from bacteria to humans. *Cell*, 100, 391-398.

[61] Selkoe, D. and Kopan, R. (2003) Notch and Presenilin: regulated intramembrane proteolysis links development and degeneration. *Annu Rev.Neurosci*, 26, 565-597.

[62] Roehl, H., Bosenberg, M., Blelloch, R. and Kimble, J. (1996) Roles of the RAM and ANK domains in signaling by the C. elegans GLP-1 receptor. *EMBO J.*, 15, 7002-7012.

[63] Wu, L., Aster, J.C., Blacklow, S.C., Lake, R., Artavanis-Tsakonas, S. and Griffin, J.D. (2000) MAML1, a human homologue of Drosophila mastermind, is a transcriptional co-activator for NOTCH receptors. *Nat. Genet*, 26, 484-489.

[64] Wallberg, A.E., Pedersen, K., Lendahl, U. and Roeder, R.G. (2002) p300 and PCAF act cooperatively to mediate transcriptional activation from chromatin templates by notch intracellular domains in vitro. *Mol. Cell Biol*, 22, 7812-7819.

[65] Kageyama, R., Ohtsuka, T. and Kobayashi, T. (2007) The Hes gene family: repressors and oscillators that orchestrate embryogenesis. *Development*, 134, 1243-1251.

[66] Logeat, F., Bessia, C., Brou, C., LeBail, O., Jarriault, S., Seidah, N.G. and Israel, A. (1998) The Notch1 receptor is cleaved constitutively by a furin-like convertase. *Proc. Natl. Acad. Sci. USA*, 95, 8108-8112.

[67] Pan, D. and Rubin, G.M. (1997) Kuzbanian controls proteolytic processing of Notch and mediates lateral inhibition during Drosophila and vertebrate neurogenesis. *Cell*, 90, 271-280.

[68] Brou, C., Logeat, F., Gupta, N., Bessia, C., LeBail, O., Doedens, J.R., Cumano, A., Roux, P., Black, R.A. and Israel, A. (2000) A novel proteolytic cleavage involved in Notch signaling: the role of the disintegrin-metalloprotease TACE. *Mol. Cell*, 5, 207-216.

[69] Kopan, R., Schroeter, E.H., Weintraub, H. and Nye, J.S. (1996) Signal transduction by activated mNotch: importance of proteolytic processing and its regulation by the extracellular domain. *Proc. Natl. Acad. Sci. USA*, 93, 1683-1688.

[70] Schroeter, E.H., Kisslinger, J.A. and Kopan, R. (1998) Notch-1 signalling requires ligand-induced proteolytic release of intracellular domain. *Nature*, 393, 382-386.

[71] Okochi, M., Steiner, H., Fukumori, A., Tanii, H., Tomita, T., Tanaka, T., Iwatsubo, T., Kudo, T., Takeda, M. and Haass, C. (2002) Presenilins mediate a dual intramembranous gamma-secretase cleavage of Notch-1. *EMBO J.*, 21, 5408-5416.

[72] Esch, F.S., Keim, P.S., Beattie, E.C., Blacher, R.W., Culwell, A.R., Oltersdorf, T., McClure, D. and Ward, P.J. (1990) Cleavage of amyloid beta peptide during constitutive processing of its precursor. *Science*, 248, 1122-1124.

[73] Vassar, R., Bennett, B.D., Babu-Khan, S., Kahn, S., Mendiaz, E.A., Denis, P., Teplow, D.B., Ross, S., Amarante, P., Loeloff, R. *et al.* (1999) Beta-secretase cleavage of Alzheimer's amyloid precursor protein by the transmembrane aspartic protease BACE. *Science*, 286, 735-741.

[74] Buxbaum, J.D., Liu, K.N., Luo, Y., Slack, J.L., Stocking, K.L., Peschon, J.J., Johnson, R.S., Castner, B.J., Cerretti, D.P. and Black, R.A. (1998) Evidence that tumor necrosis factor alpha converting enzyme is involved in regulated alpha-secretase cleavage of the Alzheimer amyloid protein precursor. *J. Biol.Chem*, 273, 27765-27767.

[75] Lammich, S., Kojro, E., Postina, R., Gilbert, S., Pfeiffer, R., Jasionowski, M., Haass, C. and Fahrenholz, F. (1999) Constitutive and regulated alpha-secretase cleavage of Alzheimer's amyloid precursor protein by a disintegrinmetalloprotease. *Proc. Natl. Acad. Sci. USA*, 96, 3922-3927.

[76] Farzan, M., Schnitzler, C.E., Vasilieva, N., Leung, D. and Choe, H. (2000) BACE2, a beta -secretase homolog, cleaves at the beta site and within the amyloid-beta region of the amyloid-beta precursor protein. *Proc. Natl. Acad. Sci. USA*, 97, 9712-9717.

[77] Lee, M.K., Slunt, H.H., Martin, L.J., Thinakaran, G., Kim, G., Gandy, S.E., Seeger, M., Koo, E., Price, D.L. and Sisodia, S.S. (1996) Expression of presenilin 1 and 2 (PS1 and PS2) in human and murine tissues. *J.Neurosci*, 16, 7513-7525.

[78] Doan, A., Thinakaran, G., Borchelt, D.R., Slunt, H.H., Ratovitsky, T., Podlisny, M., Selkoe, D.J., Seeger, M., Gandy, S.E., Price, D.L. *et al.* (1996) Protein topology of presenilin 1. *Neuron*, 17, 1023-1030.

[79] Thinakaran, G., Borchelt, D.R., Lee, M.K., Slunt, H.H., Spitzer, L., Kim, G., Ratovitsky, T., Davenport, F., Nordstedt, C., Seeger, M. *et al.* (1996) Endoproteolysis of presenilin 1 and accumulation of processed derivatives in vivo. *Neuron*, 17, 181-190.

[80] Yu, G., Nishimura, M., Arawaka, S., Levitan, D., Zhang, L., Tandon, A., Song, Y.Q., Rogaeva, E., Chen, F., Kawarai, T. *et al.* (2000) Nicastrin modulates presenilin-mediated notch/glp-1 signal transduction and betaAPP processing. *Nature*, 407, 48-54.

[81] Shah, S., Lee, S.F., Tabuchi, K., Hao, Y.H., Yu, C., LaPlant, Q., Ball, H., Dann, C.E., 3rd, Sudhof, T. and Yu, G. (2005) Nicastrin functions as a gamma-secretase-substrate receptor. *Cell*, 122, 435-447.

[82] Struhl, G. and Adachi, A. (2000) Requirements for presenilin-dependent cleavage of notch and other transmembrane proteins. *Mol. Cell*, 6, 625-636.

[83] Ikeuchi, T. and Sisodia, S.S. (2003) The Notch ligands, Delta1 and Jagged2, are substrates for presenilin-dependent "gamma-secretase" cleavage. *J. Biol. Chem.*, 278, 7751-7754.

[84] LaVoie, M.J. and Selkoe, D.J. (2003) The Notch ligands, Jagged and Delta, are sequentially processed by alpha-secretase and presenilin/gamma-secretase and release signaling fragments. *J. Biol. Chem.*, 278, 34427-34437.

[85] Hiratochi, M., Nagase, H., Kuramochi, Y., Koh, C.S., Ohkawara, T. and Nakayama, K. (2007) The Delta intracellular domain mediates TGF-beta/Activin signaling through binding to Smads and has an important bi-directional function in the Notch-Delta signaling pathway. *Nucleic Acids Res.*, 35, 912-922.

[86] Edbauer, D., Willem, M., Lammich, S., Steiner, H. and Haass, C. (2002) Insulin-degrading enzyme rapidly removes the beta-amyloid precursor protein intracellular domain (AICD). *J. Biol.Chem*, 277, 13389-13393.

[87] Schulz, J.G., Annaert, W., Vandekerckhove, J., Zimmermann, P., De Strooper, B. and David, G. (2003) Syndecan 3 intramembrane proteolysis is presenilin/gamma-secretase-dependent and modulates cytosolic signaling. *J. Biol. Chem.*, 278, 48651-48657.

[88] Ni, C.Y., Murphy, M.P., Golde, T.E. and Carpenter, G. (2001) gamma -Secretase cleavage and nuclear localization of ErbB-4 receptor tyrosine kinase. *Science*, 294, 2179-2181.

[89] Buoso, E., Lanni, C., Schettini, G., Govoni, S. and Racchi, M. beta-Amyloid precursor protein metabolism: focus on the functions and degradation of its intracellular domain. *Pharmacol. Res*, 62, 308-317.

[90] Kimberly, W.T., Zheng, J.B., Town, T., Flavell, R.A. and Selkoe, D.J. (2005) Physiological regulation of the beta-amyloid precursor protein signaling domain by c-Jun N-terminal kinase JNK3 during neuronal differentiation. *J.Neurosci*, 25, 5533-5543.

[91] Lott, I.T. and Head, E. (2005) Alzheimer disease and Down syndrome: factors in pathogenesis. *Neurobiol. Aging*, 26, 383-389.

[92] Goate, A., Chartier-Harlin, M.C., Mullan, M., Brown, J., Crawford, F., Fidani, L., Giuffra, L., Haynes, A., Irving, N., James, L. *et al.* (1991) Segregation of a missense mutation in the amyloid precursor protein gene with familial Alzheimer's disease. *Nature*, 349, 704-706.

[93] Hardy, J. and Selkoe, D.J. (2002) The amyloid hypothesis of Alzheimer's disease: progress and problems on the road to therapeutics. *Science*, 297, 353-356.

[94] Selkoe, D.J. (2002) Alzheimer's disease is a synaptic failure. *Science*, 298, 789-791.

[95] Klein, W.L., Krafft, G.A. and Finch, C.E. (2001) Targeting small Abeta oligomers: the solution to an Alzheimer's disease conundrum? *Trends Neurosci*, 24, 219-224.

[96] Terry RD, K.R., Bick KL, Sisodia SS. (1999) *Alzheimer's disease.* Lippincott, Williams and Wilkins, Philadelphia.

[97] A, A. (2008) The plaque plan. *Nature*, 456, 161-164.

[98] Schnabel, J. (2009) Alzheimer's theory makes a splash. *Nature*, 459, 310.

[99] Nikolaev, A., McLaughlin, T., O'Leary, D.D. and Tessier-Lavigne, M. (2009) APP binds DR6 to trigger axon pruning and neuron death via distinct caspases. *Nature*, 457, 981-989.

[100] Ghosal, K., Vogt, D.L., Liang, M., Shen, Y., Lamb, B.T. and Pimplikar, S.W. (2009) Alzheimer's disease-like pathological features in transgenic mice expressing the APP intracellular domain. *Proc. Natl. Acad. Sci. USA*, 106, 18367-18372.

[101] Ryan, K.A. and Pimplikar, S.W. (2005) Activation of GSK-3 and phosphorylation of CRMP2 in transgenic mice expressing APP intracellular domain. *J. Cell Biol*, 171, 327-335.

[102] Vogt, D.L., Thomas, D., Galvan, V., Bredesen, D.E., Lamb, B.T. and Pimplikar, S.W. (2009) Abnormal neuronal networks and seizure susceptibility in mice overexpressing the APP intracellular domain. *Neurobiol. Aging*.

[103] Vazquez, M.C., Vargas, L.M., Inestrosa, N.C. and Alvarez, A.R. (2009) c-Abl modulates AICD dependent cellular responses: transcriptional induction and apoptosis. *J. Cell Physiol.*, 220, 136-143.

[104] Blalock, E.M., Geddes, J.W., Chen, K.C., Porter, N.M., Markesbery, W.R. and Landfield, P.W. (2004) Incipient Alzheimer's disease: microarray correlation analyses reveal major transcriptional and tumor suppressor responses. *Proc. Natl.Acad.Sci. USA*, 101, 2173-2178.

[105] Liang, W.S., Dunckley, T., Beach, T.G., Grover, A., Mastroeni, D., Ramsey, K., Caselli, R.J., Kukull, W.A., McKeel, D., Morris, J.C. *et al.* (2008) Altered neuronal gene expression in brain regions differentially affected by Alzheimer's disease: a reference data set. *Physiol. Genomics*, 33, 240-256.

In: Amyloids
Editors: I. P. Halcheck et al.
ISBN: 978-1-62100-538-4

Chapter 3

STUDY OF AMYLOID PRECURSOR PROTEIN EXPRESSION AFTER TRAUMATIC BRAIN INJURY IN THE RAT

Tatsuki Itoh[1*], Motohiro Imano[2], Shozo Nishida[3], Masahiro Tsubaki[3], Shigeo Hashimoto[4], Akihiko Ito[1] and Takao Satou[1,5]

[1]Department of Pathology, Kinki University Faculty of Medicine, Osaka, Japan
[2]Department of Surgery, Kinki University Faculty of Medicine, Osaka, Japan
[3]Kinki University Faculty of Pharmaceutical Sciences, Osaka, Japan
[4]Department of Pathology, PL Hospital, Osaka, Japan
[5]Division of Hospital Pathology, Hospital of Kinki University Faculty of Medicine, Osaka, Japan

ABSTRACT

The β-amyloid protein derived from amyloid precursor protein (APP) is a major component of senile plaque, and β-amyloid protein deposits in Alzheimer disease (AD) are thought to be produced by alternative processing of APP. Therefore, the overexpression of APP is an important factor in the neuronal degeneration of AD. The expression of APP has also been shown to increase in swollen axons and neuronal perikarya after neuronal injury, including ischemia and stab injury. As the pathological findings in dementia pugilistica mimic those in AD, it has been posited that traumatic brain injury (TBI) is an epidemiological risk factor for AD, although other genetic factors, which include the apoE genotype, might act additively. These findings suggest that the overexpression of APP after TBI may potentiate Alzheimer disease pathology. However, the chronological changes in APP expression and cerebral function of APP have not been evaluated after TBI.

In this study, we immunohistochemically investigated chronogical changes in cellular sources and levels of APP production compared to that of mRNA for APPas assessed by RT-PCR in the rat brain following TBI. Traumatic brain injury increased the

* Corresponding author: Tatsuki Itoh, Address: 377-2, Ohno-higashi, Osakasayama-city, Osaka 589-8511, Japan. Email: tatsuki@med.kindai.ac.jp. Tel: +81-72-366-0221. Fax: +81-72-360-2028

expression of APP in the neuronal perikarya and in damaged dystrophic neurites from 1 day to 90 days after injury. Moreover, at 7 days after injury, some macrophages/microglia also were co-localized with APP, which was overproduced by the neuronal perikarya and APP-positive dystrophic neurites after injury and then APP were phagocytosed by macrophages/microglia during this phase. However, astroglia did not express APP immunopositivity after brain injury. These results suggested that long-term overexpression of APP was confirmed by immunohistochemical and biological technique after TBI.

Subsequently, we investigated the effects of direct infusion of an anti-APP antibody into the damaged brain region on cerebral function and morphologically changes following TBI in the rat. At three days after TBI, there were many TUNEL-positive neurons and astrocytes around damaged area and a significant increase in numbers of TUNEL-positive cells in the PBS group versus the anti-APP group. At seven days after TBI, there was a significant increase in the numbers of large GFAP-positive cells, long elongated projections, and MAP-2-positive cells around damaged area in the anti-APP group versus the PBS group. At seven days after TBI, both the area of the damaged brain region and the arrival time to platform were significantly decreased in the anti-APP group versus the PBS group. Further, after TBI the arrival time to platform in the anti-APP group recovered to that observed in uninjured sham operation group animals.

These data suggest that overproduction of APP after TBI inhibits astrocyte activity and reduces neural cell survival around the damaged brain region.

On the basis of these findings, we speculate that overexpression of APP after TBI is related to Alzheimer type dementia, and is an important risk factor for this disease.

INTRODUCTION

The β-amyloid protein derived from amyloid precursor protein (APP) [1,2] is a major component of senile plaque, and β-amyloid protein deposits in Alzheimer disease (AD) [3,4] are thought to be produced by alternative processing of APP [5]. Therefore, the overexpression of APP is an important factor in the neuronal degeneration of AD [6]. The expression of APP has also been shown to increase in swollen axons and neuronal perikarya after neuronal injury, including ischemia [7] and stab injury [8]. As the pathological findings in dementia pugilistica mimic those in AD [9], it has been posited that traumatic brain injury (TBI) is an epidemiological risk factor for AD [10,11], although other genetic factors, which include the apoE genotype, might act additively [11,12]. These findings suggest that the overexpression of APP after TBI may potentiate Alzheimer disease pathology. However, the chronological changes in APP expression have not been evaluated after TBI. Furthermore, the function of APP has not been evaluated after TBI.

In this study,

(1) We immunohistochemically investigated chronogical changes in cellular sources and levels of APP production compared to that of mRNA for APP as assessed by RT-PCR in the rat brain following traumatic brain injury.
(2) In addition, we investigated the function of APP by assessing water maze and morphological changes following direct infusion of the anti-APP antibody into the damaged brain region following TBI.

Materials and Methods

Animals and Surgical Procedures

Male Wistar rats (8–9 weeks old, 200–250g in weight) were anesthetized by intraperitoneal pentobarbital (50mg/kg) injection. The scalp was incised on the midline and the skull was exposed. A 2–2.5mm hole was drilled (1 mm posterior, + 1 mm right lateral to bregma) in the right parietal calvaria. Brain injury above the dura mater was then induced with a pneumatic control injury device [13-15] at an impact velocity of 4m/sec (impact tip diameter of 1mm; fixed impact deformation of 2mm depth from the cerebral surface) [13-15]. Control rats were subjected to a sham operation, but no cortical penetration injury was inflicted. The contralateral hemisphere was not used as a control, since that area may have been affected by the impact. Rats were placed in a heated cage to maintain their body temperature at 37°C during the recovery from anesthesia.

Drug Infusion

The drug infusion protocol was as previously described [16]. Immediately, after TBI, the animals were placed in a stereotactic frame and fitted with a brain infusion cannula (Alzet brain infusion kit 3, 1mm, Alzet, San Diego, CA, USA). The cannula was implanted with the tip inserted into the damaged brain region (1 mm posterior, + 1 mm right lateral to bregma). The cannula was secured with dental cement. A micro-osmotic pump (Model 2001; rate of 1μl/h, Alzet), filled with 300 μl infusion volume kept 37°C, was implanted subcutaneously in the neck and connected to the infusion cannula. Rats received an infusion of either anti-APP antibody (IgG2a isoform, 1μg/ml; anti-APP antibody group; Chemicon, Temecula, CA, USA) diluted in sterilized phosphate buffer saline (0.1M PBS, pH 7.4–7.5; n=20) or sterilized PBS only (PBS group; n=20) for seven days immediately after TBI. For the control of the Morris water maze experiments, ten animals with no operation (sham operation group, no injury) were also collected. In sham operation group, a micro-osmotic pump was implanted subcutaneously in the neck without infusion cannula. Moreover, sham operation group was used only the behavioural experiments. As probe and the extent of APP antibody infusion test, although data not shown in this study, we investigated that the extent of APP antibody diffusion with Evans blue using the osmotic pump and confirmed Evans blue extended whole brain from center of infusion area.

Morris Water Maze Experiments

A circular, thermostatically regulated, dark gray PVC-plastic water tank (180cm wide, 45cm deep, filled with tap water at 22±1°C), located in the center of the testing room and surrounded by extra-maze cues, was used in the spatial learning task. A constant asymmetrical array of lamps and pictures served as cues for spatial orientation. A circular dark gray platform (15cm wide) submerged 1cm below the water surface served as a platform. The platform was placed in the center of one of the quadrants, i.e., the target quadrant, of the water maze [17]. The experiments were monitored using a digital TV system connected to a computer [17]. Training took place between 8:00a.m. and 3:00p.m. during seven consecutive days. Each daily training session consisted of four trials with a 120s cutoff time, followed by 30s rest on the platform. Memory was tested in fifteen animals from each of the anti-APP antibody group and the PBS group, and in ten sham operation group animals, starting at day one and continued up to seven days after TBI/sham operation.

Reverse Transcription-Polymerase Chain Reaction (RT-PCR)

The mRNA expression levels were investigated by RT-PCR at 1 to 90 days after TBI. Five rats were used for each time point. Briefly, total RNA was isolated from an area of the cerebral cortex (without the corpus callosum and hippocampus) with a diameter of 2 mm from the center of the lesion using RNA-Bee (Tel-Test Inc., Friendswood, TX) and redissolved in water, before the concentration was determined photometrically using the wavelength ratio of 260 nm/280 nm.

The two oligonucleotide primers used for APPmRNA were: (accession No X07648): sense primer: 5 '-GGA TGC GGA GTT CGG ACA TG -3 '; antisense primer: 5 '-GTT CTG CAT CTG CTC AAA G -3 '; Rat glyceraldehyde-3-phosphate dehydrogenase (GAPDH) (CLONTECH, No5507, BD Biosciences Pharmingen, San Diego, CA) was used as an internal control.

RT-PCR was performed using an mRNA PCR Kit (TaKaRa Biochemicals Co., Kyoto, Japan) according to the manufacturer's instructions. RT was performed at 42°C for 25 min, followed by heat inactivation at 99°C for 5 min. PCR amplification was carried out in a PCR EXPRESS (Hybaid US, Franklin, MA) for 30 cycles of 2 min at 94°C, 30 s at 94°C, 30 s at 55°C and 1.5 min at 72°C.

The final step was extended to 5 min at 72°C. After separation by electrophoresis in 1.5% agarose gels, the products were stained with ethidium bromide and located by fluorescence using UV light (3UV Transilluminator; UVP, Upland, CA). Bio Max 1D TM1.5.1 (Kodak, Tokyo, Japan) was used to evaluate the band intensities of the PCR bands. The signals were normalized to the corresponding GAPDH signal in the same RNA sample.

Enzyme-Linked Immunosorbent Assay (ELISA) for APP

The APP protein levels after TBI were investigated by ELISA using extracts of the TBI area. Five rats were used for each time point. Briefly, brain tissue from the above-described TBI area was isolated at each time point, and homogenized (10% w/v) in homogenizing buffer (PBS containing 0.25 mol/L sucrose, 5 mmol/L ethylene glycol bis(2-aminoethyl ether)-tetraacetic acid (EGTA), 25 μg/ml leupeptin, 25 μg/ml aprotinin) at 4°C using a hand homogenizer.

After treatment of the homogenates with ultrasonic waves at 4°C for 100 s and centrifugation at 20,000 *g* for 30 min at 4°C, the resulting supernatants were collected. The protein concentrations of the supernatants were determined using a DC Protein Assay Kit (Bio-Rad Laboratories, Hercules, CA) according to the manufacturer's instructions, and then adjusted to 1 μg/μl. The brain tissue samples were stored at -80°C until analysis by ELISA.

ELISAs were performed as previously described. Briefly, flat-bottomed microtiter plates (type H for ELISA; Sumitomo Bakelite Co., Tokyo, Japan) were coated with a rabbit polyclonal anti-APP antibody (diluted to 0.5 μg/ml in PBS containing 3% BSA (PBS-B); Chemicon) and blocked with 300 μl of PBS containing 3% BSA, 1% sucrose and 0.05% Tween 20 for 120 min at room temperature. The plates were then washed three times with PBS containing 0.05% Tween 20 (PBS-T). Recombinant APP (Chemicon) was serially diluted with PBS-B (0.001-12 ng/ml) and used to construct a standard curve. Aliquots (100 μl) of the diluted samples were incubated in the wells at 4°C overnight. After washing, 100 μl of a mouse monoclonal anti-APP antibody (1:10000 dilution in PBS-B; Chemicon) was added to each well and incubated for 2 h at room temperature. After washing, 100 μl of simple-stain MAX-PO (mouse)-kit (Nichirei) consisting of a peroxidase-conjugated anti-mouse IgG was added to each well. Color development was carried out using an Enzyme Reaction Kit Type T (Sumitomo Bakelite Co., Tokyo, Japan), and the absorbances at 490 nm were measured using a microplate reader (Model 3550UV; Bio-Rad Laboratories). The detection limit of the assay for APP was 0.005 ng/ml. The measurements were made in duplicate.

Immunohistochemistry

1, 3, 7, 30 and 90 days after TBI and after the last the Morris water maze test at seven days, rats from each group were perfused intracardially with 300ml of PBS followed by 300ml of 4% paraformaldehyde (PFA; pH 7.4–7.5) in PBS. The brains were then removed and stored in PFA for three days, before the maximum size of the lesion was sliced into serial coronal sections (50μm thick) using a microslicer (Dousaka EM, Kyoto, Japan).

Sections were treated with 3% H_2O_2 in Tris-buffered saline (TBS; 0.1M Tris-HCl, pH 7.5, 0.15M NaCl) containing 0.1% Triton X-100 (TBS-T) for 30min. Next, the sections were washed three times with TBS-T, blocked with 3% bovine serum albumin (BSA; Sigma-Aldrich, St. Louis, MO, USA) in TBS-T for 30min, and incubated with a mouse monoclonal anti- microtubule associate protein-2 (MAP-2) antibody as a marker of neuronal cell (1:10000 dilution; Sigma-Aldrich), a mouse monoclonal anti-neurofilament (NF) antibody as a marker of neuronal cell (1:10000 dilution; DAKO, Glostrup, Denmark), a rabbit polyclonal anti-Glial Fibrillary Acidic Protein (GFAP) antibody (1:10000 dilution; DAKO, Glostrup, Denmark), a mouse monoclonal anti-APP antibody (1:1000 dilution; Chemicon, Temecula, CA, USA) or a mouse monoclonal anti-CD11b antibody as a marker of microglia/ macrophage (1:1000 dilution; Chemicon, Temecula, CA, USA) overnight at room temperature. Following extensive washing, the sections were further incubated with a HISTIFINE Rat-PO (multi)-kit (Nichirei, Osaka, Japan), consisting of a mixed solution of peroxidase-conjugated anti-mouse and rabbit IgG as the secondary antibody, for 60min at room temperature. The HISTIFINE Rat-PO kits contained preabsorbed rat serum, and showed very minimal non-specific binding by rat serum in injured rat tissues. Labeling was visualized using diaminobenzidine (DAB; Vector Peroxidase Substrate Kit; Vector Laboratories, Burlingame, CA, USA) for 5min, and the sections were counterstained with hematoxylin.

For TUNEL-staining at three days after TBI, the brains from five rats in each of the anti-APP antibody group and the PBS group were collected as above. Sections were used for TUNEL-staining, a marker of apoptosis. TUNEL-staining was performed using an *in situ*

Apoptosis Detection Kit (TaKaRa Biochemicals Co., Kyoto, Japan) according to the manufacturer's instructions. Sections were treated with Proteinase K for 10min, followed by treatment with 3% H_2O_2 in TBS-T for 30min. Next, the sections were washed three times with TBS-T, and incubated with a TdT Enzyme Labeling solution for 90min at room temperature. Labeling was visualized using DAB. The controls for the TUNEL stains were performed without enzyme. Furthermore, to investigate TUNEL-positive cells phenotype, MAP-2 or GFAP immunofluorescence staining was performed following TUNEL-staining. Sections were treated with Proteinase K for 10min. Next, the sections incubated with a fluorescein isothiocyanate (FITC)-conjugated TdT Enzyme Labeling solution for 90min at room temperature. Next, the sections were washed extensively and incubated with an anti-MAP-2 antibody (1:300 dilution; Sigma) or anti-GFAP antibody (1:300 dilution; DAKO) overnight at room temperature. Following extensive washing, the sections were further incubated with Alexa Flour 555 anti-mouse IgG (for anti-MAP-2 antibody) or rabbit IgG (for anti-GFAP antibody, 1:300 dilution; BD Biosciences Pharmingen, San Diego, CA) for 80 min at room temperature. Subsequently, the sections were observed using a confocal laser-scanning microscope (LSM5 PASCAL; Carl Zeiss Jena GmbH, Jena, Germany).

Double-Immunofluorescence Staining

Each serial section was washed with TBS-T, blocked with 3% BSA in TBS-T for 30 min, and incubated with a polyclonal rabbit anti-APP antibody (1:300 dilution; Chemicon) overnight at room temperature. Following extensive washing, the sections were further incubated with a polyclonal rhodamine-conjugated anti-rabbit IgG antibody (1:300 dilution; DAKO) for 80 min at room temperature. Next, the APP-stained sections were washed extensively and incubated with a monoclonal mouse anti-NF antibody (1:300 dilution; DAKO), a marker for neurons and nerve fiver; a monoclonal mouse anti-MAP-2 antibody (1:300 dilution; Sigma), a marker for neurons; a monoclonal mouse anti-CD11b antibody (1:300 dilution; AbD Serotec, Raleigh, NC), a marker for macrophages/microglia or a monoclonal mouse anti-GFAP antibody (1:300 dilution; DAKO), a marker for astroglia overnight at room temperature. Following extensive washing, the sections were further incubated with a polyclonal fluorescein isothiocyanate (FITC) -conjugated anti-mouse IgG antibody (1:300 dilution; DAKO) for 80 min at room temperature. Subsequently, the sections were observed by fluorescence microscopy (Nikon E-800; Nikon, Tokyo, Japan).

Area Measurement of Damaged Brain Region

To determine the area of the damage region at seven days after TBI, after the last the Morris water maze five rats per group were collected as above, and the anteroposterior of the maximum size of the lesion was sliced into forty serial coronal sections (50μm thick). Anteroposterior sections 1–40 were stained with hematoxylin and eosin (HE). Images of the HE stain in the forty anteroposterior serial sections were captured using a Nikon E 1000M microscope at 1X magnification and a CCD camera. The area of the damaged region in each image was traced and measured by computer, and the average area of the damaged region calculated over the forty serial sections.

Quantification

To determined the number of APP -positive neurites and cells, MAP-2-, GFAP-, or TUNEL-positive cells, each DAB-positive cells around the damaged area (cortex) without white matter after TBI were observed and DAB-labeled cells were counted in three serial sections (each section was 50 µm in thickness) under a Nikon E 1000M microscope (Nikon Corporation, Tokyo, Japan) using a 20x objective. To determine the measured area of DAB-positive cells, an image of the measured area was captured under the Nikon E 1000M microscope at 1x magnification using a CCD camera (ACT-2U; Nikon Corporation). The measured area in each image was traced and measured using a computer (Power Macintosh G3; Apple Computers, Cupertino, CA) and the NIH Image 1.6 software (NIH, Bethesda, MD). The number of DAB-positive cells was expressed as positive cell number/100μm^3.

Statistical Analysis

Water maze data were expressed as mean±SE. Data were analyzed using ANOVA and Fisher's PLSD-test (Stat View®; SAS Institute Inc, Cary, NC. USA). Other data were expressed as mean±SD and analyzed using ANOVA. $p<0.05$ was considered statistically significant.

RESULTS

RT-PCR analysis of APP

As determined by RT-PCR, the mRNA levels for APP were elevated after cortical injury. The APP transcript level displayed an increase in the cortex as early as 1 day following injury, with the elevation becoming significant between 1 day and 30 days and returning to the control level by 90 days ($p<0.05$, Figure 1A) . One peak of elevation was determined between 1 day and 30 days after injury. There were no such changes in the sham group (Figure 1A).

APP Protein in Brain Tissue after TBI

As determined by ELISA, the levels of APP were elevated after cortical injury. ELISA demonstrated a pattern similar to that showing by RT-PCR. APP levels displayed an increase in the injured side as early as 1 day following injury, with the elevation becoming significant between 1 day and 90 days ($p<0.001$, Figure 1B). One peak of elevation was determined between 1 day and 90 days after injury (Figure 1B). There were no such changes in the sham group (Figure 1B).

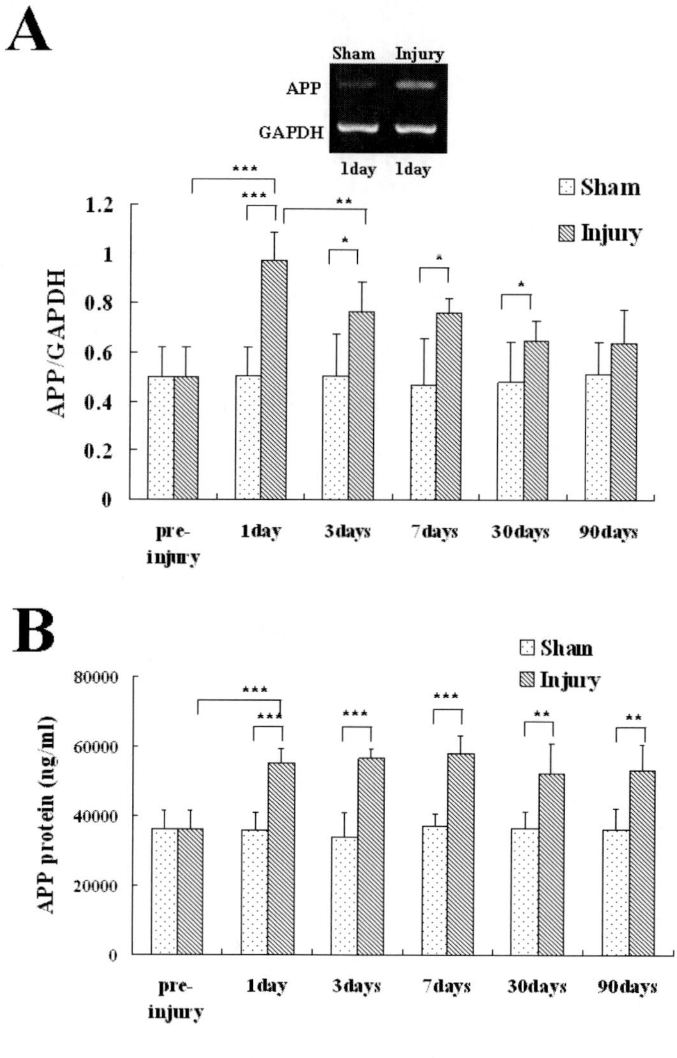

Figure 1. Graphs show the optical density units of ethidium bromide-stained RT-PCR products for APPmRNA expression (A) and the level of APP protein (B) after traumatic brain injury. The results are shown as the mean ± SD. $*P < 0.05$, $**P < 0.01$, $***P < 0.001$, $n = 5$.

Immunostaining and APP-Immunopositive Cell and Neurites Count after TBI

There were many APP-positive damaged swollen and dystrophic neurites at 1 day after TBI (Figure 2A).

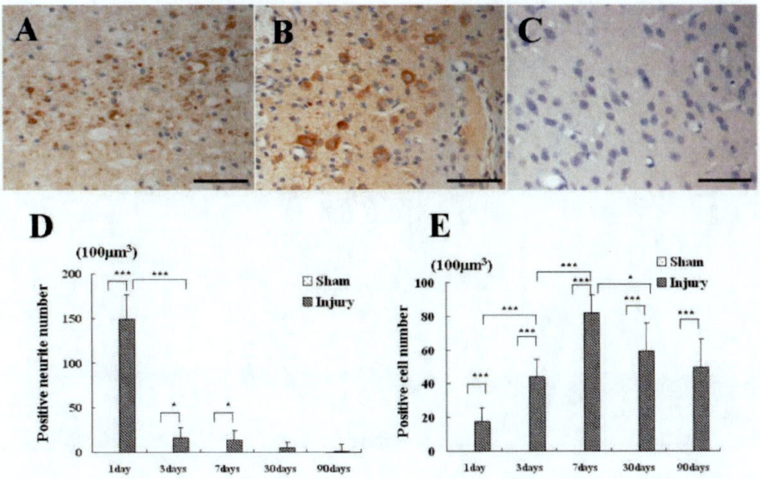

Figure 2. Immunostaining for APP around the damaged cerebral cortex after traumatic brain injury. At 1 day after injury, there are many APP-positive damaged dystrophic and swollen neurites (A). At 7 days after injury, there are many APP-positive neurons and a few APP-positive damaged dystrophic and swollen neurites (B). However, the sham-operated control cortex does not contain any APP staining or APP-immunopositive cells (C). Scale bar = 50 μm. Graph showing the numbers of APP-immunopositive neurites (D) and cells (E) around the damaged cerebral cortex after traumatic rat brain injury. The results are shown as the mean ± SD. $*P < 0.05$, $***P < 0.001$, $n = 5$.

At 7 days after TBI, there were many APP-positive neurons and a few APP-positive damaged swollen and dystrophic neurites (Figure 2B). However, there were no APP-immunopositive cells and fivers in the sham-operated cerebral cortex (Figure 2C). The number of APP-positive neurites (p<0.001, Figure 2D) and cells (p<0.001, Figure 2E) was increased in the cortex as early as 1 day following injury, with the elevation becoming significant from 1 day to 90 days(p<0.001, Figure 2E). There were no such changes in the sham group (Figure 2DandE).

Double-Immunofluorescence of Amyloid Precursor Protein (APP) and Neurofilament (NF), Microtubule-Associated Protein-2 (MAP-2), Glial Fibrillary Acidic Protein (GFAP) or CD11b

The neuronal or microglial localizations of APP protein in the cortical traumatic injured rats at 1 day, 3 days and 7 days post-lesion are illustrated (Figure 3 and 4).

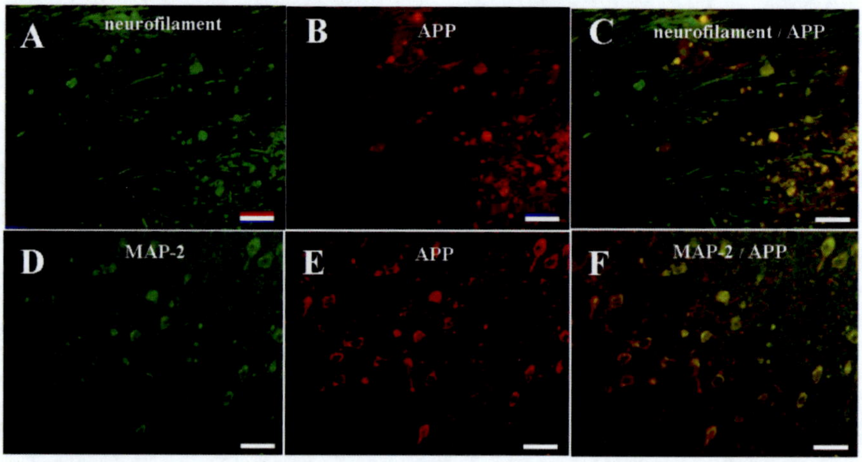

Figure 3. Double-immunofluorescence staining of localized neurofilaments (A) or MAP-2 (D) and APP (B, E) around the damaged area after traumatic brain injury. At 1 day after injury, damaged dystrophic and swollen neurites (A, green) in the cortex were APP-positive (B, red). The merged image (C) of panels (A) and (B) reveals colocalization of these proteins (yellow). At 7 days after injury, neurons near the injured portion were positive for MAP-2 (D, green) and APP (E, red). The merged image (F) of panels (D) and (E) reveals colocalization of these proteins (yellow). Scale bar = 50 μm.

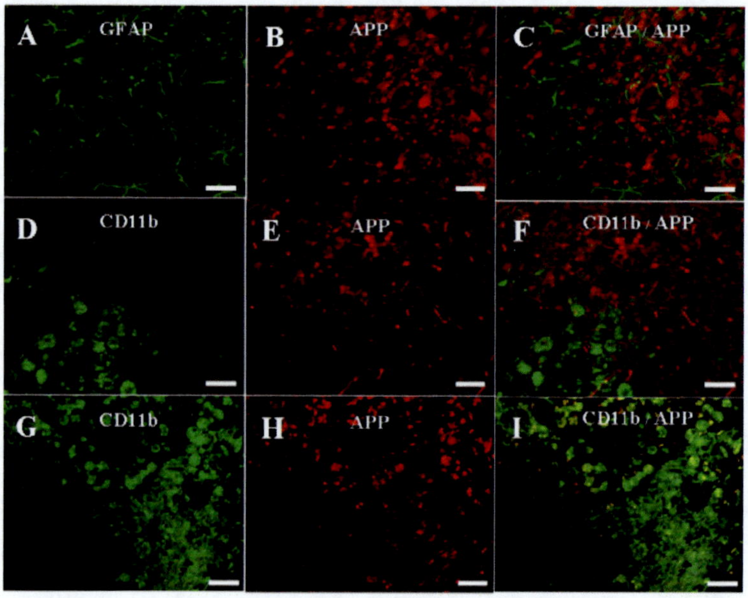

Figure 4. Double-immunofluorescence staining for GFAP (A) or CD11b (D, G) and APP (B, E, H) around the damaged area after traumatic brain injury. At 3 days after the injury, GFAP-immunopositive (A, green) and APP-immunopositive (B, red) cells are observed. The merged image (C) of panels (A) and (B) does not show any colocalization. At 3 days after the injury, CD11b-immunopositive (D, green) and APP-immunopositive (E, red) cells are observed. The merged image (F) of panels (D) and (E) does not show any colocalization. However, at 7 days after the injury, CD11b-immunopositive (G, green) and APP-immunopositive (H, red) cells are colocalization (I, yellow). Scale bar = 50 μm.

Neurons, astroglia or microglia were identified by immunohistochemical staining using anti-NF, anti-MAP-2, and anti-GFAP or anti-CD11b, respectively. Double labeling immunohistochemistry with anti-NF (Figure 3A) and anti-APP (Figure 3B) antibodies clearly demonstrated the co-localization of NF and APP proteins in swollen and dystrophic nerve fibers at 1day after TBI (Figure 3C). At 7 days after cortical traumatic injury, MAP-2-positive nerve cell cytoplasm (Figure 3D) was positive for APP (Figure 3E and F).

Anti-GFAP (Figure 4A) or anti-CD11b (Figure 4D) and anti-APP (Figure 4B and E) antibodies did not demonstrate co-localization in the same cells at 3 days after cortical traumatic injury (Figure 4C and F). At 7 days after traumatic injury, a few CD11b-positive cells (Figure 4G) expressed weakly positive APP immunoreactivity (Figure 4H and I).

Area of the Damaged Brain Region after TBI

HE images of the damaged brain region at seven days after TBI can be seen in Figure 5A-B. The area of the damaged brain region was significantly decreased in the anti-APP antibody group versus the PBS group ($p<0.001$, Figure 5C).

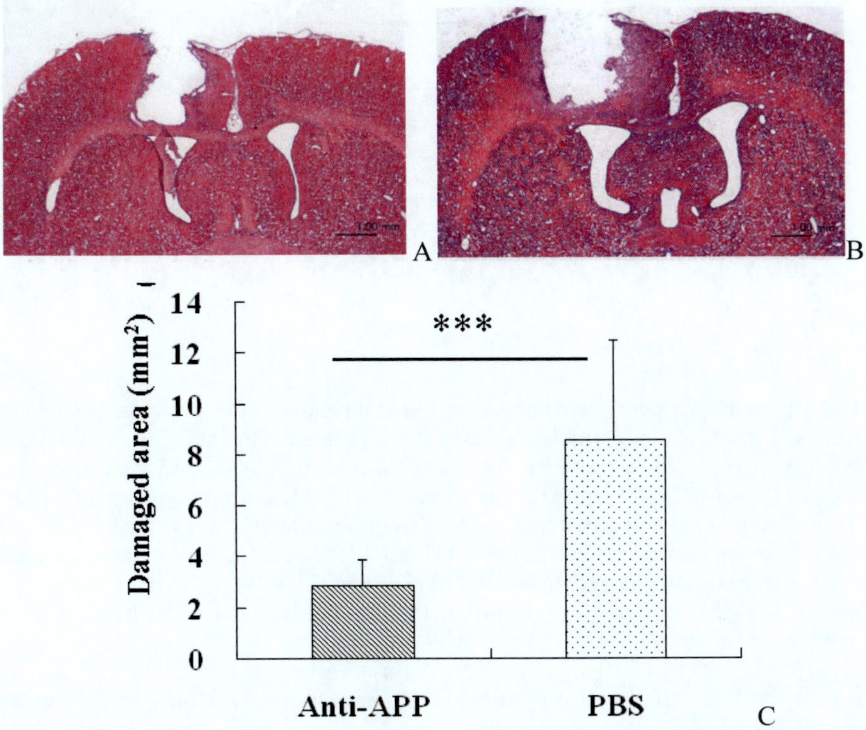

Figure 5. The size of the damaged brain region at seven days after TBI. HE stains show the anti-APP antibody group (1μg/ml) (A) and the PBS group (B). Scale bar = 1mm. (C) Measurement results of each group (n=5/group). Anti-APP antibody group *vs*. PBS group. ***$p<0.001$.

TUNEL-staining and TUNEL-positive cells counts after TBI

TUNEL-staining results and the numbers of TUNEL-positive cells around the damaged region at three days after TBI can be seen in Figure 6A-E.

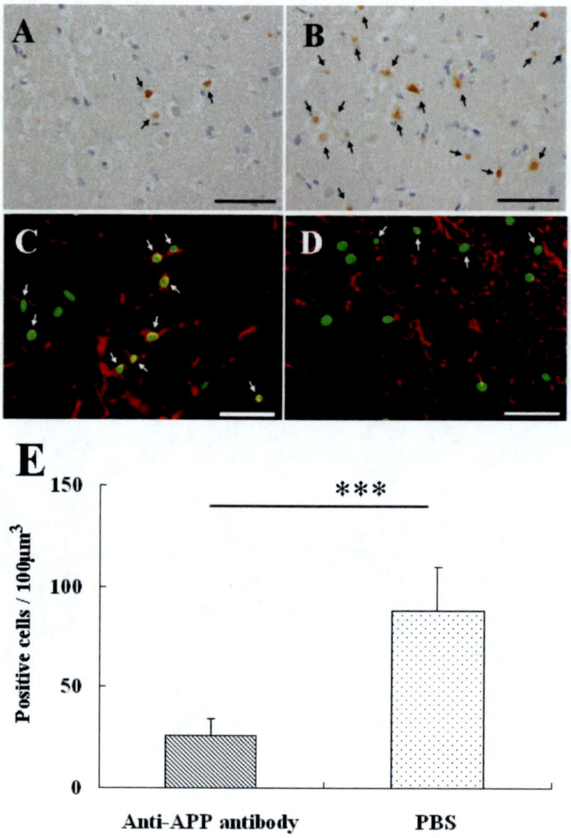

Figure 6. TUNEL-staining around the damaged region at three days after TBI. Note that there were only a few TUNEL-positive cells, which possessed small or large cytoplasms, in the anti-APP antibody group (1μg/ml; A, arrows), while many small or large cytoplasm TUNEL-positive cells were observed in the PBS group (B, arrows). (C) and (D) were indicated TUNEL-staining (green) and MAP-2 (C, red) or GFAP (D, red) immunofluorescence staining in PBS group. Many TUNEL-positive (C, green) cells showed MAP-2-immunopositive (C, red). A few TUNEL-positive (D, green) cells showed GFAP-immunopositive (D, red). The merged image of TUNEL- and MAP-2- or GFAP-positive cells showed arrows (CandD). Scale bar = 50μm. (E) Number of TUNEL-positive cells in each group (n=5/group). Anti-APP antibody group *vs*. PBS group. ***p<0.001.

In the anti-APP antibody group there were only a few TUNEL-positive cells which possessed a small cytoplasm, such as glial cells, or a large cytoplasm, such as neural cells (Figure 6A).

In contrast, in the PBS group there were many TUNEL-positive cells which possessed small or large cytoplasm's (Figure 6B). Furthermore, Almost TUNEL-positive cells in PBS group colocalize with the MAP-2 staining (Figure 6C). A few TUNEL-positive cells in PBS group colocalize with the GFAP staining (Figure 6D). In addition, the number of TUNEL-positive cells in the anti-APP antibody group was significantly decreased versus the PBS group ($p<0.001$, Figure 6E).

Immunostaining and GFAP-Immunopositive Cell Counts after TBI

Immunostaining results for GFAP expression and the numbers of GFAP-positive cells around the damaged brain region at seven days after TBI can be seen in Figure 7A-E. In the anti-APP antibody group there were many larger GFAP-positive cells with GFAP staining in their cytoplasm and long elongated projections around the damaged region (Figure 7. A-B).

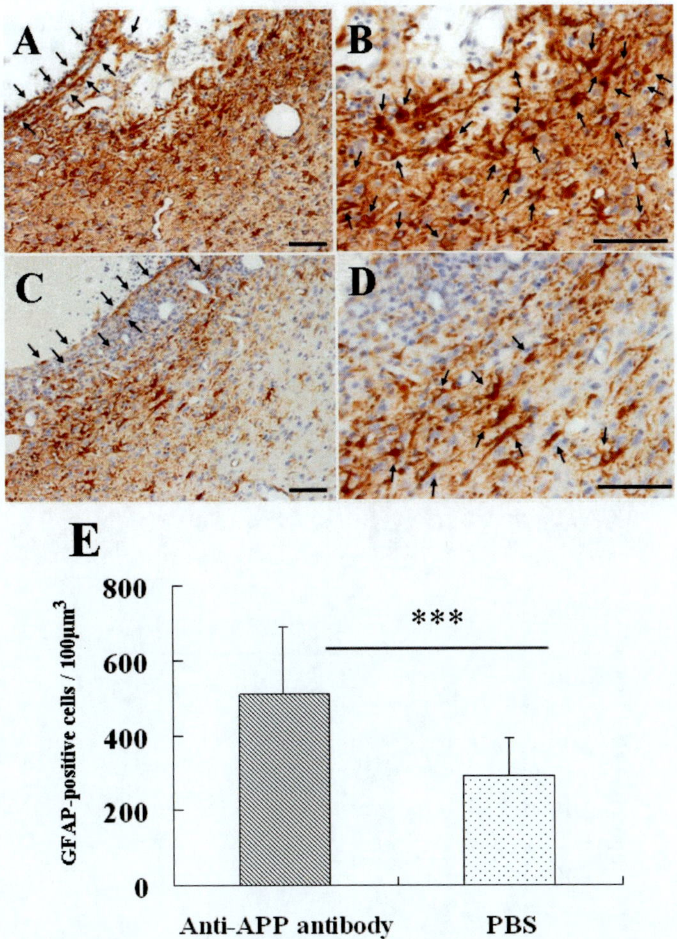

Figure 7. GFAP expression around the damaged region at seven days after TBI. Note that there were many glial scars (A, arrows) and many large GFAP-positive cells and projections (B, arrows) in the anti-APP antibody group (1μg/ml; A: low power, B: high power), while glial scars (C, arrows), only a few small GFAP-positive cells, and a few projections (D, arrows) were observed in the PBS group (C: low power, D: high power). Scale bar = 50μm. (E) Number of GFAP-positive cells in each group (n=10/group). Anti-APP antibody group *vs*. PBS group. ***$p<0.001$.

Furthermore, there were many GFAP-positive fibers that formed glial scars after TBI, and these fibers were enriched at the damaged brain region (Figure 7A). In contrast, in the PBS group there were only a few small GFAP-positive cells which possessed a small GFAP-positive cytoplasm and a few projections (Figure 7C-D). There was a significant increase in the number of GFAP-positive cells in the anti-APP antibody group versus the PBS group (p<0.001, Figure 7E).

Immunostaining and MAP-2-Immunopositive Cell Counts after TBI

Immunostaining results for MAP-2 expression and the numbers of MAP-2-positive cells around the damaged region at seven days after TBI can be seen in Figure 8A-C. In the anti-APP antibody group there were many large MAP-2-positive nerve cells (cytoplasm stained) and MAP-2-positive fibers (Figure 8A). In contrast, in the PBS group there were only a few small MAP-2-positive cells which possessed a small MAP-2-immunopositive cytoplasm and a few MAP-2-positive fibers (Figure 8B). There was a significant increase in the number of MAP-2-positive cells in the anti-APP antibody group versus the PBS group (p<0.001, Figure 8C).

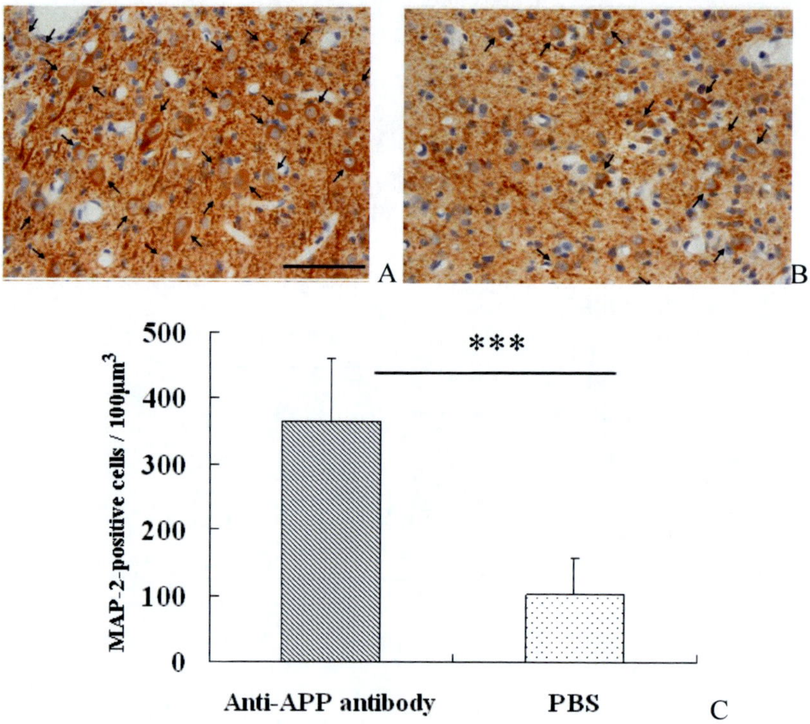

Figure 8. MAP-2 expression around the damaged region at seven days after TBI. Note that numerous large MAP-2-positive cells and fibers were observed in the anti-APP antibody group (1μg/ml; A, arrows), while only a few small MAP-2-positive cells and a few fibers were observed in the PBS group (B, arrows). Scale bar = 50μm. (C) The number of MAP-2-positive cells in each group (n=10/group). Anti-APP antibody group *vs*. PBS group. ***p<0.001.

Water Maze Experiments

When compared with the sham operation group, the arrival time to platform was significantly increased at seven days after TBI in the PBS group ($p<0.001$, Figure 9). However, at seven days after TBI, the arrival time to platform was significantly decreased in the anti-APP antibody group versus the PBS group ($p<0.001$, Figure 9), and was not different from the sham operation group at seven days after TBI (Figure 9).

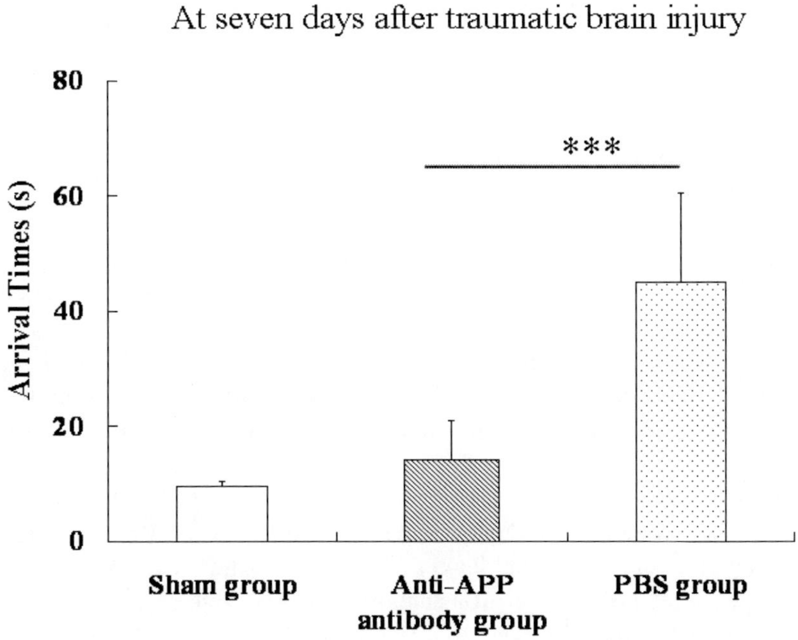

Figure 9. The effect of anti-APP antibody on cerebral function at seven days after TBI. Anti-APP antibody (1μg/ml) or PBS was infused over seven days by osmotic pump (1μl/hr). The sham operation group had no injury. The effects of treatment group on arrival time to platform are shown. Values represent mean±SE (n=10/group). PBS group *vs*. anti-APP antibody group *vs*. sham operation group. ***$p<0.001$.

Discussion

The hypothesis that head trauma is a risk factor associated with subsequent development of Alzheimer disease is based on the neuropathology of dementia [18,19]. Some reports have shown that the expression of APP is enhanced by various forms of brain injury, including traumatic [8], chemical [20] , and ischemic injury [7]. Further, the overexpression of APP suggests the possibility of an Alzheimer disease-like pathology after traumatic brain injury [21]. Previously, the appearance of APP in the brain during pathological events was usually attributed to its synthesis by neurons, macrophages, microglia or astrocytes [8]. However, APP was synthesized by neurons, as showing by immunohistochemical, ELISA and RT-PCR obtained performed in this study. In the present report, double labeling immunohistochemistry with anti-NF or anti-MAP-2 and anti-APP antibodies demonstrated the co-localization. Furthermore, we showed that the expression of APP in neurons and neurites was significantly increased in the cerebral cortex after traumatic brain injury. It has been pointed out that increased APP immunoreactivity occurrs in damaged axons due to disturbance of fast anterograde axonal transport [22]by destroyed cytoskeletal proteins, including neurofilaments [23] and MAP-2 [24]. These results suggested that APP synthesis was increased from the early phase after brain injury and it lasted for a long period during the experiment.

In the present study, the mRNA and protein levels for APP were elevated after TBI. It was previously reported that APP mRNA and protein levels were increased in neurons after TBI [25,26]. Furthermore, expression of APP was increased in neurons after ischemia and axonal injury [7,27]. Taken together, these results suggest that APP is synthesized by neurons, rather than glial cells and macrophages/microglia, after TBI. In future studies, it will be necessary to further clarify the cell type producing APP after TBI.

TBI occurs as the result of a direct mechanical insult to the brain, and induces degeneration and death in the central nervous system (CNS) [28,29]. Following the initial mechanical insult, secondary pathways are activated that contribute to the ischemic damage induced by circulatory disturbance, blood-brain barrier disruption and excitotoxic damage [30-32]. The expression of APP has also been shown to increase in swollen axons and neuronal perikarya after neuronal injuries, including ischemia [7] and stab injuries [8]. CNS disorders can be caused by the widespread neuronal and axonal degeneration induced by TBI [28,29]. Taken together, these results suggest that APP leaks out from damaged and necrotic axons and neuronal cytoplasm after TBI. In the present study, CD11b-positive cells were colocalized with APP protein at 7 days after TBI. It appeared that macrophages/microglia engulfed and digested the APP that leaked from axons and neurons during this phase. Therefore, at 7 days after TBI, both neurons and macrophages/microglia expressed APP.

The overexpression of APP suggests the possibility of an Alzheimer disease-like pathology after TBI [14,21]. Previously, the appearance of APP in the brain during pathological events was usually attributed to its synthesis by neurons, macrophages, microglia, or astrocytes [8]. Sun *et al.* reported that overproduction of APP induced expression of the apoptosis-related Fas antigen in cultured neural, astrocytes, and microglia [33]. Furthermore, APP induced nitric oxide synthetase (iNOS) and nitric oxide (NOS) production, and induced neural and glial cell apoptosis[33]. In the present study, a continuous infusion of the anti-APP antibody infusion inhibited neural and glial apoptotic cell death at three days after TBI. This neuroprotection was associated with numerous GFAP-positive cells and glial scars, suggesting that the anti-APP antibody inhibited APP-induced glial cell apoptosis.

It was previously reported that APP which was expressed on the cell membrane of cultured cortical neural cells controlled intracellular Ca^{2+} entry [34]. Intracellular entry of Ca^{2+} activates Ca^{2+}-dependent signal pathways and can induce neural cell death. Additionally, choline acetyltransferase (ChAT) activity was decreased in the frontal cortex and hippocampus of APP 695 transgenic mice, an Alzheimer model overexpressing APP, and increased the number of apoptotic neurons[35]. These APP 695 transgenic mice exhibit increased learning and memory impairment [35]. In the present study, in the anti-APP antibody group there was an increase in the number of MAP-2-positive cells and a reduction in the arrival time to platform versus the PBS group. These data suggested that the anti-APP antibody increased ChAT activity and inhibited neural degeneration induced by overproduced APP in the frontal cortex and hippocampus after TBI, while cerebral function may be improved. As such, measurement of ChAT activity in the anti-APP antibody and the PBS groups after TBI will form the basis of future research.

Recent evidence suggests that APP, a ubiquitously expressed, highly conserved integral membrane glycoprotein, is not only a sensitive marker of axonal injury [36,37], but may also have an important functional role following TBI. APP is upregulated acutely in injured neurons and reactive astrocytes following TBI [25,38], and this upregulation has been associated with increase hippocampal cell death [6]. Although the mechanisms leading to cell death are unknown, as the precursor to the neurotoxic Aβ protein, the conversion of APP to Aβ may have detrimental effects including an increased risk for development of Alzheimer's disease [39]. Indeed, there are numerous clinical studies demonstrating substantial Aβ deposition and the formation of amyloid plaques [40,41]. Furthermore, following TBI in the rat there is an increased β-secretase enzyme expression (BACE 11) that could potentially facilitate Aβ production [42], a direct link between increase APP levels and increased Aβ deposition and toxicity [43].

In contrast to the potential deleterious effects of APP, an early acute rise in APP during the reparative phase of injury has resulted in the hypothesis that APP may actually serve a neuroprotective function [44]. APP has been shown to be both beneficial and detrimental depending on its method of posttranslational processing within cells. The beneficial, secreted α form of APP (sAPPα) is generated by α-secretase cleavage, whereas the secreted APPβ (sAPPβ) and deleterious Aβ are generated from cleavage by β- and γ-secretases [45]. sAPPα has been reported to have many neuroprotective and neurotrophic functions within the central nervous system (CNS) [46]; for instance, sAPPα administration reduced neuronal injury and improved functional outcome following rat TBI [47]. The α-secretase pathway is a non-amyloidogenic pathway in which the majority of APP is normally processed [48], while after brain injury the β- and γ-secretase pathways are activated and are responsible for producing the toxic sAPPβ [49,50] and Aβ [42,43,51]. For instance, sAPPβ was shown to induce apoptotic cell death in PC 12 cells and neuronal degeneration in rat hippocampal neurons [49,50]. Nakagawa et al. reported that the APP overexpression of ST6Gal-1 in Neuro2 enhanced producing sAPP and the level of the extracellular sAPP β form is increased the level of sAPPα form[52]. In AD brain, increase of the extracellular sAPP beta level could affect the pathology of AD[52]. In the present study, anti-APP antibody treatment resulted in a reduction of neuronal and glial apoptotic cell death and recovery of memory and learning function to sham operation group levels, after TBI. These data suggest that toxic sAPPβ is produced more than sAPPα after TBI, and that toxic sAPPβ may be responsible for the neuronal and glial cell degeneration and death. Additionally, overproduction of sAPPβ after brain injury may facilitate Aβ production and deposition.

In the adult CNS, TBI results in a rapid response from resident astrocytes, a process often referred to as reactive astrocytosis or glial scarring [53,54]. Glial scars have been reported to inhibit neurite elongation of damaged neurons and axonal regeneration, and thus prevent functional recovery [53,54]. Furthermore, neurite outgrowth of cultured rat hippocampal neurons was found to be inhibited by glial scars *in vitro* [55]. However, gliosis and glial scars protect against secondary insults that contribute to the ischemic damage induced by circulatory disturbance, blood-brain barrier disruption, excitotoxic damage, and free radicals [56]. Moreover, reactive astrocytes secrete neurotrophic factor (NTF), nerve growth factor

(NGF), and extracellular matrix which induce axonal outgrowth and regeneration of the neural network [57,58]. In addition, the formation of glial scars prevents leakage of secreted factors, and separates non-damaged areas from damaged areas, thereby maintaining normal CNS homeostasis [59,60]. Therefore, although the role of gliosis and glial scars in neuronal regeneration after brain injury remains controversial, the results from the present study suggest that gliosis and glial scars after TBI might be beneficial.

The results from the present study also demonstrated that infusion of the anti-APP antibody into the damaged region following TBI inhibited degeneration of neuronal and glial cells induced by overproduced APP after brain injury, with a significant reduction in injury size. This may be attributed a protective effect of the anti-APP antibody against neural and glial cell death induced by overproduced APP after TBI.

Conclusion

In conclusion, the results of the present study have demonstrated that TBI induces long-term increases in APP overexpression in the neuronal perikarya and neurites. In addition, endogenous overproduced APP after TBI inhibited astrocyte activity around the damaged brain region and induced neural cell degeneration. On the basis of these findings, we speculate that overexpression of APP after TBI is related to Alzheimer type dementia, and is an important risk factor for this disease.

References

[1] Goldgaber D., Lerman M.I., McBride O.W., Saffiotti U. and Gajdusek D.C. (1987). Characterization and chromosomal localization of a cDNA encoding brain amyloid of Alzheimer's disease. *Science*, 235, 877-880.

[2] Robakis N.K., Ramakrishna N., Wolfe G. and Wisniewski H.M. (1987). Molecular cloning and characterization of a cDNA encoding the cerebrovascular and the neuritic plaque amyloid peptides. *Proc. Natl. Acad. Sci. USA*, 84, 4190-4194.

[3] Masters C.L., Multhaup G., Simms G., Pottgiesser J., Martins R.N. and Beyreuther K. (1985). Neuronal origin of a cerebral amyloid: neurofibrillary tangles of Alzheimer's disease contain the same protein as the amyloid of plaque cores and blood vessels. *Embo J,* 4, 2757-2763.

[4] Selkoe D.J., Abraham C.R., Podlisny M.B. and Duffy L.K. (1986). Isolation of low-molecular-weight proteins from amyloid plaque fibers in Alzheimer's disease. *J. Neurochem*., 46, 1820-1834.

[5] Golde T.E., Estus S., Younkin L.H., Selkoe D.J. and Younkin S.G. (1992). Processing of the amyloid protein precursor to potentially amyloidogenic derivatives. *Science*, 255, 728-730.

[6] Murakami N., Yamaki T., Iwamoto Y., Sakakibara T., Kobori N., Fushiki S. and Ueda S. (1998). Experimental brain injury induces expression of amyloid precursor protein, which may be related to neuronal loss in the hippocampus. *J. Neurotrauma*, 15, 993-1003.

[7] Stephenson D.T., Rash K. and Clemens J.A. (1992). Amyloid precursor protein accumulates in regions of neurodegeneration following focal cerebral ischemia in the rat. *Brain Res*, 593, 128-135.
[8] Otsuka N., Tomonaga M. and Ikeda K. (1991). Rapid appearance of beta-amyloid precursor protein immunoreactivity in damaged axons and reactive glial cells in rat brain following needle stab injury. *Brain Res,* 568, 335-338.
[9] Roberts G.W., Allsop D. and Bruton C. (1990). The occult aftermath of boxing. *J. Neurol. Neurosurg Psychiatry*, 53, 373-378.
[10] Schofield P.W., Tang M., Marder K., Bell K., Dooneief G., Chun M., Sano M., Stern Y. and Mayeux R. (1997). Alzheimer's disease after remote head injury: an incidence study. *J. Neurol. Neurosurg Psychiatry,* 62, 119-124.
[11] DeKosky S.T., Abrahamson E.E., Ciallella J.R., Paljug W.R., Wisniewski S.R., Clark R.S. and Ikonomovic M.D. (2007). Association of increased cortical soluble abeta42 levels with diffuse plaques after severe brain injury in humans. *Arch. Neurol*, 64, 541-544.
[12] Nicoll J.A., Roberts G.W. and Graham D.I. (1995). Apolipoprotein E epsilon 4 allele is associated with deposition of amyloid beta-protein following head injury. *Nat. Med.*, 1, 135-137.
[13] Itoh T., Satou T., Hashimoto S. and Ito H. (2005). Isolation of neural stem cells from damaged rat cerebral cortex after TBI. *Neuroreport,* 16, 1687-1691.
[14] Itoh T., Satou T., Nishida S., Tsubaki M., Hashimoto S. and Ito H. (2009). Expression of amyloid precursor protein after rat traumatic brain injury. *Neurol. Res.*, 31, 103-109.
[15] Itoh T., Satou T., Nishida S., Hashimoto S. and Ito H. (2007). Immature and mature neurons coexist among glial scars after rat traumatic brain injury. Neurol Res, 29, 734-742.
[16] Green-Sadan T., Kinor N., Roth-Deri I., Geffen-Aricha R., Schindler C.J. and Yadid G. (2003). Transplantation of glial cell line-derived neurotrophic factor-expressing cells into the striatum and nucleus accumbens attenuates acquisition of cocaine self-administration in rats. *Eur. J. Neurosci.*, 18, 2093-2098.
[17] Elvander E., Schott P.A., Sandin J., Bjelke B., Kehr J., Yoshitake T. and Ogren S.O. (2004). Intraseptal muscarinic ligands and galanin: influence on hippocampal acetylcholine and cognition. *Neuroscience,* 126, 541-557.
[18] Roberts G.W. (1988). Immunocytochemistry of neurofibrillary tangles in dementia pugilistica and Alzheimer's disease: evidence for common genesis. *Lancet,* 2, 1456-1458.
[19] Roberts G.W., Gentleman S.M., Lynch A., Murray L., Landon M. and Graham D.I. (1994). Beta amyloid protein deposition in the brain after severe head injury: implications for the pathogenesis of Alzheimer's disease. *J. Neurol. Neurosurg. Psychiatry*, 57, 419-425.
[20] Nakamura Y., Takeda M., Niigawa H., Hariguchi S. and Nishimura T. (1992). Amyloid beta-protein precursor deposition in rat hippocampus lesioned by ibotenic acid injection. *Neurosci. Lett*, 136, 95-98.
[21] Rumble B., Retallack R., Hilbich C., Simms G., Multhaup G., Martins R., Hockey A., Montgomery P., Beyreuther K. and Masters C.L. (1989). Amyloid A4 protein and its precursor in Down's syndrome and Alzheimer's disease. *N. Engl. J. Med.*, 320, 1446-1452.

[22] Koo E.H., Sisodia S.S., Archer D.R., Martin L.J., Weidemann A., Beyreuther K., Fischer P., Masters C.L. and Price D.L. (1990). Precursor of amyloid protein in Alzheimer disease undergoes fast anterograde axonal transport. *Proc. Natl. Acad. Sci. USA*, 87, 1561-1565.

[23] Posmantur R., Hayes R.L., Dixon C.E. and Taft W.C. (1994). Neurofilament 68 and neurofilament 200 protein levels decrease after traumatic brain injury. *J. Neurotrauma*, 11, 533-545.

[24] Taft W.C., Yang K., Dixon C.E. and Hayes R.L. (1992). Microtubule-associated protein 2 levels decrease in hippocampus following traumatic brain injury. *J. Neurotrauma,* 9, 281-290.

[25] Van den Heuvel C., Blumbergs P.C., Finnie J.W., Manavis J., Jones N.R., Reilly P.L. and Pereira R.A. (1999). Upregulation of amyloid precursor protein messenger RNA in response to traumatic brain injury: an ovine head impact model. *Exp. Neurol*, 159, 441-450.

[26] Ciallella J.R., Ikonomovic M.D., Paljug W.R., Wilbur Y.I., Dixon C.E., Kochanek P.M., Marion D.W. and DeKosky S.T. (2002). Changes in expression of amyloid precursor protein and interleukin-1beta after experimental traumatic brain injury in rats. *J. Neurotrauma*, 19, 1555-1567.

[27] Xie Y., Yao Z., Chai H., Wong W.M. and Wu W. (2003). Potential roles of Alzheimer precursor protein A4 and beta-amyloid in survival and function of aged spinal motor neurons after axonal injury. *J. Neurosci. Res.,* 73, 557-564.

[28] Chirumamilla S., Sun D., Bullock M.R. and Colello R.J. (2002). Traumatic brain injury induced cell proliferation in the adult mammalian central nervous system. *J. Neurotrauma*, 19, 693-703.

[29] Rice A.C., Khaldi A., Harvey H.B., Salman N.J., White F., Fillmore H. and Bullock M.R. (2003). Proliferation and neuronal differentiation of mitotically active cells following traumatic brain injury. *Exp. Neurol.,* 183, 406-417.

[30] Azbill R.D., Mu X., Bruce-Keller A.J., Mattson M.P. and Springer J.E. (1997). Impaired mitochondrial function, oxidative stress and altered antioxidant enzyme activities following traumatic spinal cord injury. *Brain Res*., 765, 283-290.

[31] Kawamata T., Katayama Y., Hovda D.A., Yoshino A. and Becker D.P. (1995). Lactate accumulation following concussive brain injury: the role of ionic fluxes induced by excitatory amino acids. *Brain Res*., 674, 196-204.

[32] Xiong Y., Gu Q., Peterson P.L., Muizelaar J.P. and Lee C.P. (1997). Mitochondrial dysfunction and calcium perturbation induced by traumatic brain injury. *J. Neurotrauma*, 14, 23-34.

[33] Sun K.H., Sun G.H., Su Y., Chang C.I., Chuang M.J., Wu W.L., Chu C.Y. and Tang S.J. (2004). Acidic-rich region of amyloid precursor protein induces glial cell apoptosis. *Apoptosis,* 9, 833-841.

[34] Bouron A., Mbebi C., Loeffler J.P. and De W.M. (2004). The beta-amyloid precursor protein controls a store-operated Ca2+ entry in cortical neurons. *Eur. J. Neurosci*., 20, 2071-2078.

[35] Feng Z., Chang Y., Cheng Y., Zhang B.L., Qu Z.W., Qin C. and Zhang J.T. (2004). Melatonin alleviates behavioral deficits associated with apoptosis and cholinergic system dysfunction in the APP 695 transgenic mouse model of Alzheimer's disease. *J. Pineal Res*., 37, 129-136.

[36] Gentleman S.M., Nash M.J., Sweeting C.J., Graham D.I. and Roberts G.W. (1993). Beta-amyloid precursor protein (beta APP) as a marker for axonal injury after head injury. *Neurosci. Lett*, 160, 139-144.

[37] Blumbergs P.C., Scott G., Manavis J., Wainwright H., Simpson D.A. and McLean A.J. (1995). Topography of axonal injury as defined by amyloid precursor protein and the sector scoring method in mild and severe closed head injury. *J. Neurotrauma,* 12, 565-572.

[38] Pierce J.E., Trojanowski J.Q., Graham D.I., Smith D.H. and McIntosh T.K. (1996). Immunohistochemical characterization of alterations in the distribution of amyloid precursor proteins and beta-amyloid peptide after experimental brain injury in the rat. *J. Neurosci, 16*, 1083-1090.

[39] Chen X.H., Siman R., Iwata A., Meaney D.F., Trojanowski J.Q. and Smith D.H. (2004). Long-term accumulation of amyloid-beta, beta-secretase, presenilin-1, and caspase-3 in damaged axons following brain trauma. *Am. J. Pathol*, 165, 357-371.

[40] Smith D.H., Uryu K., Saatman K.E., Trojanowski J.Q. and McIntosh T.K. (2003). Protein accumulation in traumatic brain injury. *Neuromolecular Med.*, 4, 59-72.

[41] Ikonomovic M.D., Uryu K., Abrahamson E.E., Ciallella J.R., Trojanowski J.Q., Lee V.M., Clark R.S., Marion D.W., Wisniewski S.R. and DeKosky S.T. (2004). Alzheimer's pathology in human temporal cortex surgically excised after severe brain injury. *Exp. Neurol*, 190, 192-203.

[42] Blasko I., Beer R., Bigl M., Apelt J., Franz G., Rudzki D., Ransmayr G., Kampfl A. and Schliebs R. (2004). Experimental traumatic brain injury in rats stimulates the expression, production and activity of Alzheimer's disease beta-secretase (BACE-1). *J. Neural. Transm*, 111, 523-536.

[43] Stone J.R., Okonkwo D.O., Singleton R.H., Mutlu L.K., Helm G.A. and Povlishock J.T. (2002). Caspase-3-mediated cleavage of amyloid precursor protein and formation of amyloid Beta peptide in traumatic axonal injury. *J. Neurotrauma*, 19, 601-614.

[44] Van Den Heuvel C., Donkin J.J., Finnie J.W., Blumbergs P.C., Kuchel T., Koszyca B., Manavis J., Jones N.R., Reilly P.L. and Vink R. (2004). Downregulation of amyloid precursor protein (APP) expression following post-traumatic cyclosporin-A administration. *J. Neurotrauma*, 21, 1562-1572.

[45] Hardy J. (1997). Amyloid, the presenilins and Alzheimer's disease. *Trends Neurosci*, 20, 154-159.

[46] Mattson M.P., Cheng B., Culwell A.R., Esch F.S., Lieberburg I. and Rydel R.E. (1993). Evidence for excitoprotective and intraneuronal calcium-regulating roles for secreted forms of the beta-amyloid precursor protein. *Neuron*, 10, 243-254.

[47] Thornton E., Vink R., Blumbergs P.C. and Van Den Heuvel C. (2006). Soluble amyloid precursor protein alpha reduces neuronal injury and improves functional outcome following diffuse traumatic brain injury in rats. *Brain Res.*, 1094, 38-46.

[48] Suh Y.H. and Checler F. (2002). Amyloid precursor protein, presenilins, and alpha-synuclein: molecular pathogenesis and pharmacological applications in Alzheimer's disease. *Pharmacol. Rev.*, 54, 469-525.

[49] Sola Vigo F., Kedikian G., Heredia L., Heredia F., Anel A.D., Rosa A.L. and Lorenzo A. (2009). Amyloid-beta precursor protein mediates neuronal toxicity of amyloid beta through Go protein activation. *Neurobiol Aging*, 30, 1379-1392.

[50] Matrone C., Di Luzio A., Meli G., D'Aguanno S., Severini C., Ciotti M.T., Cattaneo A. and Calissano P. (2008). Activation of the amyloidogenic route by NGF deprivation induces apoptotic death in PC12 cells. *J. Alzheimers Dis.*, 13, 81-96.

[51] Mills J. and Reiner P.B. (1999). Regulation of amyloid precursor protein cleavage. *J. Neurochem*, 72, 443-460.

[52] Nakagawa K., Kitazume S., Oka R., Maruyama K., Saido T.C., Sato Y., Endo T. and Hashimoto Y. (2006). Sialylation enhances the secretion of neurotoxic amyloid-beta peptides. *J. Neurochem*, 96, 924-933.

[53] Jurynec M.J., Riley C.P., Gupta D.K., Nguyen T.D., McKeon R.J. and Buck C.R. (2003). TIGR is upregulated in the chronic glial scar in response to central nervous system injury and inhibits neurite outgrowth. *Mol. Cell Neurosci.,* 23, 69-80.

[54] Davies S.J., Goucher D.R., Doller C. and Silver J. (1999). Robust regeneration of adult sensory axons in degenerating white matter of the adult rat spinal cord. *J. Neurosci,* 19, 5810-5822.

[55] Rudge J.S. and Silver J. (1990). Inhibition of neurite outgrowth on astroglial scars in vitro. *J. Neurosci.,* 10, 3594-3603.

[56] Pekny M. and Nilsson M. (2005). Astrocyte activation and reactive gliosis. *Glia*, 50, 427-434.

[57] Bechmann I. and Nitsch R. (2000). Involvement of non-neuronal cells in entorhinal-hippocampal reorganization following lesions. *Ann. N Y Acad. Sci.,* 911, 192-206.

[58] Deller T., Haas C.A. and Frotscher M. (2000). Reorganization of the rat fascia dentata after a unilateral entorhinal cortex lesion. Role of the extracellular matrix. *Ann. N Y Acad. Sci.*, 911, 207-220.

[59] Gallo V. and Chittajallu R. (2001). Neuroscience. Unwrapping glial cells from the synapse: what lies inside? *Science,* 292, 872-873.

[60] Silver I.A., Deas J. and Erecinska M. (1997). Ion homeostasis in brain cells: differences in intracellular ion responses to energy limitation between cultured neurons and glial cells. *Neuroscience,* 78, 589-601.

In: Amyloids
Editors: I. P. Halcheck et al.
ISBN: 978-1-62100-538-4

Chapter 4

Mechanism of Neurotoxicity of Amyloid-Beta Peptide and the Protective Effects of Docosahexaenoic Acid

***Shahdat Hossain*[1, 2] *and Michio Hashimoto*[2]**
[1]Department of Biochemistry and Molecular Biology,
Jahangirnagar University, Savar, Dhaka, Bangladesh,
[2]Department of Environmental Physiology,
Shimane University Faculty of Medicine, Izumo,
Shimane 693-8501, Japan

Abstract

The plasma membrane-bound amyloid precursor protein is cleaved by sequential actions of two proteases β- and γ-secretases into a short amyloid-beta peptide (Aβ) fragment. The fragment passes through a dynamic process to precipitate into insoluble fibrous aggregates. These aggregates interact with brain neurons and cause neurotoxicity, leading to neurodegeneration and cognitive and behavioral decline, characteristics of Alzheimer's disease (AD).

In this review, we focus on how Aβ behaves during self-association and causes neurotoxicity while emphasizing the protective effect of docosahexaenoic acid (DHA; C22:6*n*-3), a highly polyunsaturated fatty acid required for incessant maintenance of neuronal functions, against amyloidogenesis. DHA inhibits in vitro fibrillation of Aβ with concomitant inhibition of Aβ-induced cytotoxic effects in neuronal cells and learning-related memory impairments in AD model rats infused with the peptide in the cerebral ventricle.

Introduction

Since November 4, 1906, when Alois Alzheimer delivered his remarkable lecture on a form of dementia and cognitive impairment in his patient Auguste D [1], research on the mechanism of the disease, namely Alzheimer's disease (AD), has been an interesting field for neuroscientists. For the last decade, the extra cellular plaques and intracellular neurofibrillary tangles described by Alois Alzheimer have been known to be accumulations of insoluble fibrillar amyloid-beta peptide (Aβ) derived from the proteolytic processing of the plasma membrane-bound amyloid precursor protein (APP) [2-4]. AD pathology, neurotoxicity, and the related neurobehavioral problems are attributed to mature fibrillar deposits of abnormal and non-native protein folding of Aβ [5,6]. The early studies demonstrated that amyloid plaque numbers are directly related to quantitative measures of cognitive decline in the aged population [7]. However, this relationship appears to be misleading because plaque formation is a common feature of the aging human brain and can occur in the absence of cognitive decline [8]. Although substantial evidence indicates that insoluble Aβ deposits are associated with AD pathology, the recent studies suggest that soluble Aβ oligomers, the so-called "on-pathway" intermediates of mature amyloid plaques, correlate more strongly with neuronal dysfunction, damage, and AD symptoms [9] than mature fibers, and they are now considered the ultimate perpetrator for the pathogenesis of AD.

Aβ aggregates interact with neuronal cells and cause toxicity in the brains of patients with AD. However, the toxicity can be inhibited, at least partially, by enriching neuronal cells with docosahexaenoic acid (DHA, C22:6*n*-3), an important component of fish oil. DHA received enthusiastic attention following reports that chronic dietary administration of DHA, besides its beneficial role in cardiovascular diseases, improves memory-related learning ability in rats and the level of DHA is severely depleted in the brains of patients with AD [10, 11]. Dietary administration of DHA also protects against [12] and ameliorates [13] the impairment of memory-related learning induced by the infusion of $A\beta_{1-40}$ into the rat cerebral ventricle, suggesting that DHA plays an important role in the neurochemical basis of memory. Epidemiological studies suggest that AD is less frequent in populations that consume DHA [14]; the mechanisms, however, remain largely unknown. Dietary supplementation of DHA alters the risk of cognitive impairment with aging and/or developing AD over the long term and low concentrations of blood DHA are a critical risk factor for cognitive dysfunction [15]. The mechanism(s) of amyloid fibrillation and DHA-induced inhibition of fibrillation are reviewed here.

$A\beta_{1-42}$, $A\beta_{1-40}$, and $A\beta_{25-35}$ Exhibit Differential Self-Association Behaviors

AD is characterized by the presence of stamps of misfolded amyloid peptides in the affected brain areas. The genetic, biochemical and histopathological data from both human and animal studies strongly suggest that Aβ plays the crucial role in the pathogenesis of AD. Aβ is derived from the membrane-bound APP by the sequential action of two proteases called β- and γ-secretase [2-4]. APP is first cleaved by β-secretase shedding its large ectodomain and leaving a membrane-bound C-terminal end [16, 17]. The 99 amino acid segment is subsequently cleaved by γ-secretase and Aβ is released [18]. Depending on the exact point of cleavage by γ-secretase, two main forms of Aβ (comprising either 40- or 42-amino acid residues) are produced intracellularly and extracellularly. After their proteolytic release from the membrane, they undergo fibrillation (Figure 1).

Despite numerous studies, the mechanism of fibrillation and mechanistic association of fibers with neuromembranes and AD pathology are still not clear. Depending on various factors, including peptide length [19], concentration [20], and pH [21] of the solvent, Aβ passes through different stages, including α-helix to β-sheet transformation, nucleation, beading or elongation of oligomers (fibrillation) and coalescence of fibers into larger aggregates [22]. These molecular transitions are typically associated with neuronal loss and cognitive deficits in AD. Microscopic examinations reveal that amyloid plaques consist of morphologically different amyloid proteins, mainly $A\beta_{1\text{-}40}$ and $A\beta_{1\text{-}42}$. Although the latter is only two amino acids longer than the former, its fibrillation kinetics is quite different: its self-association occurs more rapidly than the shorter and less hydrophobic $A\beta_{1\text{-}40}$ form [19, 23] $A\beta_{1\text{-}40}$ is the major Aβ sequence in cerebrospinal fluid, whereas $A\beta_{1\text{-}42}$ and other sequences are minor components [2, 3, 24].

$A\beta_{1\text{-}42}$ is the major amyloid component in the neuritic plaques of the affected brain [25], indicating the key role of $A\beta_{1\text{-}42}$ in amyloidogenesis and the associated neurobehavioral impairments in AD [26]. Although $A\beta_{1\text{-}40}$ and $A\beta_{1\text{-}42}$ are the predominant forms, other fragments can be present as well: $A\beta_{25\text{-}35}$, a stretch of 11-amino acid residues from position 25 to position 35 (Figure 2), is produced in the brains of aged patients by the proteolysis of full-length $A\beta_{1\text{-}40}$ [27].

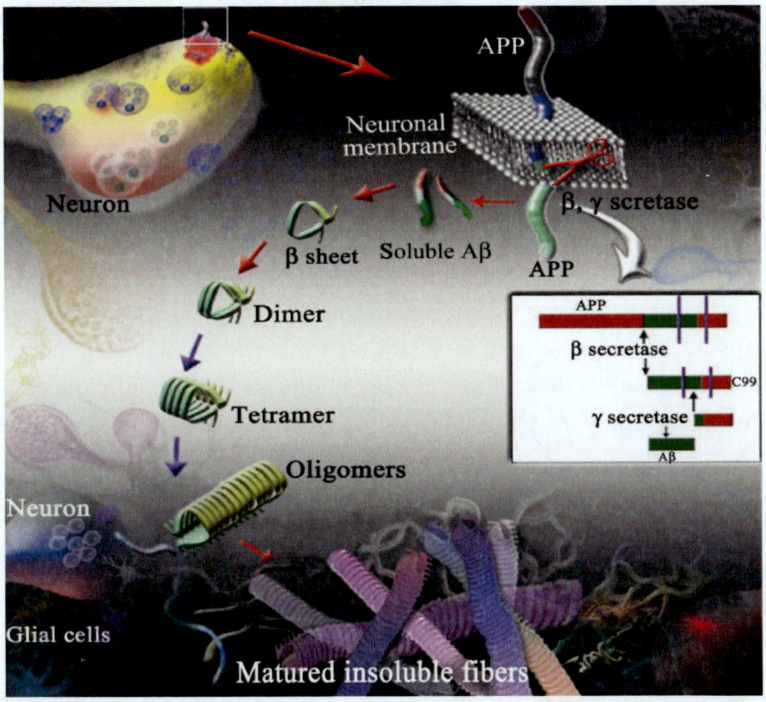

Figure 1. Production of Aβ by proteolytic cleavage from APP followed by association of Aβ to form oligomers and fibers. Aβ is cleaved from APP by the sequential action of two proteases: β-secretase carries out the initial cleavage to form the N-terminus of Aβ and γ-secretase then cleaves the C99 stub to produce the C-terminus of Aβ. The secretase action leads to the formation of Aβ monomers. Monomers self-associate to form dimers/tetramers and oligomers, which increase in size and eventually lead to fibrillation. The fibrillation starts typically from a lag phase, progresses to an exponential growth phase, and then declines to a plateau phase (see Figure 6).

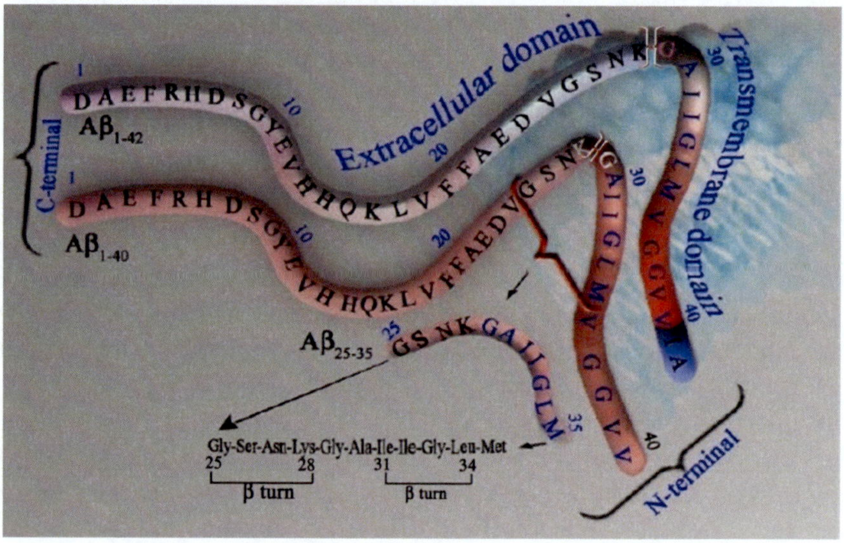

Figure 2. Schematic Diagrams of the Short Sequence of $A\beta_{25-35}$ and its Parent $A\beta_{1-40}$ or $A\beta_{1-42}$.

The $A\beta_{25-35}$ fragment is biologically active and analogous to its full-length Aβ progenitor [28]. The capacity of $A\beta_{25-35}$ to form fibers and cause toxicity analogous to that of the full-length peptide has been proved by in vitro studies of neurons [29]. These findings are supported by in vivo studies demonstrating that direct cerebral infusion of $A\beta_{25-35}$ into the ventricle impairs memory in AD model mice [30, 31].

In Vitro Fibrillation of $A\beta_{1-40}$, $A\beta_{1-42}$, and $A\beta_{25-35}$

Amyloid fibers clogging the brains of patients with AD have confused scientists with their robust structures. They can withstand the actions of powerful detergents that cripple most other proteins. The fibers are, in fact, so tough that researchers have not yet found a way to attack these resilient molecules. There are continuing endeavors to explore ways to control them for the management of AD and many other neurological conditions. Compact, neuritic amyloid plaques in patients with AD contain thioflavin S- and Congo red-positive fibrillar deposits of both $A\beta_{1-40}$ and $A\beta_{1-42}$ [25]. Fluorescence spectroscopic measurements with thioflavin T (ThT) also clearly indicate that polymerization of $A\beta_{1-40}$, $A\beta_{1-42}$, and $A\beta_{25-35}$ occurs in the assembly buffer. However, equimolar concentrations of $A\beta_{1-40}$, $A\beta_{1-42}$, and $A\beta_{25-35}$ samples differed significantly in their degree of fibrillation in identical assembly buffer over 24 h (Figure 3A–C).

The ability of $A\beta_{1-40}$, $A\beta_{1-42}$ and $A\beta_{25-35}$ fibers to withstand the detergent action of SDS also differ significantly. As shown in Figure 3D, the $A\beta_{1-42}$ fibers showed the highest resistance to the detergent, $A\beta_{25-35}$ showed moderate resistance, and $A\beta_{1-40}$ fibers exhibited no resistance. This may be related to the differences in the accessibility of SDS molecules to the core structures of the fibers: SDS can solubilize the fibers if it can form electrostatic bonds with positively charged amino acid residues in the core structures of the fibers. With regard to the resistance of amyloid fibers to SDS, an important piece of information derived from our experiments is that the 25 to 35 amino acid residues of $A\beta_{1-42}$ was possibly constitute a critical sequence of the peptide that confers significant resistance to the SDS-mediated solubilization. However, the same sequence of $A\beta_{1-40}$ fiber is accessible to SDS, rendering it soluble.

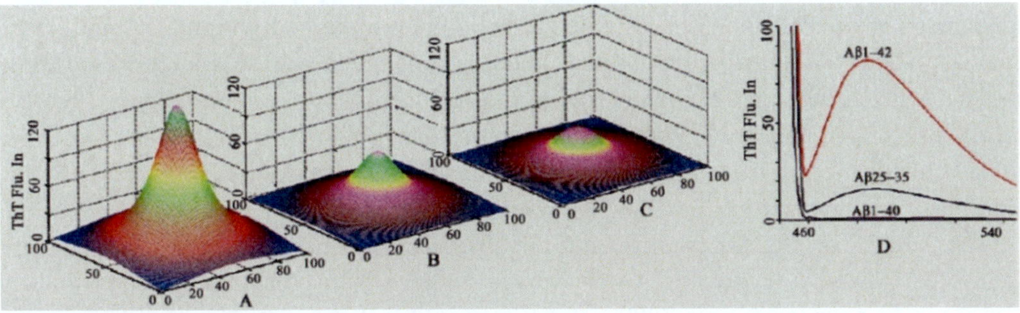

Figure 3. Equimolar concentrations (50 μM) of $A\beta_{1-42}$ (A), $A\beta_{1-40}$ (B), and $A\beta_{25-35}$ (C) were incubated in identical assembly buffer for 24 h at 37 °C. After complete fibrillation, 40-μL aliquots of the samples were added to 210 μL of 5 μM ThT in glycine buffer (pH 8.5) and analyzed the ThT fluorescence intensity at the excitation wavelength of 448 nm and emission wavelength of 488 nm. The $A\beta_{1-42}$ fibers showed the highest ThT intensity, indicating greater fibrillation than those of the $A\beta_{1-40}$ and $A\beta_{25-35}$ fibers. The results are representative of three determinations on different days. D: Effect of sodium dodecylsulphate (SDS) on amyloid fibrillation. To test the sensitivity to SDS, $A\beta_{1-42}$, $A\beta_{1-40}$, and $A\beta_{25-35}$ (50 μM) fibrillations were carried out in the presence of 1% SDS and the ThT fluorescence intensity was measured. The $A\beta_{1-42}$ fibers exhibited the highest ThT fluorescence intensity and the $A\beta_{25-35}$ fibers showed moderate intensity, but no intensity was detected in the $A\beta_{1-40}$ samples. Postfibrillated $A\beta_{1-42}$, $A\beta_{1-40}$, and $A\beta_{25-35}$ were also treated with 1% SDS and incubated for a further 24 h. The ThT intensity profile was almost similar in both the SDS-pretreated and the SDS-posttreated amyloid samples. These results indicate that $A\beta_{1-42}$ and $A\beta_{25-35}$ fibers are SDS (detergent) stable whereas $A\beta_{1-40}$ fibers are either unable to form in the presence of SDS or dissolved by the posttreatment.

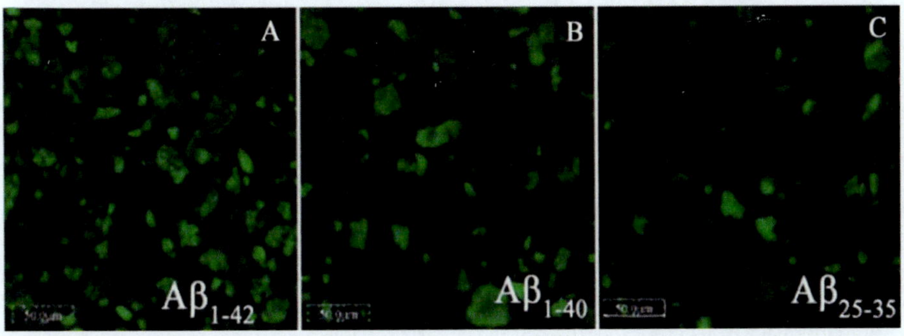

Figure 4. ThT staining of polymerized $A\beta_{1-40}$ (A), $A\beta_{1-42}$, (B) and $A\beta_{25-35}$ (C) aggregates. Under fluorescence illumination with ThT staining, the amyloid fibrillar aggregates appear green. A 2.5-μl aliquot of the fibrillated Aβ peptide (50 μM) sample was diluted 2X with 5 μM ThT in 50 mM glycine–NaOH buffer, pH 8.5, transferred onto slides and photographed by the confocal laser microscope system (CLSM FV300; Olympus, Tokyo, Japan).

In vitro $A\beta_{1-40}$, $A\beta_{1-42}$ or $A\beta_{25-35}$ aggregates also showed characteristic green fluorescence after their binding with thioflavin T (Figure 4 A-C). Microfluorescence study revealed differences in the fibrillation with the highest level of green fluorescence from the $A\beta_{1-42}$ samples.

Transmission electron microscopy revealed that the Aβ samples contained abundant aggregated amyloid fibers in the assembly buffer. The $A\beta_{1-42}$ fibers appeared as diffuse beaded structures forming necklace-like decorative filaments as well as showed extensive branching. They also demonstrated some projections from the body (Figure 5A and its inset, lower panel) and periodically intertwined strands around the growth axis (Figure 5A inset).

The $A\beta_{1-40}$ fibers predominantly had round to flat ribbon-like morphology (Figure 5 B and its inset). Flat ribbons were the primary morphology of the $A\beta_{25-35}$ fibers, and ribbons with a helical pattern were not observed in the transmission electron micrograph (TEM) grids of the $A\beta_{25-35}$ samples (Figure 5C and its inset).

The lengths of the fibers differed from grid to grid and were almost inconspicuous because of extensive branching. Their average length (from branching point to branching point) and diameter were 100–2000 and 5–20 nm, respectively. In plaques of the brain of patients with AD, Aβ fibers measure 6–10 nm in diameter [32, 33], which is within the range of the in vitro fibers.

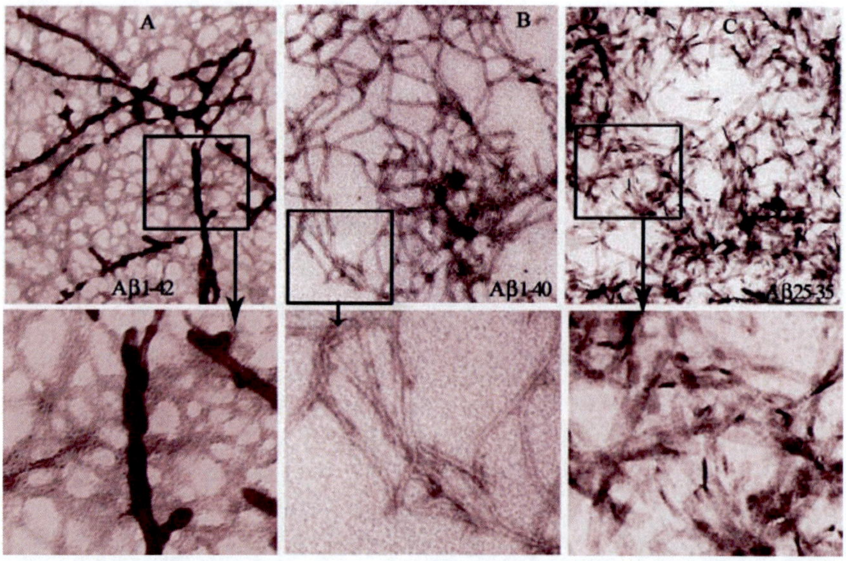

Figure 5. Representative TEMs of $A\beta_{1-42}$ (A), $A\beta_{1-40}$ (B), and $A\beta_{25-35}$ (C) fibers. Six microliters of each sample was applied to a 400-mesh grid stained with 1% uranyl acetate and the excess stain was removed. Afterwards, the stained samples were dried for 1 min and visualized under a transmission electron microscope. The amyloid fibers displayed heterogeneous morphology: both round, beaded-necklace and ribbon-like structures were seen in the TEM grids.

DRIVERS OF Aβ FIBRILLATION

Despite numerous studies, the molecular event of α-helix to β-sheet transformation is not clearly understood. Figures 1 and 6 schematically show the morphological changes from monomers to amyloid fibers. First, monomeric α-helical units are transformed into U-turn bent β-sheets, or the so-called "β-units." These basic units are stabilized by hydrophobic interactions between nonpolar amino acid residues and electrostatic interactions (derived from the so-called "salt bridging").

The fibrillation is a dynamic process that increases as a function of the concentration of Aβ [34, 35]. Depending on the concentration of the peptide, both nucleation-dependent and nucleation-independent fibrillations of $A\beta_{1-40}$ and $A\beta_{1-42}$ occur. Fibrillation at a high concentration (50 μM) of $A\beta_{1-42}$ does not exhibit any lag time; instead, an exponential growth phase is prominent after a short time lag (<10 min).

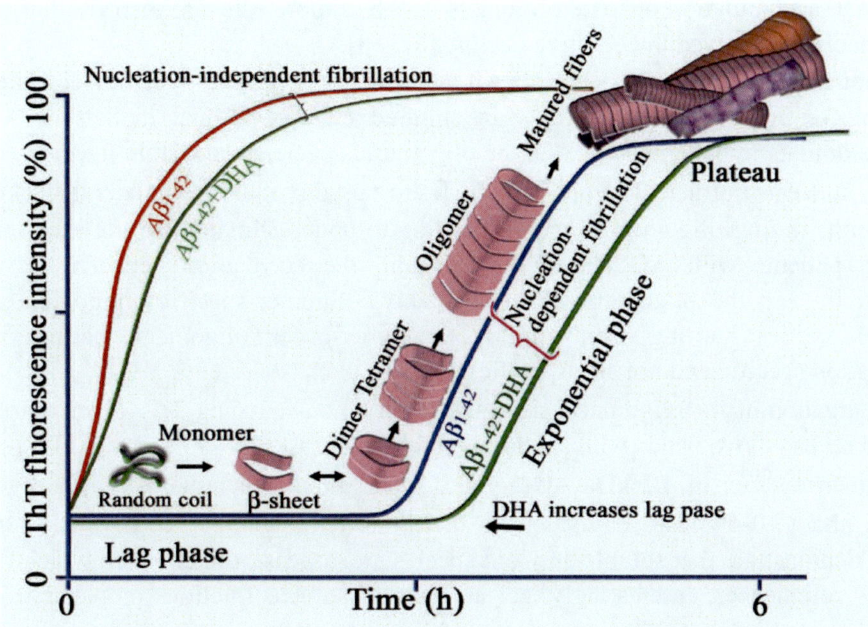

Figure 6. The hierarchical structure of amyloid fibers, from nanostructure to macrostructure. Nucleation-dependent fibrillation, which occurs at low amyloid concentration (e.g. 5, 10 μM), consists of a lag phase, during which a thermodynamically stable nucleus needs to accumulate, which is then followed by a rapid elongation phase that sees the formation of mature amyloid fibrils. At lag phase the amyloids are transformed into beta sheets, which by hexagonal stacking form oligomers. Oligomers act as the seeds of amyloidogenesis. Then exponential growth of the oligomers occurs leading to matured fibrils and finally reaches a plateau phase. At high concentration of amyloid (~50 μM) nucleation-independent fibrillation occurs with a short or no lag phase (3-7 min).

The fibrillation of $A\beta_{1-42}$ commences at a faster rate than that of $A\beta_{1-40}$ (data not shown), indicating the strong fibrillation potential of the former. Indeed, fibrillation at low concentrations of $A\beta_{1-42}$ (5 and 10 μM) starts typically from a lag phase, progresses to an exponential growth phase, and then declines to a plateau phase (Figure 6). The kinetics at lower concentrations appears as a characteristic sigmoid curve.

OLIGOMERS OF Aβ

Although the hypothesis that aggregation of Aβ into fibers is essential for toxicity was maintained until recently, current observations cast doubt on this notion. Many patients with

AD do not show dementia and progressive cognitive impairment despite the presence of mature fibrillar deposits in their brain [36-38], indicating a very weak correlation between the severity of AD dementia and the density of amyloid fibrillar aggregates [38]. Mature amyloid fibers are not cytotoxic or are negligibly so, suggesting that the presence of fibers is not necessarily associated with toxicity. Undoubtedly, a strong correlation exists between the oligomeric levels and the extent of axodendritic loss and decline in neuronal populations [36 [35]. These soluble oligomeric species can also be isolated from brain homogenates of patients with AD and correlate excellently with the cognitive impairments, demonstrating clearly that accumulation of toxic oligomers of Aβ is more relevant than fibrillar plaques to the mechanisms involved in cognitive decline [37, 38].

Although the oligomeric species of Aβ rather than the mature fibers are considered more important for imparting neurotoxicity in cultured cells or animal models of AD and/or cognitive deficits in patients with AD, the oligomeric species responsible for the toxic effects is largely unclear. Further, the epitopes of Aβ are possibly masked by lateral stacking of the β-sheet units during oligomer formation, leading to underestimation of the exact oligomeric levels in patients with AD and defying, again, the correlation between the extent of neurotoxicity and the oligomeric levels. Notably, oligomer-specific antibodies estimate a range of species starting from dimers to higher-order oligomers, because they are conformation specific and not size specific.

Investigation of the oligomeric size of Aβs on a non-reducing Tris–glycine gradient (4–20%) gel revealed two bands with apparent molecular masses of ~15 and ~20 kDa (equivalent to the dimer/tetramer of 4.2 kDa $A\beta_{1-42}$) [35], and soluble fractions of $A\beta_{1-40}$ revealed two bands of about 10–14 kDa (equivalent to the dimer/tetramer of 4 kDa $A\beta_{1-40}$) [34]. These analyses demonstrate that soluble amyloid species are, at least, dimers from which fibrillation of $A\beta_{1-42}$ commences. Interestingly, gel analysis of soluble fractions prepared from 1 kDa $A\beta_{25-35}$ revealed a band of 10 kDa (equivalent to the decamer of $A\beta_{25-35}$), indicating that $A\beta_{25-35}$ decamers act as the seed of $A\beta_{25-35}$ amyloidogenesis. Our observations of the size of $A\beta_{1-40}$ or $A\beta_{1-42}$ oligomers [34,35] are consistent with those in other studies showing SDS-stable dimers and trimers (the so-called "low-*n* oligomers") of Aβ in the buffer-soluble fraction of the human cerebral cortex and CSF [39, 40]. Similar oligomers are also formed by a fibroblast cell line genetically manipulated to express mutant human *APP* (7PA2 cells). These cells produce and secrete significant amounts of SDS-stable low-*n* oligomers of Aβ which migrate on denaturing SDS–polyacrylamide gels with molecular masses consistent with dimers, trimers, and occasionally tetramers [41, 42].

The size of Aβ oligomers as toxic species is, therefore, consistent with the discovery that these oligomers can inhibit hippocampal long-term potentiation (a measure of synaptic plasticity) [43, 44], impair complex learned behavior in rats [45], and reduce the density of dendritic spines in cultured hippocampal neurons [46, 47].

Interaction of Aβ Aggregates with DHA Hinders Amyloid-Induced Toxicity

Why DHA?

DHA is an essential *n*-3 polyunsaturated fatty acid in the brain (Figure 7). It remains highly concentrated in synaptic plasma membranes and increases neurotransmission and related brain functions. It also has strong antioxidant properties [48, 49, 50]. DHA supplementation imparts numerous beneficial effects on the health of the human brain through the modulation of genes associated with synaptic plasticity, neuroinflammation, and energy metabolism [51].

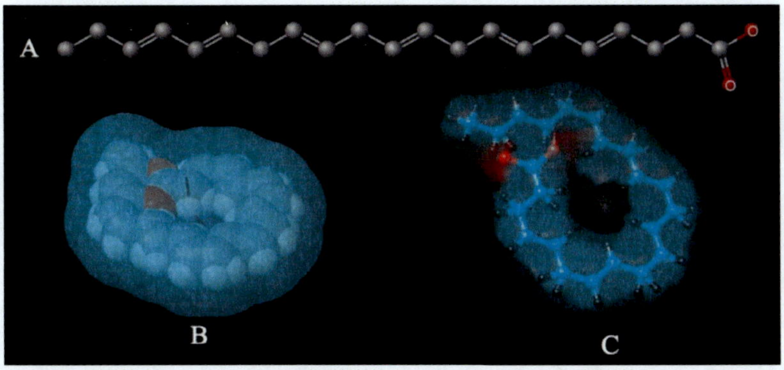

Figure 7. Structure of DHA. **A**: The straight chain DHA (C22:6, n3) was sketched by Marvin Sketch. 5.5.0.1. **B**: Space fill model with surface of straight chain DHA was constructed by Jmol-12.2.2 (an open-source Java viewer for chemical structures in 3D. http://www.jmol.org) **C**: Ball-Stick model of straight chain DHA with carbon labelling was produced by using the UCSF Chimera package from the Resource for Biocomputing, Visualization, and Informatics at the University of California, San Francisco (supported by NIH P41 RR-01081).

Chronic administration of DHA increases the DHA content in the hippocampus and cerebral cortex and concurrently improves memory-related learning ability [52]. A high serum level of total cholesterol is considered an independent risk factor for AD [53]. High-cholesterol diets induce AD-like Aβ deposition in the rabbit brain [54]. An elevated cholesterol level, which causes the formation of the so-called "lipid rafts" in the membrane bilayer, also facilitates the deposition of neurotoxic Aβ [55, 56]. The hypocholesterolemic effect of DHA on lipid rafts is well documented [34]. Therefore, the initial localization of amyloid peptides in these cholesterol-rich lipid rafts, their subsequent removal therefrom, and the underlying mechanisms are the focus of many investigations. At least up to 1500 mg/kg body weight, DHA causes no acute toxicity [57]. Moreover, the blood–brain barrier does not block fatty acids [58].

To date, there is no effective treatment to prevent the progression of AD; the available drugs only delay the worsening of symptoms. Therefore, therapies that alter the progression of AD are urgently needed. DHA boosts neurogenesis both in vitro and in vivo [59]. It decreases the proportion of S-phase cells and increases the proportion of G_o/G_1-phase cells, indicating that DHA promotes differentiation of neural stem cells. Therefore, it might help to restore the injured neurons in neurodegenerative diseases, including AD, by controlling the fate of the neuronal cycle [60]. Because of these beneficial effects, we investigated whether dietary administration of DHA inhibits the memory-related learning deficits of Aβ-infused AD rats: DHA protected against and ameliorated the impairment of learning ability in the AD rats with a concurrent decrease in the amyloid burden in their brains [12,13]. These results led us to explore the mechanism of action of DHA. We examined whether DHA inhibits in vitro Aβ fibrillation and whether such an anti-Aβ outcome affects neurotoxicity in a neuronal line.

DHA Inhibits *In Vitro* Amyloid Fibrillation

The decrease in the amyloid burden in the brains of the $A\beta_{1-40}$-infused AD rats [12] indicates that DHA affects the process of fibrillation. DHA significantly decreased in vitro amyloid fibrillation [35, 36] (Figure 8), indicating that the dietary DHA-induced protection against memory impairment of AD rats [12,13] was probably mediated by the inhibition of amyloid fibrillation. DHA not only inhibited the formation of amyloid fibers but also destabilized SDS-sensitive and SDS-resistant Aβ fibers.

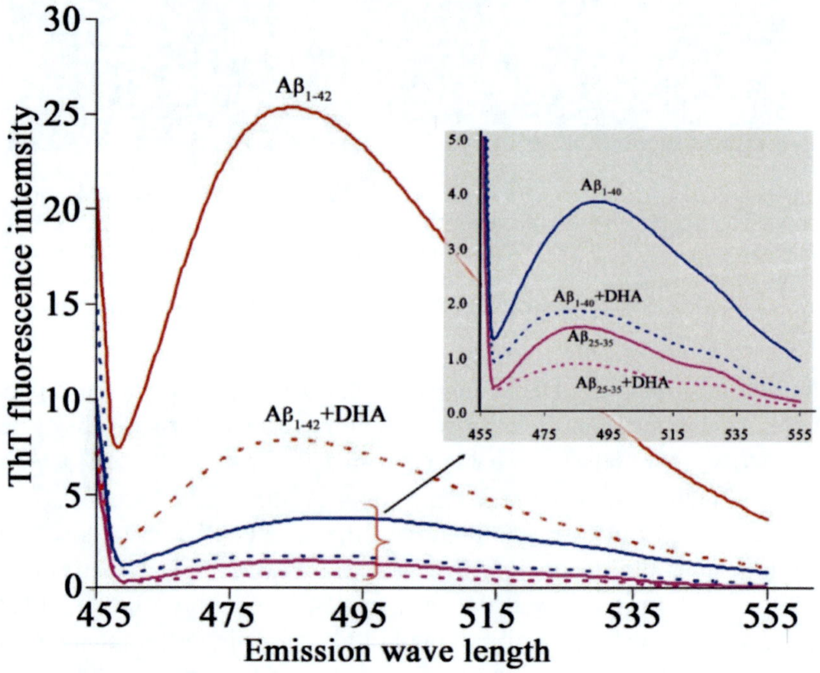

Figure 8. Effect of DHA on amyloid fibrillation. DHA reduced the ThT fluorescence intensity, demonstrating its anti-Aβ fibrillation (50 μM) property.

The Aβ+DHA samples had a lower number of green fluorescent areas than the Aβ controls. The fluorescence intensity of the green illuminated areas was significantly lower in the DHA-treated samples than in the controls, again indicating the clear inhibitory effect of DHA on the extent of Aβ fibrillation (Figure 9).

Fatty acids at concentrations that favor lipid vesicle formation in the assembly buffer stimulate nucleation and polymerization of α-synuclein [61] and tau [62]. DHA at concentrations higher than 40 μM showed an increased tendency to enhance $A\beta_{1-42}$ fibrillation, suggesting that DHA at high concentrations (>40 μM) increases oligomer formation by favoring Aβ nucleation; DHA ethyl ester and 1-palmitoyl-2-docosahexaenoyl phosphatidylcholine increased the fibrillation to an even greater extent (unpublished data).

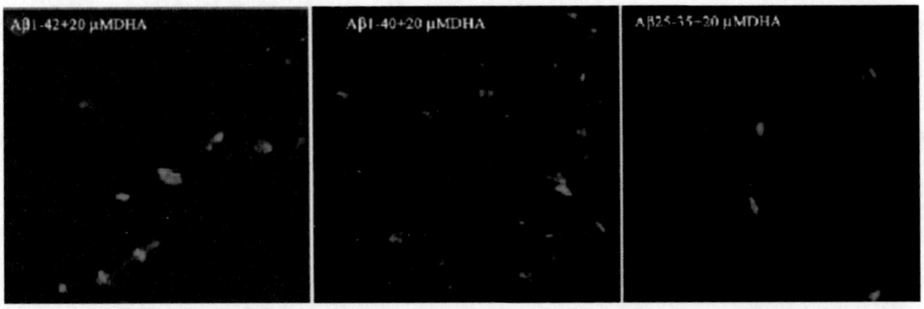

Figure 9. Effect of DHA on the ThT fluorescence intensity of Aβ. After 24 h of fibrillation at 37 °C, an aliquot from each sample was subjected to fluorescence microscopy. DHA clearly inhibited fibrillation, as indicated by the smaller numbers of illuminated Aβ deposits than those of the control samples without DHA (see Figure 4 for larger number of illuminated deposits).

Therefore, the presence of DHA in either free or micellar form in the face of a bulk of individual amyloid molecules is critical to affect the fibrillation of Aβ. Kinetically, DHA delays the lag phase and decelerates the exponential growth of Aβ fibers (Figure 6), demonstrating that DHA inhibits fibrillation by influencing nucleus formation (oligomerization) as well as the elongation of oligomers into mature fibers. The decreased oligomeric levels indicate that DHA decreases both the formation and the stabilization of oligomers, and hence their propagation into fibers [34, 35].

To evaluate the ability of DHA in inhibiting amyloid fibrillation, Aβ+DHA samples were directly visualized by transmission electron microscopy (Figure 10). TEM grids of the DHA-pretreated Aβ samples were almost devoid of fibers or contained, if at all, only a minute number of fibers with diffused, shorter, and narrower morphology. Because DHA kinetically delays and decelerates fibrillation (Figure 6), the morphological transformation of Aβ might be attributable to the effect of DHA.

The DHA-pretreated Aβ preparations had lower levels of oligomers, as determined by the use of oligomer-specific antibodies [35]. When isolated oligomers were posttreated with DHA, the extent of fibrillation was minimal. This finding demonstrates that DHA prevents not only the formation of oligomers but also the propagation of oligomers into fibers [35]. Accordingly, gel electrophoresis analyses of the samples incubated with DHA showed significantly decreased density of oligomers (10–14 kDa bands for $A\beta_{1-40}$, 15–20 kDa bands for $A\beta_{1-42}$, and ~10 kDa band for $A\beta_{25-35}$), indicating that these protofibrillar on-pathway intermediate oligomers are inhibited by DHA [34, 35].

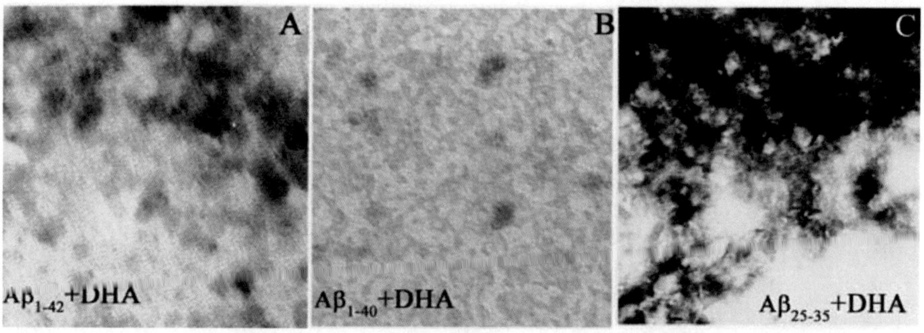

Figure 10. Effect of DHA on Aβ fibrillation. Aβ samples were incubated with 20 μM DHA for 24 h at 37 °C and the extent of fibrillation was visualized by transmission electron microscopy. DHA significantly reduced the extent of fibrillation.

Therefore, DHA might affect net fibrillation by inhibiting α-monomer to oligomer transformation and perturbing the stacking of oligomers into mature fibers. Nonetheless, the exact mechanism remains to be clarified. The amino acids of Aβ involved β-sheet constructs are not known. In amyloid fibrillation, hydrophobic amino acid residues are embedded in the core of the fibers [63]. The quicker fibrillation rate of $A\beta_{1-42}$ than that of $A\beta_{1-40}$ is attributable to increased hydrophobicity [64]. Associations of the C-terminal amino acid R-groups and those of the central hydrophobic core are important determinants of amyloid polymerization [65]. Therefore, unlike hydrophilic substances, hydrophobic or surface-active substances tend to penetrate into and act at the core of the fibers.

DHA may inhibit the intrachain and/or interchain amino acid interactions (crucial for the formation of β-sheets, leading to fibrillation). To reveal the DHA-induced inhibition of amyloid fibrillation, the effect of DHA on the emission intensity of intrinsic fluorophore Tyr10 was studied. Nonpolar aromatic residues, particularly tyrosine and phenylalanine, promote molecular recognition during amyloid fibrillation and stabilize the resulting fibers. We speculate that the Tyr10 residue in the $A\beta_{1-40}$ or $A\beta_{1-42}$ backbone is entangled in a region that acts as a site for intramolecular bending (during α- to β-transformation) of Aβ segments and/or intermolecular lateral stacking along the long axis during fibrillation. Further, the "hexagonal-stacking-of-nucleus" and/or "beading-to-string" of oligomers may be prevented by hydrophobic DHA from the π-stacking of planar aromatic rings. In so doing, DHA shields Tyr10 in a hydrophobic pocket different from that of the matrix of $A\beta_{1-40}$ or $A\beta_{1-42}$, spatially affects its conformation and emission, and consequently reduces fibrillation [34, 35].

This proposition was further supported by determining Tyr10 fluorescence quenching with nonpolar quencher acrylamide [34, 35]. DHA dose-dependently protected Tyr10 from acrylamide collisional quenching, demonstrating that Tyr10 was less accessible to the quencher in the presence of DHA and suggesting the interaction between hydrophobic DHA and Tyr10. The interactions between Tyr10 and hydrophobic DHA are, therefore, considered to affect Tyr10 conformation and emission, and hence, fibrillation. The levels of dityrosine (formed by Tyr10–Tyr10 residues of two adjacent Aβ monomers) also decreased in the $A\beta_{1-40}$ and $A\beta_{1-42}$ fibers in the presence of DHA, again suggesting that the DHA-induced inhibition of $A\beta_{1-40}$ or $A\beta_{1-42}$ fibrillation is, at least partially, related to its flanking ability between two neighboring Tyr residues, or the anti-dityrosine (DT) property of DHA.

On the other hand, DHA affects $A\beta_{25-35}$ differently. The fibrillation of $A\beta_{25-35}$ must also be independent of salt bridges between Asp/Glu and His/Lys residues because of the lack of these residues in this Aβ form. These bridges are crucial for amyloidogenesis of full-length $A\beta_{1-40}$ or $A\beta_{1-42}$ [21, 66, 67]. $A\beta_{25-35}$ contains only one ionizable residue, namely Lys28, and the ε-amino group of Lys28 is positively charged at pH 7.0, the pH at which fibrillation was conducted. Therefore, the presence of net ε-positive charges of Lys28 very likely leads native $A\beta_{25-35}$, at least partially, to a putative orientation facilitating the formation of β-sheets and the subsequent fibrillation. $A\beta_{25-35}$ fibers are stabilized by the electrostatic interactions between the Lys28 ε-amino group and the C-terminus of a neighboring peptide [68].

Amino acid residues 31–35 (Ile31-Ile32-Gly33-Leu34-Met35) of $A\beta_{25-35}$ adopt a reverse conformation, inducing short anti-parallel strands in the surrounding residues, which in turn promote fibrillation [69]. Both Ile31 and Leu34 residues of $A\beta_{25-35}$ show elements of the β-sheet structure [70]. $A\beta_{25-35}$ contains a type I β-turn motif in residues 25–28, analogous to $A\beta_{1-42}$, indicating that this Aβ form retains the structural features of its parent peptide in similar environments [34]. These speculations are also consistent with the fact that Aβ forms of different lengths aggregate through different modes of action; therefore, the targeting of Aβ sequences by drugs is a difficult task.

DHA itself has six double bonds with an ultimate formation of kinks along its axis. In addition, it has a strong anti-order (disordering) property because of its large hydrophobic volume, higher average area per molecule, and spring-like efficacy attributed to rapid interconversion between extended and bent conformations [71]. In contrast, Aβ fibers are sterically stacked and highly ordered structures. The high anti-order or hydrophobic property of DHA may, therefore, be related to its anti-fibrillar properties.

Effects of DHA on Aβ-Induced Toxicity in Neuronal Cell Culture

No significant neurotoxicity was observed in the cells directly treated with preformed matured fibers (data not shown), indicating that oligomers but not mature fibers are the toxic species. Although oligomers caused toxicity in cultured cells [72], DHA significantly decreased the toxicity in neuronal cells, as evaluated by MTT reduction [35, 73] and morphological examinations (Figure 11), probably by reducing the number of toxic oligomers, the “on-pathway” intermediates of amyloid fibers.

Aβ caused severe axodendritic loss, with cell bodies having a shrunken appearance. However, co-incubation of DHA with Aβ+SH-S5Y5 cells clearly improved the morphological features of the cells (Figure 11).

These results corroborate the effects of DHA on in vitro fibrillation, oligomeric levels, and amelioration of cognitive decline in $A\beta_{1-40}$-infused AD rats.

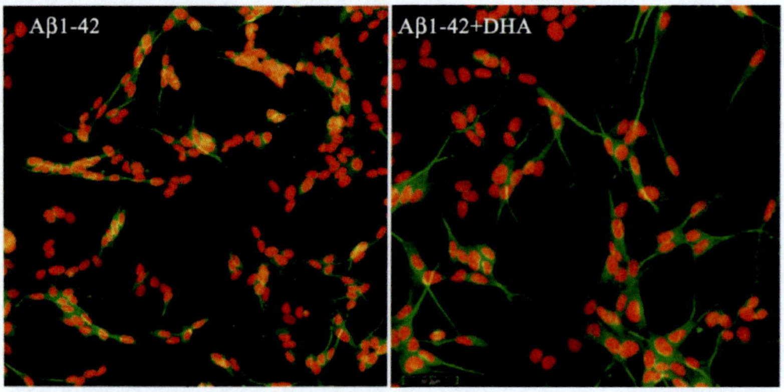

Figure 11. Neuronally differentiated SH-S5Y5 cells were incubated with $A\beta_{1-42}$ in the absence (A) or presence (B) of DHA for 24 h to commence fibrillation in the medium. The qualitative effect of DHA is shown by the representative morphological changes. Altered neuritic sprouting with dystrophic axodendritic systems were clearly observed after the treatment with $A\beta_{1-42}$ for 24 h. DHA inhibited the toxicity, as indicated by the appearance of well-defined axodendritic sprouting processes, good viability, and full spherical somas.

Conclusion

The results so far provide compelling evidence that soluble nonfibrillar forms of Aβ are potent neurotoxins. Removal or neutralization of such toxic species could, therefore, be an attractive therapeutic strategy. However, the lack of a unique mechanism to explain the complex process of amyloidogenesis hinders anti-Aβ therapy. Inhibition of Aβ fibrillation either at the monomeric or oligomeric level is the most likely method to decrease the incidence of AD. Easily accessible agents to the brain are also a strategy to treat AD. Treatment of AD with immunotherapy (using an anti-oligomer antibody) slows the disease progression. The incidences of encephalitis and edema with immunotherapy, however, discouraged the clinical trials [74]. and the incidences of encephalitis and edema; however, clinical trials are restrained [74]. On the other hand, considering the urgency for disease-modifying therapeutics to treat AD, the present data could provide insights into innovative strategies that interfere with the toxic processes initiated on the neuronal membrane. Finally, DHA may inhibit the in vivo deposition of Aβ and destabilize preformed Aβ fibers. Therefore, dietary DHA could be an effective means to prevent AD pathogenesis and the related cognitive decline.

Acknowledgments

This work was supported in part by a grant-in-Aid for Scientific Research from the Ministry of Education, Science and Culture of Japan (19500324 to MH).

REFERENCE

[1] Alzheimer A. (1906) Über einen eigenartigen schweren Erkrankungsprozeß der Hirnrinde. *Neurologisches Centralblatt* 23, 1129–36.

[2] Haass, C., Schlossmacher, M.G., Hung, A.Y., Vigo-Pelfrey, C., Mellon, A., Ostaszewski, B.L., Lieberburg, I., Koo, E.H., Schenk, D., Teplow, D.B., Selkoe, D.J. (1992) Amyloid beta-peptide is produced by cultured cells during normal metabolism. *Nature* 359, 322–325.

[3] Seubert, P., Vigo-Pelfrey, C., Esch, F., Lee, M., Dovey, H., Davis, D., Sinha, S., Michael Schiossmacher, M., Whaley, J., Swindlehurst, C., McCormack, R., Wolfert, R., Selkoe, D., Lieberburg, I., Dale Schenk, D (1992) Isolation and quantification of soluble Alzheimer's beta-peptide from biological fluids. *Nature* 359, 325–327.

[4] Shoji, M., Golde, TE., Ghiso, J., Cheung, TT., Estus, S., LM Shaffer, L.M., XD Cai, XD., DM McKay, DM., R Tintner, R., Frangione, B and et al. (1992) Production of the Alzheimer amyloid beta protein by normal proteolytic processing. *Science* 258,126–129.

[5] Glenner, G.G., Wong, C.W. (1984) Alzheimer's disease: initial report of the purification and characterization of a novel cerebrovascular amyloid protein. *Biochem. Biophys. Res. Commun.* 120, 885–890.

[6] Wischik, C.M., Novak, M., Edwards, P.C., Klug, A., Tichelaar, W., Crowther, R.A. (1988) Structural characterization of the core of the paired helical filament of Alzheimer disease. *Proc. Natl. Acad. Sci. USA* 85, 4884-4888.

[7] Roth, M., Tomlinson, B.E., Blessed, G. (1966) Correlation between scores for dementia and counts of 'senile plaques' in cerebral grey matter of elderly subjects. *Nature* 209, 109–110.

[8] Yankner, B.A., Lu, T., and Loerch, P. (2008) The aging brain. *Annu. Rev. Pathol* 3, 41–66.

[9] Haass, C., and Selko, D. J. (2007) Soluble protein oligomers in neurodegeneration: lessons from the Alzheimer's amyloid b-peptide. *Nat Rev Mol Cell Biol* 8, 101–112.

[10] Prasad, M.R., Lovell, M.A., Yatin, M., Dhillon, H., and Markesbery W. R. (1998) Regional membrane phospholipid alterations in Alzheimer's disease. *Neurochem. Res.* 23, 81-88.

[11] So¨derberg, M., Edlund, C., Kristensson, K., and Dallner, G. (1991) Fattyacid composition of brain phospholipids in aging and in Alzheimer's disease. *Lipids* 26, 421–425.

[12] Hashimoto, Hossain, M.S., Shimada, T., Sugioka, K., Yamasaki, H., Fujii, Y. Ishibashi, Oka, J., and Shido, O. (2002) Docosahexaenoic acid provides protection from impairment of learning ability in Alzheimer's disease model rats. *J. Neurochem.* 81, 1084–1091.

[13] Hashimoto, M., Tanabe, Y., Fujii, Y., Kikuta T., Shibata, H., Shido, O. (2005) Chronic administration of docosahexaenoic acid ameliorates the impairment of spatial cognition learning ability in amyloid β-infused rats, *J. Nutr.* 135, 549–555.

[14] Morris, M.C., Evans, D.A., Bienias, J.L., Tangney, C.C., Bennett, D.A., Wilson, R.S., Aggarwal, N., and Schneider, J. (2003) Consumption of fish and n-3 fatty acids and risk of incident Alzheimer disease. *Arch. Neurol.* 60, 940–946.

[15] Schaefer, E.J., Bongard, V., Beiser, A.S., Lamon-Fava, S., Robins, S.J., Au, R., Tucker, K.L., Kyle, D.J., Wilson, P.W., and Wolf, P.A. (2006) Plasma phosphatidylcholine docosahexaenoic acid content and risk of dementia and Alzheimer disease: the Framingham Heart Study *Arch. Neurol.* 63, 1545–1550.

[16] Cai,., H, Wang Y., McCarthy, D., Wen, H., Borchelt, D.R., Price, D.L., Wong, P.C. (2001) BACE1 is the major beta-secretase for generation of Abeta peptides by neurons. *Nat. Neurosci.* 4, 233–234.

[17] Vassar R; Bennett BD; Babu-Khan S; Kahn S; Mendiaz EA; Denis P; Teplow DB; Ross S; Amarante P; Loeloff R; Luo Y; Fisher S; Fuller J; Edenson S; Lile J; Jarosinski MA; Biere AL; Curran E; Burgess T; Louis JC; Collins F; Treanor J; Rogers G; Citron M. (1999) Beta-secretase cleavage of Alzheimer's amyloid precursor protein by the transmembrane aspartic protease BACE. *Science* 286, 735–41.

[18] Schroeter, E.H., Ilagan, M.X., Brunkan, A.L., Hecimovic, S., Li, Y.M., Xu, M., Lewis, H.D., Saxena, M.T., De Strooper, B., Coonrod, A., Tomita, T., Iwatsubo, T., Moore, C.L., Goate, A., Wolfe, M.S., Shearman, M., Kopan, R. (2003) A presenilin dimer at the core of the gamma-secretase enzyme: insights from parallel analysis of Notch 1 and APP proteolysis. *Proc. Natl. Acad. Sci. USA*. 100, 13075–80.

[19] Jarrett, J.T., Berger, E.P,. and Lansbury, P.T., Jr. (1993) The carboxy terminus of the b amyloid protein is critical for the seeding of amyloid formation: implications for the pathogenesis of Alzheimer's disease. *Biochemistry* 32, 4693–4697.

[20] Burdick, D., Soreghan, B., Kwon, M., Kosmoski, J., Knauer, M., Henschen, A., Yates, J., Cotman, C., and Glabe, C. (1992) Assembly and aggregation properties of synthetic Alzheimer's A4/b amyloid peptide analogs. *J. Biol. Chem.* 267, 546–554.

[21] Fraser, P.E., Nguyen, J.T., Surewicz, W.K., and Kirschner, D.A. (1991) pH-dependent structural transitions of Alzheimer's amyloid peptides. *Biophys. J.* 60, 1190–1201.

[22] Serpell, L.C. (2000) Alzheimer's amyloid fibrils: structure and assembly. *Biochim. Biophys. Acta Mol. Basis Dis.* 1502, 16–30.

[23] Snyder S. W., Ladror U. S., Wade W. S., Wang G. T., Barrett L. W., Matayoshi E. D., Huffaker H. J., Kraft G. A. and Holzman T. F. (1994) Amyloid-b aggregation: selective inhibition of aggregation in mixtures of amyloid with different chain lengths. *Biophys. J.* 67, 1216–1228.

[24] Skovronsky, D.M., Doms, R.W., and Lee, V.M. (1998) Detection of a novel intraneuronal pool of insoluble amyloid b protein that accumulates with time in culture. *J. Cell Biol.* 141, 1031–1039.

[25] Iwatsubo, T., Odaka, A., Suzuki, N., Mizusawa, H., Nukina, N., and Ihara, Y. (1994) Visualization of Ab 42(43) and Ab 40 in senile plaques with end-specific Ab monoclonals: evidence that an initially deposited species is Aβ 42(43). *Neuron* 13, 45–53.

[26] Berman, D.E., Dall'Armi, C., Voronov, S.V., McIntire, L.B., Zhang, H., Moore, A.Z., Staniszewski, A., Arancio, O., Kim, T.W., and Di Paolo, G. (2008) Oligomeric amyloid-b peptide disrupts phosphatidylinositol 4,5-bisphosphate metabolism. *Nat. Neurosci.* 11, 547–554.

[27] Kubo, T., Nishimura, S., Kumagae, Y., and Kaneko, I. (2002) In vivo conversion of racemized β- amyloid ([D-Ser 26]Aβ 1-40) to truncated and toxic fragments ([D-Ser 26]A 25– 35/40) and fragment presence in the brains of Alzheimer's patients, *J. Neurosci. Res.* 70, 474–483.

[28] Yankner, B.A., Duffy, L.K., and Kirschner, D.A. (1990) Neurotrophic and neurotoxic effects of amyloid β protein: reversal by tachykinin neuropeptides. *Science* 250 (4978) 279–282.

[29] Pike, C.J., Burdick, D., Walencewicz, A.J., Glabe, C.G., Cotman, C.W. (1993) Neurodegeneration induced by β-amyloid peptide in vitro: the role of peptide assembly state. *J. Neurosci.* 16, 1676–1687.

[30] Tohda, C., Tamura, T., Komatsu, K. (2003) Repair of amyloid β(25–35)-induced memory impairment and synaptic loss by a Kampo formula, Zokumei-to. *Brain Res.* 990, 141–147.

[31] Yamada, M., Chiba, T., Sasabe, J., Nawa, M., Tajima, H., Niikura, T., Terashita, K., Aiso, S., Kita, Y., Matsuoka, M., Nishimoto, I. (2005) Implanted cannula-mediated repetitive administration of Aβ25–35 into the mouse cerebral ventricle effectively impairs spatial working memory. *Behav. Brain Res.* 164, 139–146.

[32] Terry, R.D., Gonatas, N.K., Weiss, M. (1964) Ultrastructural Studies In Alzheimer's Presenile Dementia. *Am. J. Pathol.* 44, 269–297.

[33] Kidd, M. (1963) Paired helical filaments in electron microscopy of Alzheimer's disease. *Nature* 197, 192–193.

[34] Hashimoto M., Shahdat H.M., Yamashita, S., Katakura, M., Tanabe, T., Fujiwara, H., Ganoh, S., Miyazawa, T., Arai, H., Shimada T., Shido, O. (2008) Docosahexaenoic acid disrupts in vitro amyloid b fibrillation andconcomitantly inhibits amyloid levels in cerebral cortex of Alzheimer's disease model rats. *J. Neurochem.* 107, 1634–1646.

[35] Hossain, S., Hashimoto, M., Katakura, M., Miwa, K., Shimada, T., Shido O. (2009). Mechanism of docosahexaenoic acid-induced inhibition of in vitro Aβ1-42 fibrillation and Aβ1-42-induced toxicity in SH-S5Y5 cells. *Neurochem.* 111, 568-579, 2009.

[36] Terry, R.D., Masliah, E., Salmon, D.P., Butters, N., DeTeresa, R., Hill, R., Hansen, L.A., Katzman, R. (1991) Physical basis of cognitive alterations in Alzheimer's disease: synapse loss is the major correlate of cognitive impairment. *Ann. Neurol.* 30, 572-580.

[37] Lue, L.F., Kuo, Y.M., Roher, A.E., Brachova, L., Shen, Y., Sue, L., Beach, T., Kurth, J.H., Rydel, R.E., Rogers, J. (1999) Soluble amyloid beta peptide concentration as a predictor of synaptic change in Alzheimer's disease. *Am. J. Pathol.* 155, 853–862.

[38] McLean, C.A., Cherny, R.A., Fraser, F.W., Fuller, S.J., Smith, M.J., Beyreuther, K., Bush, A.I., Masters, C.L. (1999) Soluble pool of Abeta amyloid as a determinant of severity of neurodegeneration in Alzheimer's disease. *Ann. Neurol.* 46:860–6.

[39] Walsh, D.M., Tseng, B.P., Rydel, R.E., Podlisny, M.B., Selkoe, D.J. (2000) The oligomerization of amyloid beta-protein begins intracellularly in cells derived from human brain. *Biochemistry* 39, 10831–9.

[40] Enya M, Morishima-Kawashima, M., Yoshimura, M., †Yasuhisa Shinkai, Y., Kusui, K., Karen K., Games, D., Schenk, D., Sugihara, S., Yamaguchi, H., Ihara, Y. Appearance of sodium dodecyl sulfate-stable amyloid beta-protein (Abeta) dimer in the cortex during aging. *Am J Pathol* 154, 271–279.

[41] Podlisny, M.B., Ostaszewski, B.L., Squazzo, S.L., Koo, E.H., Rydell, R.E., Teplow, D.B., Selkoe, D.J. (1995) Aggregation of secreted amyloid beta-protein into sodium dodecyl sulfate-stable oligomers in cell culture. *J Biol Chem* 270, 9564–9570.

[42] Walsh, D.M., Klyubin, I., Fadeeva, J.V., Rowan, M.J,, Selkoe, D,.J. (2002) Amyloid-beta oligomers: their production, toxicity and therapeutic inhibition. *Biochem Soc Trans* 30, 552–557.

[43] (2002) Walsh, D.M., Klyubin, I., Fadeeva, J.V., Cullen, W.K., Anwyl, R., Wolfe, M.S., Rowan, M.J., Selkoe, D.J. (**2002**) Naturally secreted oligomers of amyloid beta protein potently inhibit hippocampal long-term potentiation in vivo. *Nature* 416, 535–539.

[44] Wang, Q., Walsh, D.M., Rowan, M.J., Selkoe, D.J., Anwyl, R. (2004) Block of long-term potentiation by naturally secreted and synthetic amyloid beta-peptide in hippocampal slices is mediated via activation of the kinases c-Jun N-terminal kinase, cyclin-dependent kinase 5, and p38 mitogen-activated protein kinase as well as metabotropic glutamate receptor type 5. *J. Neurosci.,* 24, 3370-3374.

[45] Cleary, J.P., Walsh, D.M., Hofmeister, J.J., Shankar, G.M., Kuskowski, M.A., Selkoe, D.J., Ashe, K.H. (2005) Natural oligomers of the amyloid-beta protein specifically disrupt cognitive function. *Nat. Neurosci.* 8, 79–84.

[46] Calabrese, B., Shaked, G.M., Tabarean, IV., Braga, J., Koo, E.H., Halpain, S. (2007) Rapid, concurrent alterations in pre- and postsynaptic structure induced by naturally-secreted amyloid-beta protein. *Mol. Cell Neurosci.* 35, 183–93.

[47] Shankar, G.M., Bloodgood, B.L., Townsend, M., Walsh, D.M., Selkoe, D.J., Sabatini, B.L. (2007) Natural oligomers of the Alzheimer amyloid-beta protein induce reversible synapse loss by modulating an NMDA-type glutamate receptordependent signaling pathway. *J. Neurosci.* 27, 2866–75.

[48] Hossain, M.S., Hashimoto, M., Gamoh, S., Masumura, S. (1999) Antioxidative effects of docosahexaenoic acid in the cerebrum versus cerebellum and brainstem of aged hypercholesterolemic rats. *J. Neurochem.* 72, 1133–1138.

[49] Green, P., and Yavin, E. (1998) Mechanisms of docosahexaenoic acid accretion in the fetal brain. *J. Neurosci. Res.* 52, 129–136.

[50] Bazan, N.G. (2005) Neuroprotectin D1 (NPD1): a DHA-derived mediator that protects brain and retina against cell injury-induced oxidative stress. *Brain Pathol.* 15,159-66.

[51] Horrocks, L.A., and Farooqui, A.A. (2004) *Docosahexaenoic acid in the diet: its importance in maintenance and restoration of neural membrane function. Prost. Leukot. Essent. Fatty Acids* 70, 361–372.

[52] Gamoh, S., Hashimoto, M., Sugioka, K., Hossain, S., Hata, N., Misawa, Y., Masumura, S. (1999) Chronic administration of docosahexaenoic acid improves reference memory-related learning ability in young rats. *Neuroscience* 93, 237–241.

[53] Notkola, I.L., Sulkava, R., Pekkanen, J., Erkinjuntti, T., Ehnholm, C., Kivinen, P., Tuomilehto, J., and Nissinen, A. (1998) Serum total cholesterol, apolipoprotein E epsilon 4 allele, and Alzheimer's disease. *Neuroepidemiology* 17, 14–20.

[54] Sparks, D.L., Scheff, S.W., Hunsaker, III J.C., Liu, H., Landers, T. and Gross, D.R. (1994) Induction of Alzheimer-like b-amyloid immunoreactivity in the brains of rabbits with dietary cholesterol. *Exp. Neurol.* 126, 88–94.

[55] Kojro, E., Gimpl, G., Lammich, S., Marz, W., and Fahrenholz, F. (1998) Low cholesterol stimulates the nonamyloidogenic pathway by its effect on the □-secretase ADAM 10. *Proc. Natl Acad. Sci. USA* 95, 6460–6464.

[56] Casserly, I., and Topol, E. (2004) Convergence of atherosclerosis and Alzheimer's disease: inflammation, cholesterol, and misfolded proteins. *Lancet* 363, 1139–1146.

[57] Hempenius, R.A., Van Delft, J.M., Prinsen, M., and Lina, B.A. (1997) Preliminary safety assessment of an arachidonic acid-enriched oil derived from Mortierella alpina: summary of toxicological data. *Food Chem Toxico.* 5, 573–581.

[58] Dhopeshwarkar, G A., and Mead J. F. (1973) Uptake and transport of fatty acids into the brain and the role of the blood-brain barrier system. *Adv. Lipid Res* 11, 109–142.

[59] Kawakita, E., Hashimoto, M., Shido, O. (2006) Docosahexaenoic acid promotes neurogenesis in vitro and in vivo. *Neuroscience* 139, 991–997.

[60] Katakura, M., Hashimoto, M., Shahdat, H.M., Gamoh, S., Okui, T., Matsuzaki, K., Shido, O. (2009) Docosahexaenoic acid promotes neuronal differentiation by regulating basic helix-loop transcription factors and cell cycle in neural stem cells. *Neuroscience* 160, 651–60.

[61] Zhu, M., and Fink, A.L. (2003) Lipid binding inhibits a-synuclein fibril formation. *J. Biol. Chem.* 278, 16873–16877.

[62] Gamblin, T.C., King, M.E., Kuret, J., Berry, R.W., and Binder, L.I. (2000) Oxidative regulation of fatty acid induced tau polymerization. *Biochemistry* 39, 14203–14210.

[63] Pellarin, R., and Caflisch, A. (2006) Interpreting the aggregation kinetics of amyloid peptides. *J. Mol. Biol.* 360, 882–892.

[64] Stine, W.B., Dahlgren, K.N., Krafft, G.A., LaDu M, J. (2003) In vitro characterization of conditions for amyloid-b peptide oligomerization and fibrillogenesis. *J. Biol. Chem.* 278, 11612–11622.

[65] Kirkitadze, M.D., Condron, M.M., and Teplow, D.B. (2001) Identification and characterization of key kinetic intermediates in amyloid bprotein fibrillogenesis. *J. Mol. Biol.* 312, 1103–1119.

[66] Khandogin, C.L., Brooks, III. (2007) Linking folding with aggregation in Alzheimer's β- amyloid peptides.*Proc. Natl. Acad. Sci. USA*. 104, 16880–16885.

[67] Lűhrs, T., Ritter, C., Adrian, M., Riek-Loher, D., Bohrmann, B., Döbeli, H., Schubert, D., Riek, R. (2005) 3D structure of Alzheimer's amyloid-β(1–42) fibrils. *Proc. Natl. Acad. Sci. USA.* 102, 17342–17347.

[68] Karsai, A., Nagy, A., Kengyel, A., Mártonfalvi, Z., Grama, L., Penke, B., Kellermayer, M.S.(2005) ffect of lysine-28 side-chain acetylation on the nanomechanical behavior of Alzheimer amyloid β25–35 fibrils. *J. Chem. Inf. Model* 45, 1641–1646.

[69] Bond J.P., Deverin, S.P., Inouye, H., El-Agnaf, O.M.A., Teeter, M.M., Kirschner, D.A. (2003) Assemblies of Alzheimer's peptides A β 25–35 and A beta 31–35: reverse-turn conformation and side-chain interactions revealed by X-ray diffraction. *J. Struct Biol.* 141, 56–710.

[70] Hughes, E., Burke, R.M., Doig, A.J. (2000) Inhibition of toxicity in the beta-amyloid peptide fragment β-(25–35) using N-methylated derivatives: a general strategy to prevent amyloid formation. *J. Biol. Chem.* 275, 25109–25115.

[71] Stillwell, W., Wassall, S.R. (2003) Docosahexaenoic acid: membrane properties of a unique fatty acid. *Chem. Phys. Lipids* 126, 1–27.

[72] Podlisny, M.B., Ostaszewski, B.L., Squazzo, S.L., Koo, E.H., Rydell, R.E., Teplow, D.B., and Selkoe, D.J. (1995) Aggregation of secreted amyloid-protein into sodium dodecyl sulfate-stable oligomers in cell culture. *J. Biol. Chem* 70, 9564–9570.

[73] Hashimoto, M., Katakura, M., Hossain, S., Rahman, A., Shimada, T., Shido, O. (2011) Docosahexaenoic acid withstands the Aβ(25-35)-induced neurotoxicity in SH-SY5Y cells. *J. Nutr. Biochem. 22,22-29.*

[74] Weksler, M.E. (2004) The immunotherapy of Alzheimer's diseaseImmunity and Ageing 2004, 1:2 doi:10.1186/1742-4933-1-2, http://www.immunityageing.com/content/1/1/2.

In: Amyloids
Editors: I. P. Halcheck et al.
ISBN: 978-1-62100-538-4

Chapter 5

PROTEIN AGGREGATION AS A MODULATORY FACTOR IN EPM1

Mira Polajnar[1], Slavko Čeru[1,$], Nataša Kopitar Jerala[1], Dejan Caglič[1,#] and Eva Žerovnik[1*]

[1]Department of Biochemistry and Molecular and Structural Biology, Jožef Stefan Institute, Ljubljana, Slovenia
[$]EN-FIST Centre of Excellence and Slovenian NMR centre, National Institute of Chemistry, Ljubljana, Slovenia,
[#]The Scribbs Research Institute, La Jolla, California

ABSTRACT

Conformational disorders, due to pathological changes in protein folding and oligomerization, comprise many neurodegenerative diseases, such as Parkinson's, Alzheimer's, various dementias as well as prion diseases. A common characteristic of these diseases are aberrantly folded proteins, rich in β secondary structure that have a high tendency to aggregate and form amyloid fibrils. The modified amyloid cascade hypothesis of neurodegeneration states that the soluble protein oligomers and prefibrillar aggregates are the species exerting cytotoxicity rather than full-length fibrils or extracellular amyloid plaques. In this paper we present a brief overview of the protein aggregation status in connection to neurodegenerative disease. Moreover, we provide an insight on how increased protein aggregation may modulate neurodegeneration in the progressive myoclonus epilepsy of type 1 (EPM1). EPM1, also known as Unverricht-Lundborg disease, is a rare type of hereditary epilepsy with features of cerebellar atrophy and increased oxidative stress. Our hypothesis from 2005 states that some of the mutants of human stefin B found in EPM1 patients may undergo changed aggregation in cells, thus exerting additional gain in toxic function (apart from loss of normal function as an endogenous cysteine proteases inhibitor). This in turn may influence the patients' phenotype. Our *in vitro* experiments thus far have shown that the fragment R68X forms amyloid-fibrils very fast even at neutral pH, whereas the G4R mutant exhibits a prolonged lag phase, during which the toxic prefibrillar aggregates accumulate. The

* Corresponding author: e-mail eva.zerovnik@ijs.si

experiments using cell culture have demonstrated that the fragment R68X forms cytotoxic inclusions of the aggresome type, whereas the G4R mutant forms such aggregates only in conditions of proteasome inhibition. Results on two more missense EPM1 mutants, G50E and Q71P are being collected.

1. INTRODUCTION

An active protein conformation is essential for normal cell functioning and thus important in sustaining and maintaining viability of cells and most importantly, health. Several types of disease stem from protein misfolding and aggregation. The so called conformational disorders involve protein conformational changes leading to aberrantly folded states, rich in β secondary structure that have a high tendency to form fibrous aggregates that are quite resistant to proteolysis. Among the conformational disorders systemic amyloidoses and neurodegenerative diseases such as Parkinson's, Alzheimer's, Huntington's, various other dementias, amyotropic lateral sclerosis (ALS) and prion related diseases can be counted. All of them have similar molecular backgrounds: changes in protein conformation and aggregation [1-2].

A consensus has become accepted that amyloid fibril formation is a generic property of proteins [3], after discovering that most proteins, even some α helical proteins that have not been associated with any amyloid disease, can form amyloid-like fibrils. The parts with α helical structure must undergo an α to β transition and the β strands then associate into regular fibrillar structure. Fibril formation has also been observed for proteins which are initially unfolded or predominantly β sheet and which fold through an α helical intermediate. Different proteins most probably follow different pathways towards amyloid fibrils, depending on sequence detail and structural class. Folded (e.g. β2-microglobulin) or unfolded (e.g. Aβ) proteins seem to be able to transform to amyloid-fibrils [1, 4].

Stefin B is an amyloidogenic protein that has served as a good model system to study amyloid fibril formation and prefibrillar aggregation [5-8] We launched a hypothesis in 2005 [9] that some of the mutants of human stefin B found in EPM1 patients may be more prone to undergo aggregation than the wild-type protein also intracellularly, exerting additional gain in toxic function (apart from loss of normal function as an endogenous cysteine proteases inhibitor). Indeed, our *in vitro* experiments thus far have shown that the fragment R68X forms amyloid-fibrils under mild conditions and very fast even at neutral pH, whereas the G4R mutant exhibits a prolonged lag phase, during which the toxic prefibrillar aggregates accumulate [5]. The *ex vivo* (cell culture) experiments have shown that already the endogenous protein forms small oligomers in cells [10-11]. Furthermore, it has been observed that the wild type stefin B and its EPM1 mutants all aggregate upon over expression in cells. The mutant R68X showed cytotoxic inclusions of the aggresome type, whereas G4R and the wild type protein formed such aggregates only in conditions of over-expression or proteasome inhibition [6]. In accordance, patients bearing one allele of R68X and the other of the usual dodecamer repeat in cystatin B gene, show more severe pathology and clinical symptoms, such as severe learning disabilities and even psychosis [12]. Among the stefin B variants, the R68X fragment and the Y31 variant are most aggregate prone and their aggregates get collected at the nucleus, resembling aggresomes and significantly reduce cell viability [6]. These findings raise questions whether and how might the oligomers/aggregates be cytotoxic.

It has been found that the higher-order oligomers effectively bind to lipid monolayers and are internalized through the plasma membrane into the cell where they lower cell viability [13-15]. On the cellular level alternative function(s) of stefin B linking this protein to reduced oxidative stress and presumably augmented autophagy, need to get better defined. The signaling pathway(s) where stefin B affects cell physiology/pathology also remain to be clarified, which is discussed later on.

2. Neurodegenerative Diseases

Neurodegenerative diseases (ND) have become an increasing problem in the developed world due to a prolonged life span of the aging populations. Although symptomatically usually different, they share in common the feature of protein conformational changes that lead to protein aggregation.

In Alzheimer's disease (AD) the main protein involved in pathology is APP (amyloid precursor protein), a transmembrane protein of approximately 700 amino acid residues, which has important prosurvival functions in the cell. It is normally processed by membrane proteolysis (by enzymes called secretases) into an Aβ (1-40) peptide. In the case of disease development amyloid plaques and neurofibrillary tangles are formed due to prevalence of the hydrophobic Aβ (1-42) peptide and hyperphosphorylated protein τau, respectively[16]. Recently, cellular prion protein (PrP^C) was found to function as a high affinity surface receptor for soluble Aβ-oligomers in neurons and is therefore a mediator of Aβ-oligomer induced synaptic dysfunction. This links AD with infective prion diseases, such as mad cow and Creutzfeldt-Jakob disease (CJD)[17], which has been suspected previously. Infectivity of other amyloid forms connected to Parkinson's and Huntington's disease has also been studied, however, there is no direct evidence of infectious forms of Aβ, tau, huntigtin and α-synuclein [18].

Like other NDs, prion diseases (e.g. kuru, fatal familial insomnia) can also originate spontaneously or are inherited. The main component of prions is presumably a misfolded, partially protease-resistant conformer (PrP^{Sc}) of a normal cell surface protein, the cellular prion protein (PrP^C), which exerts an anti-oxidative function. Other common features between prion diseases and neurodegenerative diseases comprise astrocytosis and vacuolization [19].

The symptoms of Parkinson's disease (PD) are caused by selective and progressive degeneration of pigmented dopaminergic (DA) neurons in the substancia nigra pars compacta (SNpc). One important feature of PD is the presence of eosinophilic, cytoplasmic inclusions of fibrillar, misfolded proteins, termed Lewy bodies, in affected brain areas. The exact composition of Lewy bodies is unknown, but they contain proteins associated with the genetic form of PD such as (ubiquinated) α-synuclein, parkin and synphilin[20]. These and other proteins linked to PD (e.g. redox-regulated chaperone DJ-1, leucine-rich repeat kinase 2 LRRK2, serine/threonine kinase PINK-1) provide insight into the pathogenesis of PD. The main mechanism underlying selective dopaminergic neuron death is believed to be severe mitochondrial dysfunction [21].

Symptoms in Huntington's disease (HD) include a characteristic movement disorder (Huntington's chorea), cognitive dysfunction, and psychiatric symptoms, such as depression. Symptoms result from the selective loss of long projection GABAergic neurons, known as medium spiny neurons that control movement of the body and limbs [22]. The disease is caused by a mutation encoding an abnormal expansion of polyglutamine repeats in a protein called huntingtin. Disease severity depends on the length of the polyQ stretch, with over 37 repeats clearly linked to HD. The expanded polyQ stretch in mutant huntingtin induces a conformational change resulting in aggregates in dendrites and nuclei and is thought to confer a gain in toxic function via transcriptional dysregulation, binding and sequestering selective transcription factors, protein aggregation, and excitotoxicity [22-23].

Table 1. A selection of aAmyloidogenic proteins and the corresponding diseases [24]

clinical syndrome	precursor protein	fibril component
Alzheimer's disease	APP	β-peptide 1-40 to 1-43
primary systemic amyloidosis	immunoglobulin light chain	intact light chain or fragment
secondary systemic amyloidosis	serum amyloid A	amyloid A (76 residue fragment)
senile systemic amyloidosis	transthyretin	transthyretin or fragment
familial amyloid polyneuropathy I	transthyretin	over 45 transthyretin variants
hereditary cerebral amyloid angiopathy	cystatin C	cystatin C minus 10 residues
hemodialysis-related amyloidosis	β_2-microglobulin	β_2-microglobulin
familial amyloid polyneropathy III	apolipoprotein A1	fragments of apolipoprotein A1
Finnish hereditary systemic amyloidosis	gelsolin	71 amino acid fragment of gelsolin
type II diabetes	islet amyloid polypeptide (IAPP)	fragment of IAPP
medullary carcinoma of the thyroid	calcitonin	fragments of calcitonin
spongiform encephalopathies	prion	prion
atrial amyloidosis	atrial natriuretic factor (ANF)	ANF
hereditary non-neurophatic systemic amyloidosis	lysozyme	lysozyme
injection-localized amyloidosis	insulin	insulin
hereditary renal amyloidosis	fibrinogen	fibrinogen fragments

3. Progressive Myoclonus Epilepsy of Type 1 (EPM1) - Unverricht-Lundborg Disease

Unverricht-Lundborg disease (EPM1) is a type of progressive myoclonus epilepsy (PME), a group of etiologically and clinically heterogeneous inherited disorders sharing a combination of myoclonus, epilepsy, and progressive neurological deterioration[25]. EPM1 is inherited in autosomal recessive manner and patients suffer from progressive myoclonic jerks, frequent tonic-clonic seizures, and progressive decline in cognition. It occurs mainly in Baltic and Mediterranean regions and the onset of disorder is between 6 and 18 years of age. Individuals are mentally alert but show emotional lability, depression, and a mild decline in intellectual performance over time. The disease can lead to dysarthria, and ataxia in later stages[26]. Pathology has shown a marked neural loss of Purkinje cells in the granular layer of the cerebellum [27]. Mutations in the gene encoding cystatin B (*CSTB*), a cysteine protease inhibitor, are responsible for the primary defect underlying EPM1. It is not currently possible to diagnose EPM1 without a genetic test, and since the early symptoms can often be mistaken for more common epilepsy types, for example, juvenile myoclonic epilepsy (JME)[28].

The most common change found in cystatin B gene is the dodecamer repeat expansion in the promoter region which leads to reduced mRNA and protein levels. Other mutations in the stefin B underlying EPM1 are shown in Table 2. The *CSTB* gene is located in chromosome 21q22.3 and altogether 10 different mutations of this gene have been reported to underlie EPM1 [29].

Table 2. Mutations in *CSTB* underlying Unverricht-Lundborg disease (EPM1) [29]

mutation	position of mutation in gene/type	predicted consequence on protein
dodecamer repeat expansion	5' UTR/expansion	reduced CSTB expression
c.10G>C	exon 1/missense	p.G4R
c.67-1G>C	intron 1/splice site	p.delV23_K56
c.149G>A	exon 2/missense	p.G50E
c.168>A	exon 2/splice site	aberrant splicing?
c.169-2A>G	intron 2/splice site	aberrant splicing?
c.168+1_18del	intron 2/deletion	p.delV23_K56 p.V57EfsX28
c.202C>T	exon 3/nonsense	p.R68X
c.218_219delTC	exon 3/deletion	p.L73FSX3
c.212A>C	exon 3/missense	p.Q71P

3.1. Disease Development and Progression

It is not clear whether the mutant stefin B protein is responsible for the disease development and neurotoxicity or, the lack of protein's protective activity – or, as we suggest here, at least for some cases bearing missense mutants – a combination of both. It also would seem important to clarify more protein's alternative function(s), such as its function in the multiprotein complex important in the regulation of cytoskeleton [30], anti-oxidative stress effect [31] and putative chaperone-like function [32-33]. Stefin B also binds to histones and indirectly regulates the cell cycle through inhibition of cathepsin L in the nucleus [6]. If such functions would be omitted in EPM1 mutants, the consequences would also likely lead to neurodegeneration.

Stefin B deficient mice (knock-out – KO mice) serve as a reasonable model to study EPM1, although they lack some symptoms typical for human patients. KO mice develop myoclonus after 1 month and ataxia after 6 months. Myoclonus occurs only during sleep with no tonic-clonic seizures, no photosensitivity nor spike-wave complexes in electroencephalogram (EEG) [34]. Of note, the hippocampal slices from *CSTB*-deficient mice are hyperexcitable and display more precocious and pronounced epileptic-like responses. GABAeric inhibitory neurons may be particularly prone to damage and show higher susceptibility to seizure-induced cell damage, which may cause the progressive nature of this disease [35].

Patients show cerebral atrophy and reduced levels of grey matter in the motor cortex and thalami and reduced cortical thickness [36]. Interestingly, the disease has been linked to obesity [37]. Although the protective role of stefin B against apoptosis seems to become apparent [38-39], it is still unclear how the reduced protein levels could lead to myoclonic and/or the epileptic phenotype. It was suggested that primary stefin B deficit triggers several secondary processes, which may be responsible for the seizure phenotype EPM1, such as overstimulation of serotoninergic transmission [40] or a defect in dopamine transmission [41].

3.2. Stefin B Effects on Innate Immunity and Neurogenesis

There are two groups of glial cells in the central nervous system (CNS): the macroglia, including astrocytes, oligodendrocytes, and ependymal cells, and the microglia. Microglia are the immunological effector cells in the CNS that participate in the innate immune defense by continuously surveying their cellular environment in the brain parenchyma [42]. In chronic neurodegenerative diseases an innate immune response within the brain is dominated by microglia [43]. Excessive reactive gliosis inhibits neurogenesis and leads to neuroinflammation and neurodegeneration through release of the S100β protein [44]. However, not only the microglia, but also the effect of systemic inflammation on brain pathology and disease progression was observed in animal models of acute neurodegeneration [45], ischemia [46] and Alzheimer disease (AD) [47]. CNS inflammation can lead to a dysfunctional uptake of glutamate in astrocytes which causes excitotoxicity [48]. The accumulation of Aβ in the brain of AD patients triggers the activation of microglia and astrocytes, which are capable of phagocytosing Aβ from the brain parenchyma. Adult astrocytes also clear Aβ efficiently both *ex vivo* [49] and *in vivo* when administered intracranially in the AD mouse brain [50]. However, microglial activation could in fact be one of the main causes of neuronal death [51] even though studies on postmortem human AD brain suggest a serious lack of effective phagocytosis by any cell type in clinical cases.

Glial cells provide support and nutrition to neurons, maintain homeostasis, form myelin, and participate in signal transmission in the nervous system. Reactive fibrillary astrocytes are capable of producing dense microglial scars, which create a physical barrier between damaged and healthy cells. This biological phenomenon seems to be a prevalent adaptation against many forms of neural damage. Glial activation has been suggested also in other neurodegenerative diseases such as scrapie, AD and CJD [52]. Both, in stefin B deficient mice and EPM1 patients signs of neuronal atrophy in the cerebellum (particularly the Purkinje cells), entorhinal cortex and hippocampus take place [34, 36]. In accordance to other neurodegenerative conditions, gliosis (e.g. production of dense glial scars) is marked in these regions [53].

Furthermore, in astrocytoma cell line T98G stefin B was found expressed in the nucleus as well as in the cytosol. It was shown that stefin B interacts with cathepsin L and and histones in the nucleus [54]. In synchronized T98G cells, stefin B co-immunoprecipitated with histones H2A.Z, H2B, and H3, predominantly in the G1 phase of the cell cycle. Stefin B-deficient mouse embryonic fibroblasts entered S phase earlier than wild type mouse embryonic fibroblasts. In contrast, increased expression of stefin B in the nucleus delayed cell cycle progression in T98G cells. The delay in cell cycle progression was associated with the inhibition of cathepsin L in the nucleus, as judged from the decreased cleavage of the CUX1 transcription factor. Cells isolated from stefin B deficent animals were more sensitive to staurosporin induced apoptosis, than control cells from wild type animals [38]. Weather nuclear or cytosolic functions of stefin B are associated with increased sensitivity to apoptosis in stefin B deficient cells is not clear yet [55].

Neurogenesis is a major cell differentiation process in the nervous system. It involves an extremely complex system of regulatory signals operating with a very precise timing along embryonic development [56]. The myristoylated alanine-rich C kinase substrate (MARCKS), a substrate of protein kinase C (PKC), plays critical functions in morphogenetic processes involved in the development of the nervous system. Experiments on MARCKS deficient mice have shown that this protein is essential for normal central nervous system and neural retina morphogenesis [57]. Besides regulation at the level of gene transcription, MARCKS concentrations within the cell are regulated by proteolytic cleavage by cysteine cathepsins and calpains. Increased cleavage of MARCKS in brain and macrophages of stefin B-deficient mice compared to wild-type mice tissue was reported. Expression and processing of cathepsin B was unaltered in the brain of stefin B-deficient mice, indicating that increased cleavage of MARCKS could be attributed to the lack of stefin B [58].

4. Cytotoxicity of Protein Aggregates and Cell Defense Systems

A consensus has been reached that soluble, prefibrillar aggregates cause most stress to the cell and are even cytotoxic [59-62]. The cell transports small protein aggregates in a microtubule-dependant manner to the centrosome where they form aggresomes [63]. The aggresome serves as a storage compartment for protein aggregates that accumulate inside the cell. One notion is that the formation of aggresomes represents a protective mechanism to sequester the aggregates of abnormally folded polypeptides.

Cells have several mechanisms that protect them against protein misfolding and aggregation. These are chaperonins, chaperons and “amateur chaperons” as the first line of defense and degradation by ubiquitin proteasome system (UPS) and autophagy as the second line. Autophagy is a process closely related to endosomal and lysosomal pathways for cell clearance. Molecular chaperones are a large and diverse group of proteins that share the property of assisting the noncovalent folding and unfolding and the assembly and disassembly of other macromolecular structures [64]. Chaperones are the main regulators of protein homeostasis (proteostasis) which is disturbed by appearance of toxic aggregates [65].

While UPS has gained much attention as it relates to neurodegenerative disease, important links between autophagy and neurodegeneration have also become evident [66]. Furthermore, recent studies have revealed interaction between UPS and autophagy, suggesting a coordinated and complementary relationship between these degradation systems that becomes critical in times of cellular stress [67]. Autophagy is a degradation and recycling process of cellular constituents – long-lived proteins and organelles. It is increased under pathological conditions that cause cell dysfunction such a starvation, oxidative stress and hypoxia. Macroautophagy, which is the most common type of autophagy, is responsible for the turnover of cellular debris which is sequestered in double-membraned vesicle termed autophagasome and later on degraded in autophagolysosomes by proteases in low pH environment. Many cytosolic proteins, as well as misfolded secretory and membrane proteins exported from endoplasmic reticulum to the cytosol, are normally degraded by UPS, however, autophagy is claimed as the predominant way of aggresomes clearance [68].

Cystatin C has a diverse roles in the cell, first being the inhibiton of cysteine proteases – cathepsins B, H, K, L, and S [69]. Its dimeric form also interacts with Aβ amd inhibits *in vitro* Aβ fibril formation [70], so, it is believed to be somehow neuroprotective in NDs by either preventing cell death, promoting cell survival, or promoting neurogenesis. Recently, it was discovered that cystatin C plays a protective role under stress conditions by inducting autophay via mTOR inhibition [71].

Stefin B belongs to the same family of cystatins as cystatin C inhibiting the same group of cathepsins [72]. Stefin B binding also prevents Aβ fibril growth, however, this binding is dependent on its oligomeric state [32]. Oligomers and aggregates of stefin B have been observed in cells [11], indicating that the endogenous protein forms smaller, occasional aggregates, which become even more abundant upon proteasome inhibition or in the presence of mutations (e.g. R68X) [54]. Recently, neuronal cytoplasmic and intranuclear inclusions, containing the lysosomal proteins cathepsin B and CD68 or FUS, respectively, were identified in one EPM1 patient [64]. We thus believe that stefin B could potentially have a positive role in autophagy. A recent report [73] by Yang et al. in Brain 2011, states that stefin B deletion improves autophagy in a mouse model of Alzheimer disease [73]. However, conflicting results have been reported thus far, which calls for more studies [74].

For many neurodegenerative diseases autophagy was found to be impaired (for review [75]). We recently proposed this also could hold for progressive myoclonus epilepsies, including EPM1 and EPM2 [76]. Impairment of autophagy can be due to a loss of function of a certain protein or a dysfunction in the lysosomal pathway and/or the susceptibility of the lysosome to oxidative stress that can originate from excessive protein aggregation, mitochondrial dysfunction or other cellular defects. Furthermore, amyloids can disrupt the membrane integrity through formation of pores resembling bacterial pore-forming toxins in their function (for review [77]). One of the biggest sources of reactive oxygen species (ROS) are (intact) mitochondria and their production can be significantly increased in various NDs due to mitochondrial defects in the respiratory and energy systems [78-80]. Indeed, the accumulated levels of H_2O_2 that are predicted to originate in mitochondria in starved cells act as a signal to initiate autophagy [81]. Dysfunctional mitochondria can finally lead to the collapse of the organelle and the release of proapoptotic molecules that lead to cell death [82].

4.1. Stefin B Aggregates Size and Cytotoxicity

Human stefin B is the most labile and amyloidogenic protein among the members of the broader family of cystatins and it serves as an appropriate model protein in studies of amyloid fibril formation. Size and morphology of the prefibrillar aggregates have been studied and are in agreement to other amyloid systems [14]. Furthermore, toxicity studies showed that the prefibrillar oligomers of stefin B are toxic [14-15]. A likely mechanism for the amyloid cytotoxicity is membrane interaction and damage [62]. As observed for a number of amyloid forming proteins, stefin B bound to and even perforated acidic phospholipid bilayers [6, 13].

As described in the Introduction, we have studied several mutations of stefin B occurring in some EPM1 patients. In order to study their stability and aggregation propensity, the mutant proteins were prepared in recombinant form. Their inhibitory activity as cysteine protease inhibitors was also probed; G4R was found not to be active, while G50E and Q71P showed lowered inhibitory activity in comparison to the wild type. This certainly holds for the fragment R68X, which is not active and not folded at all.

By measuring CD (data in preparation) it has been shown that G50E and Q71P mutants possess similar secondary structure but a more open tertiary structure compared to the wild type. Our previous studies showed that the R68X fragment was unfolded and that it transformed into fibrils already at neutral pH [5]. Interestingly and in distinction to other examples of the missense EPM1 mutants, G4R was found properly folded and compact protein of similar stability to the wild type but it showed a four times longer lag phase, during which the prefibrillar aggregates accumulated [5].

The *ex vivo* (cell culture) studies have shown that even the wild type stefin B forms small endogenous aggregates which increase in amount upon over-expression or proteasome inhibition [10]. Compared to the wild-type, all the above mentioned missense mutants formed larger cytosolic and often perinuclear aggregates. The aggregates showed some of the molecular characteristics of aggresomes [63]. In contrast to the belief that aggresomes are cytoprotective [67], in our case the cells with more abundant aggregates were less viable. Partial co-localization of stefin B aggregates with LC3 (Figure 1) is in agreement to the notion that the aggresome-like aggregates are targeted for autophagic clearance [10].

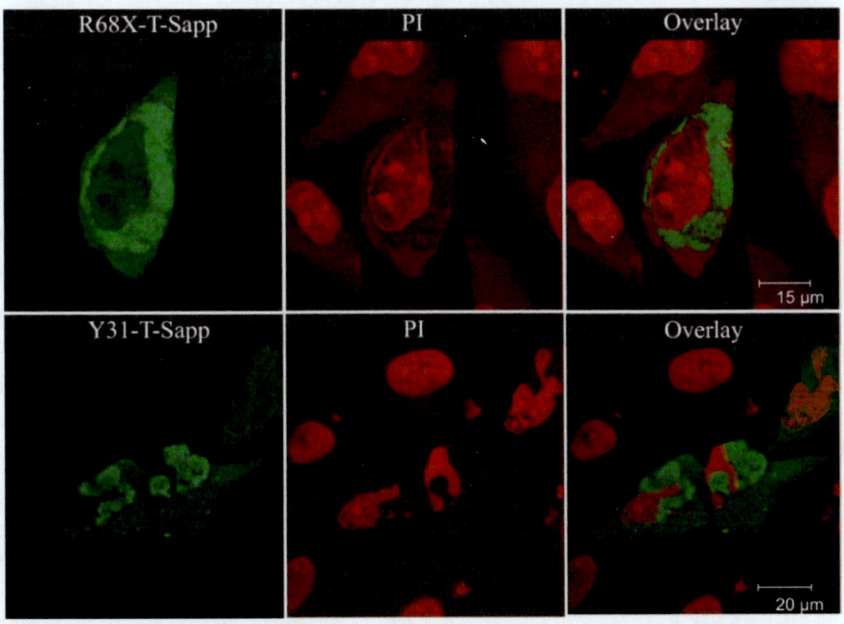

Figure 1A. Perinuclear aggregates of the EPM1 R68X mutant and of the most aggregate prone wild-type variant stefin B with Y31 (both fused with a green fluorescent protein T-Sapphire). The nuclei were labeled with propyl-iodide – PI (red color).

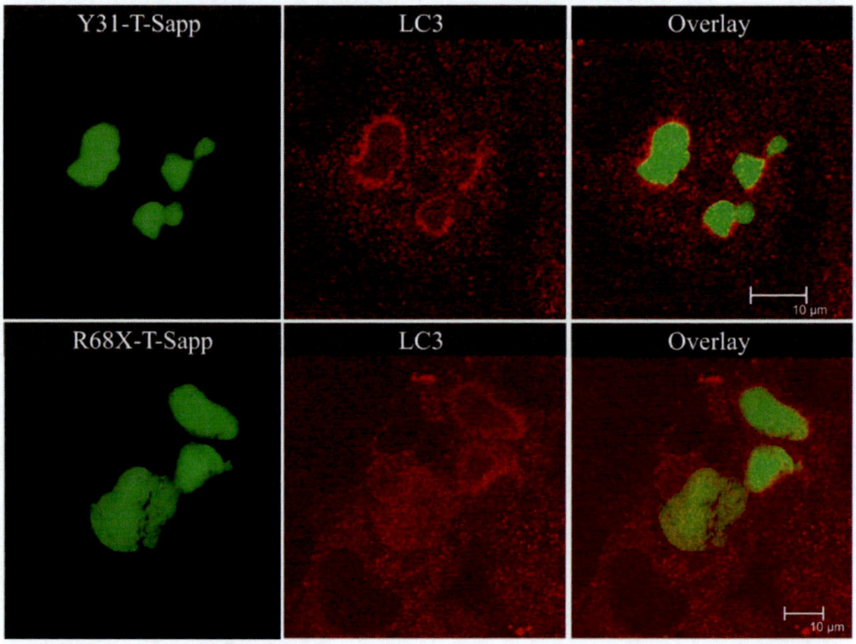

Figure 1B). Partial co-localization of stefin B - Y31 variant (top) and R68X mutant (bottom) (green color of the fused GFP, protein T-Sapphire) with LC3 a marker of autophagy (colored red by the secondary anti-bodies to LC3). A ring of LC3 at the edge of the aggregates can be seen.

Conclusion

1. Protein aggregation is at the core of many neurodegenerative diseases; the soluble oligomers and aggregates are cytotoxic, most likely acting via membrane binding and perforation.
2. Stefin B, its variants and EPM1 mutants serve as a good model for the study of protein aggregation *in vitro* and *ex vivo* as well as amyloid oligomers cytotoxicity.
3. In the study by Čeru et al., 2010 [10] we have shown that even the endogenous stefin B is prone to form smaller scattered aggregates in cells, which increase in number and size if the proteasome is inhibited. We also have shown that the missense EPM1 mutants R68X and G4R form perinuclear intracellular aggregates when expressed in cells, which are, especially in the case of R68X, more abundant and of extended size than for the wild type protein and in this way could gain a toxic function and modulate pathology of the progressive myoclonus epilepsy of type 1 (EPM1). In accordance, patients bearing one allele of R68X and the other of the usual dodecamer repeat in cystatin B gene, show more severe pathology and clinical symptoms,
4. The preliminary studies of the two additional stefin B EPM1 missense mutants: G50E and Q71P show that these two proteins possess similar secondary structure but a more open tertiary structure compared to the wild type. Undergoing are studies to characterize the aggregates in appropriate cell models and to determine the ways of their clearance.

REFERENCES

[1] Zerovnik, E. (2002) Amyloid-fibril formation. Proposed mechanisms and relevance to conformational disease, *Eur. J. Biochem.* 269, 3362-71.

[2] Irvine, G. B., El-Agnaf, O. M., Shankar, G. M. and Walsh, D. M. (2008) Protein aggregation in the brain: the molecular basis for Alzheimer's and Parkinson's diseases, *Mol. Med.* 14, 451-64.

[3] Dobson, C. M. (2002) Getting out of shape, *Nature.* 418, 729-30.

[4] Zerovnik, E., Stoka, V., Mirtic, A., Guncar, G., Grdadolnik, J., Staniforth, R. A., Turk, D. and Turk, V. (2011) Mechanisms of amyloid fibril formation - focus on domain-swapping, *FEBS J.* 278, 2263-2282.

[5] Rabzelj, S., Turk, V. and Zerovnik, E. (2005) In vitro study of stability and amyloid-fibril formation of two mutants of human stefin B (cystatin B) occurring in patients with EPM1, *Protein Sci.* 14, 2713-22.

[6] Anderluh, G., Gutierrez-Aguirre, I., Rabzelj, S., Ceru, S., Kopitar-Jerala, N., Macek, P., Turk, V. and Zerovnik, E. (2005) Interaction of human stefin B in the prefibrillar oligomeric form with membranes. Correlation with cellular toxicity, *FEBS J.* 272, 3042-51.

[7] Zerovnik, E., Skarabot, M., Skerget, K., Giannini, S., Stoka, V., Jenko-Kokalj, S. and Staniforth, R. A. (2007) Amyloid fibril formation by human stefin B: influence of pH and TFE on fibril growth and morphology, *Amyloid.* 14, 237-47.

[8] Zerovnik, E. (2009) Protein Conformational Pathology in Alzheimer's and other Neurodegenerative Diseases; New Targets for Therapy, *Cur. Alzheimer Res.*

[9] Ceru, S., Rabzelj, S., Kopitar-Jerala, N., Turk, V. and Zerovnik, E. (2005) Protein aggregation as a possible cause for pathology in a subset of familial Unverricht-Lundborg disease, *Med. Hypotheses.* 64, 955-9.

[10] Ceru, S., Layfield, R., Zavasnik-Bergant, T., Repnik, U., Kopitar-Jerala, N., Turk, V. and Zerovnik, E. (2010) Intracellular aggregation of human stefin B: confocal and electron microscopy study, *Biol. Cell.* 102, 319-34.

[11] Cipollini, E., Riccio, M., Di Giaimo, R., Dal Piaz, F., Pulice, G., Catania, S., Caldarelli, I., Dembic, M., Santi, S. and Melli, M. (2008) Cystatin B and its EPM1 mutants are polymeric and aggregate prone in vivo, *Biochim. Biophys Acta.* 1783, 312-22.

[12] Danner, N., Julkunen, P., Kononen, M., Hypponen, J., Koskenkorva, P., Vanninen, R., Lehesjoki, A. E., Kalviainen, R. and Mervaala, E. Primary motor cortex alterations in a compound heterozygous form of Unverricht-Lundborg disease (EPM1), *Seizure.* 20, 65-71.

[13] Rabzelj, S., Viero, G., Gutierrez-Aguirre, I., Turk, V., Dalla Serra, M., Anderluh, G. and Zerovnik, E. (2008) Interaction with model membranes and pore formation by human stefin B: studying the native and prefibrillar states, *FEBS J.* 275, 2455-66.

[14] Ceru, S., Kokalj, S. J., Rabzelj, S., Skarabot, M., Gutierrez-Aguirre, I., Kopitar-Jerala, N., Anderluh, G., Turk, D., Turk, V. and Zerovnik, E. (2008) Size and morphology of toxic oligomers of amyloidogenic proteins: a case study of human stefin B, *Amyloid.* 15, 147-59.

[15] Ceru, S. and Zerovnik, E. (2008) Similar toxicity of the oligomeric molten globule state and the prefibrillar oligomers, *FEBS Lett.* 582, 203-9.

[16] Querfurth, H. W. and LaFerla, F. M. Alzheimer's disease, *N. Engl. J. Med.* 362, 329-44.

[17] Lauren, J., Gimbel, D. A., Nygaard, H. B., Gilbert, J. W. and Strittmatter, S. M. (2009) Cellular prion protein mediates impairment of synaptic plasticity by amyloid-beta oligomers, *Nature.* 457, 1128-32.

[18] Cushman, M., Johnson, B. S., King, O. D., Gitler, A. D. and Shorter, J. Prion-like disorders: blurring the divide between transmissibility and infectivity, *J. Cell Sci.* 123, 1191-201.

[19] Sakudo, A. and Ikuta, K. (2009) Prion protein functions and dysfunction in prion diseases, *Curr. Med. Chem.* 16, 380-9.

[20] Bossy-Wetzel, E., Schwarzenbacher, R. and Lipton, S. A. (2004) Molecular pathways to neurodegeneration, *Nat. Med.* 10 Suppl, S2-9.

[21] Bueler, H. (2009) Impaired mitochondrial dynamics and function in the pathogenesis of Parkinson's disease, *Exp. Neurol.* 218, 235-46.

[22] Bossy-Wetzel, E., Petrilli, A. and Knott, A. B. (2008) Mutant huntingtin and mitochondrial dysfunction, *Trends Neurosci.* 31, 609-16.

[23] Friedlander, R. M. (2003) Apoptosis and caspases in neurodegenerative diseases, *N Engl. J. Med.* 348, 1365-75.

[24] Yon, J. M. (2002) Protein folding in the post-genomic era, *J. Cell Mol Med.* 6, 307-27.

[25] Genton, P. Unverricht-Lundborg disease (EPM1), *Epilepsia.* 51 Suppl 1, 37-9.

[26] Kalviainen, R., Khyuppenen, J., Koskenkorva, P., Eriksson, K., Vanninen, R. and Mervaala, E. (2008) Clinical picture of EPM1-Unverricht-Lundborg disease, *Epilepsia.* 49, 549-56.

[27] Eldridge, R., Iivanainen, M., Stern, R., Koerber, T. and Wilder, B. J. (1983) "Baltic" myoclonus epilepsy: hereditary disorder of childhood made worse by phenytoin, *Lancet.* 2, 838-42.

[28] de Haan, G. J., Halley, D. J., Doelman, J. C., Geesink, H. H., Augustijn, P. B., Jager-Jongkind, A. D., Majoie, M., Bader, A. J., Leliefeld-Ten Doeschate, L. A., Deelen, W. H., Bertram, E., Lehesjoki, A. E. and Lindhout, D. (2004) Univerricht-Lundborg disease: underdiagnosed in the Netherlands, *Epilepsia.* 45, 1061-3.

[29] Joensuu, T., Lehesjoki, A. E. and Kopra, O. (2008) Molecular background of EPM1-Unverricht-Lundborg disease, *Epilepsia.* 49, 557-63.

[30] Di Giaimo, R., Riccio, M., Santi, S., Galeotti, C., Ambrosetti, D. C. and Melli, M. (2002) New insights into the molecular basis of progressive myoclonus epilepsy: a multiprotein complex with cystatin B, *Hum. Mol. Genet.* 11, 2941-50.

[31] Lehtinen, M. K., Tegelberg, S., Schipper, H., Su, H., Zukor, H., Manninen, O., Kopra, O., Joensuu, T., Hakala, P., Bonni, A. and Lehesjoki, A. E. (2009) Cystatin B deficiency sensitizes neurons to oxidative stress in progressive myoclonus epilepsy, EPM1, *J Neurosci.* 29, 5910-5.

[32] Skerget, K., Taler-Vercic, A., Bavdek, A., Hodnik, V., Ceru, S., Tusek-Znidaric, M., Kumm, T., Pitsi, D., Pompe-Novak, M., Palumaa, P., Soriano, S., Kopitar-Jerala, N., Turk, V., Anderluh, G. and Zerovnik, E. Interaction between oligomers of stefin B and amyloid-beta in vitro and in cells, *J. Biol. Chem.* 285, 3201-10.

[33] Taler-Vercic, A. and Zerovnik, E. Binding of amyloid peptides to domain-swapped dimers of other amyloid-forming proteins may prevent their neurotoxicity, *Bioessays.* 32, 1020-4.

[34] Pennacchio, L. A., Bouley, D. M., Higgins, K. M., Scott, M. P., Noebels, J. L. and Myers, R. M. (1998) Progressive ataxia, myoclonic epilepsy and cerebellar apoptosis in cystatin B-deficient mice, *Nat. Genet.* 20, 251-8.

[35] Franceschetti, S., Sancini, G., Buzzi, A., Zucchini, S., Paradiso, B., Magnaghi, G., Frassoni, C., Chikhladze, M., Avanzini, G. and Simonato, M. (2007) A pathogenetic hypothesis of Unverricht-Lundborg disease onset and progression, *Neurobiol. Dis.* 25, 675-85.

[36] Koskenkorva, P., Khyuppenen, J., Niskanen, E., Kononen, M., Bendel, P., Mervaala, E., Lehesjoki, A. E., Kalviainen, R. and Vanninen, R. (2009) Motor cortex and thalamic atrophy in Unverricht-Lundborg disease: voxel-based morphometric study, *Neurology.* 73, 606-11.

[37] Korja, M., Ferlazzo, E., Soilu-Hanninen, M., Magaudda, A., Marttila, R., Genton, P. and Parkkola, R. T2-weighted high-intensity signals in the basal ganglia as an interesting image finding in Unverricht-Lundborg disease, *Epilepsy Res.* 88, 87-91.

[38] Kopitar-Jerala, N., Schweiger, A., Myers, R. M., Turk, V. and Turk, B. (2005) Sensitization of stefin B-deficient thymocytes towards staurosporin-induced apoptosis is independent of cysteine cathepsins, *FEBS Lett.* 579, 2149-55.

[39] Yang, F., Tay, K. H., Dong, L., Thorne, R. F., Jiang, C. C., Yang, E., Tseng, H. Y., Liu, H., Christopherson, R., Hersey, P. and Zhang, X. D. Cystatin B inhibition of TRAIL-induced apoptosis is associated with the protection of FLIP(L) from degradation by the E3 ligase itch in human melanoma cells, *Cell Death Differ.* 17, 1354-67.

[40] Vaarmann, A., Kaasik, A. and Zharkovsky, A. (2006) Altered tryptophan metabolism in the brain of cystatin B-deficient mice: a model system for progressive myoclonus epilepsy, *Epilepsia.* 47, 1650-4.

[41] Korja, M., Kaasinen, V., Lamusuo, S., Parkkola, R., Nagren, K. and Marttila, R. J. (2007) Substantial thalamostriatal dopaminergic defect in Unverricht-Lundborg disease, *Epilepsia.* 48, 1768-73.

[42] Hanisch, U. K. and Kettenmann, H. (2007) Microglia: active sensor and versatile effector cells in the normal and pathologic brain, *Nat Neurosci.* 10, 1387-94.

[43] Dheen, S. T., Kaur, C. and Ling, E. A. (2007) Microglial activation and its implications in the brain diseases, *Curr Med Chem.* 14, 1189-97.

[44] Van Eldik, L. J. and Wainwright, M. S. (2003) The Janus face of glial-derived S100B: beneficial and detrimental functions in the brain, *Restor Neurol. Neurosci.* 21, 97-108.

[45] Palin, K., Cunningham, C., Forse, P., Perry, V. H. and Platt, N. (2008) Systemic inflammation switches the inflammatory cytokine profile in CNS Wallerian degeneration, *Neurobiol. Dis.* 30, 19-29.

[46] McColl, B. W., Rothwell, N. J. and Allan, S. M. (2008) Systemic inflammation alters the kinetics of cerebrovascular tight junction disruption after experimental stroke in mice, *J. Neurosci.* 28, 9451-62.

[47] Dunn, N., Mullee, M., Perry, V. H. and Holmes, C. (2005) Association between dementia and infectious disease - Evidence from a case-control study, *Alz. Dis. Assoc. Dis.* 19, 91-94.

[48] Lobsiger, C. S. and Cleveland, D. W. (2007) Glial cells as intrinsic components of non-cell-autonomous neurodegenerative disease, *Nat. Neurosci.* 10, 1355-60.

[49] Koistinaho, M., Lin, S., Wu, X., Esterman, M., Koger, D., Hanson, J., Higgs, R., Liu, F., Malkani, S., Bales, K. R. and Paul, S. M. (2004) Apolipoprotein E promotes

astrocyte colocalization and degradation of deposited amyloid-beta peptides, *Nat. Med.* 10, 719-26.

[50] Pihlaja, R., Koistinaho, J., Malm, T., Sikkila, H., Vainio, S. and Koistinaho, M. (2008) Transplanted astrocytes internalize deposited beta-amyloid peptides in a transgenic mouse model of Alzheimer's disease, *Glia.* 56, 154-63.

[51] Herrup, K. Reimagining Alzheimer's disease--an age-based hypothesis, *J. Neurosci.* 30, 16755-62.

[52] Lieuallen, K., Pennacchio, L. A., Park, M., Myers, R. M. and Lennon, G. G. (2001) Cystatin B-deficient mice have increased expression of apoptosis and glial activation genes, *Hum. Mol. Genet.* 10, 1867-71.

[53] Shannon, P., Pennacchio, L. A., Houseweart, M. K., Minassian, B. A. and Myers, R. M. (2002) Neuropathological changes in a mouse model of progressive myoclonus epilepsy: cystatin B deficiency and Unverricht-Lundborg disease, *J. Neuropathol. Exp. Neurol.* 61, 1085-91.

[54] Ceru, S., Konjar, S., Maher, K., Repnik, U., Krizaj, I., Bencina, M., Renko, M., Nepveu, A., Zerovnik, E., Turk, B. and Kopitar-Jerala, N. (2010) Stefin B interacts with histones and cathepsin L in the nucleus, *J. Biol Chem.* 285, 10078-86.

[55] Abramov, A. Y. and Duchen, M. R. (2005) The role of an astrocytic NADPH oxidase in the neurotoxicity of amyloid beta peptides, *Philos Trans R Soc Lond B Biol Sci.* 360, 2309-14.

[56] Calof, A. L. (1995) Intrinsic and extrinsic factors regulating vertebrate neurogenesis, *Curr. Opin. Neurobiol.* 5, 19-27.

[57] Stumpo, D. J., Bock, C. B., Tuttle, J. S. and Blackshear, P. J. (1995) MARCKS deficiency in mice leads to abnormal brain development and perinatal death, *Proc. Natl. Acad. Sci. USA.* 92, 944-8.

[58] Kopitar-Jerala, N. and Turk, B. (2007) Cleavage of the myristoylated alanine-rich C kinase substrate (MARCKS) by cysteine cathepsins in cells and tissues of stefin B-deficient mice, *Biol. Chem.* 388, 847-52.

[59] Bucciantini, M., Calloni, G., Chiti, F., Formigli, L., Nosi, D., Dobson, C. M. and Stefani, M. (2004) Prefibrillar amyloid protein aggregates share common features of cytotoxicity, *J. Biol. Chem.* 279, 31374-82.

[60] Bucciantini, M., Giannoni, E., Chiti, F., Baroni, F., Formigli, L., Zurdo, J., Taddei, N., Ramponi, G., Dobson, C. M. and Stefani, M. (2002) Inherent toxicity of aggregates implies a common mechanism for protein misfolding diseases, *Nature.* 416, 507-11.

[61] Bucciantini, M., Rigacci, S., Berti, A., Pieri, L., Cecchi, C., Nosi, D., Formigli, L., Chiti, F. and Stefani, M. (2005) Patterns of cell death triggered in two different cell lines by HypF-N prefibrillar aggregates, *FASEB J.* 19, 437-9.

[62] Stefani, M. (2010) Protein aggregation diseases: toxicity of soluble prefibrillar aggregates and their clinical significance, *Methods Mol. Biol.* 648, 25-41.

[63] Kopito, R. R. (2000) Aggresomes, inclusion bodies and protein aggregation, *Trends Cell Biol.* 10, 524-30.

[64] Henderson, B. P. and Pockley, A. G. (2005) *Molecular chaperones and cell signalling*, Cambridge University Press, Cambridge.

[65] Teixeira-Castro, A., Ailion, M., Jalles, A., Brignull, H. R., Vilaca, J. L., Dias, N., Rodrigues, P., Oliveira, J. F., Neves-Carvalho, A., Morimoto, R. I. and Maciel, P.

Neuron-specific proteotoxicity of mutant ataxin-3 in C. elegans: rescue by the DAF-16 and HSF-1 pathways, *Hum. Mol. Genet.*

[66] Ventruti, A. and Cuervo, A. M. (2007) Autophagy and neurodegeneration, *Curr. Neurol. Neurosci Rep.* 7, 443-51.

[67] Nedelsky, N. B., Todd, P. K. and Taylor, J. P. (2008) Autophagy and the ubiquitin-proteasome system: collaborators in neuroprotection, *Biochim. Biophys. Acta.* 1782, 691-9.

[68] Chin, L. S., Olzmann, J. A. and Li, L. (2010) Parkin-mediated ubiquitin signalling in aggresome formation and autophagy, *Biochem. Soc. Trans.* 38, 144-9.

[69] Turk, V., Brzin, J., Kotnik, M., Lenarcic, B., Popovic, T., Ritonja, A., Trstenjak, M., Begic-Odobasic, L. and Machleidt, W. (1986) Human cysteine proteinases and their protein inhibitors stefins, cystatins and kininogens, *Biomed. Biochim. Acta.* 45, 1375-84.

[70] Sastre, M., Calero, M., Pawlik, M., Mathews, P. M., Kumar, A., Danilov, V., Schmidt, S. D., Nixon, R. A., Frangione, B. and Levy, E. (2004) Binding of cystatin C to Alzheimer's amyloid beta inhibits in vitro amyloid fibril formation, *Neurobiol Aging.* 25, 1033-43.

[71] Tizon, B., Sahoo, S., Yu, H., Gauthier, S., Kumar, A. R., Mohan, P., Figliola, M., Pawlik, M., Grubb, A., Uchiyama, Y., Bandyopadhyay, U., Cuervo, A. M., Nixon, R. A. and Levy, E. (2010) Induction of autophagy by cystatin C: a mechanism that protects murine primary cortical neurons and neuronal cell lines, *PLoS One.* 5, e9819.

[72] Turk, V. and Bode, W. (1991) The cystatins: protein inhibitors of cysteine proteinases, *FEBS Lett.* 285, 213-9.

[73] Yang, D. S., Stavrides, P., Mohan, P. S., Kaushik, S., Kumar, A., Ohno, M., Schmidt, S. D., Wesson, D., Bandyopadhyay, U., Jiang, Y., Pawlik, M., Peterhoff, C. M., Yang, A. J., Wilson, D. A., St George-Hyslop, P., Westaway, D., Mathews, P. M., Levy, E., Cuervo, A. M. and Nixon, R. A. Reversal of autophagy dysfunction in the TgCRND8 mouse model of Alzheimer's disease ameliorates amyloid pathologies and memory deficits, *Brain.* 134, 258-77.

[74] Zerovnik, E. (2009) The emerging role of cystatins in Alzheimer's disease, *Bioessays.* 31, 597-9.

[75] Wong, E. and Cuervo, A. M. Autophagy gone awry in neurodegenerative diseases, *Nat. Neurosci.* 13, 805-11.

[76] Polajnar, M. and Zerovnik, E. (2011) Impaired autophagy: a link between neurodegenerative diseases and progressive myoclonus epilepsies, *Trends Mol. Med.* 17, 293-300.

[77] Butterfield, S. M. and Lashuel, H. A. Amyloidogenic protein-membrane interactions: mechanistic insight from model systems, *Angew Chem Int Ed Engl.* 49, 5628-54.

[78] Lim, D., Fedrizzi, L., Tartari, M., Zuccato, C., Cattaneo, E., Brini, M. and Carafoli, E. (2008) Calcium homeostasis and mitochondrial dysfunction in striatal neurons of Huntington disease, *J Biol Chem.* 283, 5780-9.

[79] Rhein, V., Song, X., Wiesner, A., Ittner, L. M., Baysang, G., Meier, F., Ozmen, L., Bluethmann, H., Drose, S., Brandt, U., Savaskan, E., Czech, C., Gotz, J. and Eckert, A. (2009) Amyloid-beta and tau synergistically impair the oxidative phosphorylation system in triple transgenic Alzheimer's disease mice, *Proc. Natl. Acad. Sci. USA.* 106, 20057-62.

[80] Vila, M. and Perier, C. (2008) Molecular pathways of programmed cell death in experimental Parkinson's disease, *Parkinsonism Relat Disord.* 14 Suppl 2, S176-9.
[81] Scherz-Shouval, R. and Elazar, Z. (2007) ROS, mitochondria and the regulation of autophagy, *Trends Cell Biol.* 17, 422-7.
[82] Halestrap, A. P. (2009) Mitochondria and reperfusion injury of the heart--a holey death but not beyond salvation, *J. Bioenerg. Biomembr.* 41, 113-21.

In: Amyloids
Editors: I. P. Halcheck et al.
ISBN: 978-1-62100-538-4

Chapter 6

MOLECULAR DYNAMICS SIMULATIONS OF AMYLOID β PEPTIDES CONTAINING D-ASPARTIC ACID RESIDUES

Akifumi Oda[1,2,*], Kana Kobayashi[1] and Ohgi Takahashi[1]
[1]Faculty of Pharmaceutical Sciences,
Tohoku Pharmaceutical University, Japan
[2]Institute for Protein Research, Osaka University, Japan

ABSTRACT

Stereoinversion of aspartic acid (Asp) residues in amyloid β (Aβ) peptide is observed in the brain of Alzheimer's dementia patients. The amyloid β 1–42 ($A\beta_{1-42}$) peptide, which has 42 amino acid residues, is considered to be related to Alzheimer's dementia. This peptide includes three Asp residues—Asp1, Asp7, and Asp23. Recently, $A\beta_{1-42}$ peptides containing D-Asp residues were artificially synthesized, and their secondary structures were experimentally observed. Although the experiments indicate that stereoinversion of Asp23 causes β-sheet formation of the $A\beta_{1-42}$ peptide, the atomic-level structure of the $A\beta_{1-42}$ peptide could not be obtained experimentally. Because $A\beta_{1-42}$ containing D-Asp23 tend to aggregate, the atomic-level structural biological experiments are difficult to conduct; thus, structural bioinformatics approaches to the peptides have an important role. In this study, we performed molecular dynamics simulation, which is one of the most useful methods in structural bioinformatics, for $A\beta_{1-42}$ peptides containing D-Asp residues. In addition, the abilities of several settings of molecular dynamics methods to evaluate the structural features of Aβ peptides containing D-Asp were compared. The results indicate that stereoinversion of Asp residues induces changes in the secondary structures of the C-terminal regions of $A\beta_{1-42}$ peptides. The usefulness of some implicit solvent models for molecular dynamics simulations of Aβ peptides is also discussed.

---

* Correspondence concerning this article should be addressed to Akifumi Oda, E-mail: oda@tohoku-pharm.ac.jp, Fax: +81-22-275-2013.

1. INTRODUCTION

Amyloid β (Aβ) peptides contain 39–43 amino acid residues, and these peptides are generated from amyloid precursor proteins [1]. Aβ peptide aggregation is observed in the brains of Alzheimer's dementia (AD) patients, and a relationship between this aggregation and AD is speculated. When aggregation occurs, Aβ peptides form β-sheet structures, and interpeptide interactions are observed between these structures. The presence of apolipoprotein E4 [2], mutations of amino acid residues [3], and cyclization of the side chains of the residues [4] have been proposed as causes of aggregation. Furthermore, a relationship between Aβ peptide aggregation and amino acid residue racemization has been considered [5-7]. Because racemization of aspartic acid (Asp) residues has been observed in Aβ peptides extracted from the brains of AD patients, there is a possibility that this racemization is related to AD. Sakai-Kato et al. reported the properties of artificially synthesized $A\beta_{1-42}$ peptides, produced by solid-phase syntheses in which Asp residues were exhaustively substituted with the D-form [7]. They investigated secondary structures and aggregates of peptides and found that β-structure formation and aggregation were triggered by stereoinversion of Asp23 in the $A\beta_{1-42}$ peptide. Because a relationship between the $A\beta_{1-42}$ peptide and AD is speculated, structural bioinformatics investigations of the effects of Asp residue stereoinversion on conformations of $A\beta_{1-42}$ peptides are expected to play an important role in AD research.

Previously, all amino acids existing in living organisms were believed to be in the L-form. However, D-amino acids have recently been found in mammals [8]. In mammals, not only free D-amino acids but also D-amino acid residues that are components of proteins and peptides have been reported. D-Amino acid residues are present in many proteins and peptides, including α-crystallin [9], skin proteins [10], and Aβ peptides. Among the 20 amino acid residues, nonenzymatic stereoinversions of Asp and serine residues are frequently reported. In addition to stereoinversion, isomerization of Asp residues is frequently observed, and L- and D-β-Asp residues are found in living organisms. These stereoinversion and/or isomerization events are considered to play an important role in age-related diseases such as senile cataract, AD, and skin aging.

Although the importance of stereoinversion of amino acid residues has been reported, atomic/molecular-level investigations have not been sufficiently performed. D-Amino acid residues in proteins and peptides are considered affect the three-dimensional (3D) structures of proteins and peptides. Because denatured proteins and peptides sometimes aggregate, D-amino acid residues are speculated to play an important role in conformational diseases. However, in comparison to native proteins, the solubility of denatured proteins frequently decreases and structural biological experiments, especially atomic-level studies, become difficult to conduct. Although the secondary structural contents have been evaluated by circular dichroism (CD) spectroscopy [6, 7], atomic-level 3D structures that can be used for structure-based drug design have not been obtained experimentally. Experimental data for 3D structures of Aβ peptides have been reported [11, 12]; however, these data include only L-amino acid residues. The experimental 3D structures of Aβ peptides containing D-amino acid residues have not been obtained. Structural bioinformatics studies, such as molecular simulations and structural predictions, are frequently useful for examining structurally unknown proteins and peptides. Although some molecular dynamics (MD) simulations for wild-type Aβ peptides containing L-amino acid residues have been performed [11, 13, 14, 15], no structural bioinformatics studies for the effect of stereoinversion of amino acid residues have been conducted.

The molecular simulation technique, which is one of the structural bioinformatics methods, is used for the investigation of structural and functional modifications to proteins caused by environmental changes. In molecular simulation, the behavior of the atomic- or molecular-level structure of a protein is simulated according to physical laws. MD simulation is one of the most widely used molecular simulation techniques. In MD simulation, the behavior of the molecule can be simulated using the numerical solution of Newton's equation of motion. The outline of an MD simulation is illustrated in Figure 1.

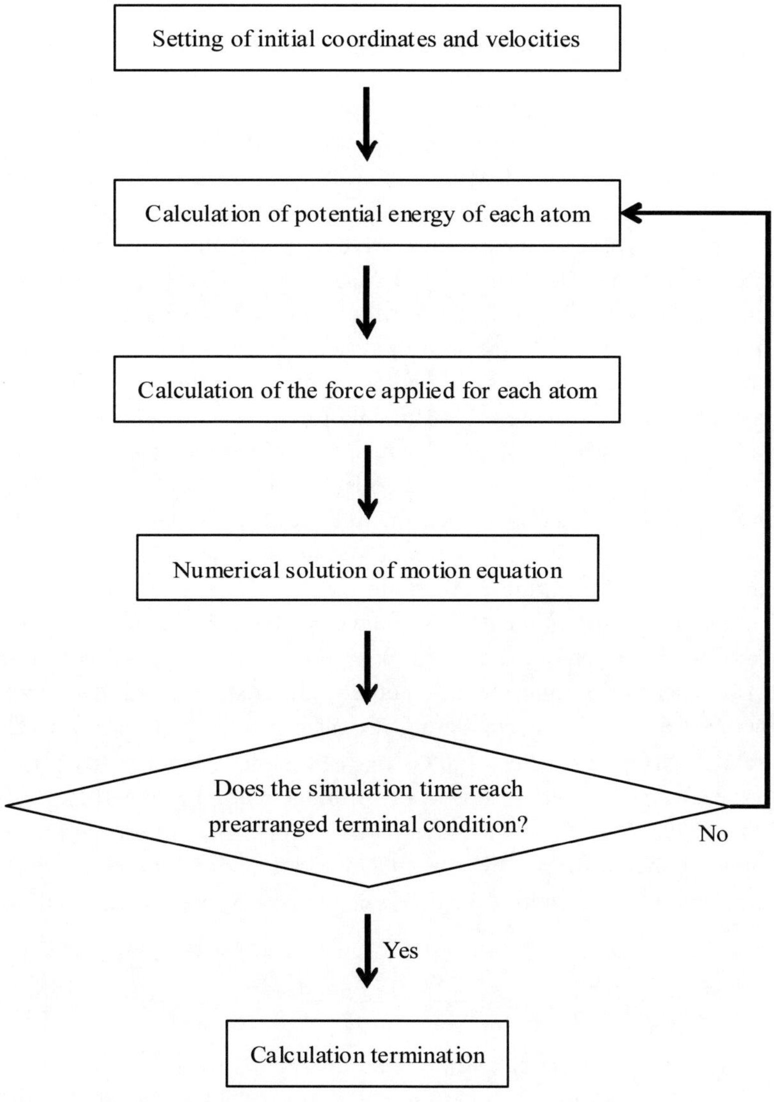

Figure 1. Flow chart of an MD simulation.

As shown in the figure, the potential energy of each atom included in the system is calculated. Next, when the potential energy is obtained, the force **F** can be calculated by differentiating with respect to the position of the atom. Using the force, positional vector **x** can be derived from Newton's equation of motion (1). In the MD simulation, equation (1) is numerically solved.

$$\mathbf{F} = m\frac{\mathrm{d}^2\mathbf{x}}{\mathrm{d}t^2} \tag{1}$$

MD simulations are used for a wide variety of biological studies, such as protein folding predictions [16], protein structure refinements [17], structural analysis of induced fit caused by ligand binding [18], and investigations into the influence of amino acid mutations on protein function [19]. MD simulations play an important role not only in basic biology but also in medical and pharmaceutical sciences. In addition to normal MD, molecular structures can be examined by generalized-ensemble algorithms such as replica exchange MD (REMD), which allows powerful sampling of molecular structures [20].

For MD simulations of biomolecules, an aqueous environment is frequently considered because many types of biomolecules in living organisms exist in a water-based solvent. When considering an aqueous solvent, an explicit solvent model in which a water molecule is represented by three or more massive point charges, e.g., TIP3P [21] and TIP4P [21,22], is sometimes used. On the other hand, because MD simulations using explicit solvent models require large numbers of molecules (several thousands or several tens of thousands of water molecules), explicit water calculations are computationally expensive. Thus, implicit solvent models are frequently used for large-scale simulations of biomolecules. The implicit solvent model is known as the continuous solvent in which the solvent environment is represented by a continuous model with the solvent dielectric constant and explicit solvent molecules are not used. Although the Poisson–Boltzmann (PB) method [23,24] is a well-known implicit solvent model, numerical solution of the PB equation is required for intricately shaped solutes, e.g., proteins and nucleic acids. Thus, the computational cost of the PB method is expensive for MD simulation in which solvation energies have to be calculated at every step of the simulation. For MD simulations, generalized Born (GB) methods [25], which are easier for calculating solvation energies than the PB method, are frequently used as implicit solvent models. GB methods are used for calculating the electrostatic portion of solvation energies (ΔG_{el}). In the AMBER [26] software, which is one of the most widely used MD programs for biopolymers, several types of GB models are implemented. In AMBER, each atom in a molecule is represented as a sphere of radius R_i with a charge q_i at its center; the interior of the atom is assumed to be uniformly filled with a material having a dielectric constant of 1. R_i are the so-called effective Born radii. ΔG_{el} is calculated by GB methods as follows:

$$\Delta G_{el} = -\frac{1}{2}\sum_{ij}\frac{q_i q_j}{f_{GB}\left(r_{ij}, R_i, R_j\right)}\left(1 - \frac{\exp\left(-\kappa f_{GB}\right)}{\varepsilon}\right) \tag{2}$$

where r_{ij} is the distance between atoms i and j, and f_{GB} is a certain smooth function of its arguments. κ is the Debye–Hückel screening parameter, and ε is the dielectric constant of the solvent. Although the effective Born radii (R_i) are equal to the van der Waals radii (ρ_i) for monoatomic iron, the effective Born radii of buried atoms in intricately shaped molecules should be calculated using some approximation functions. One of the representative methods for calculating the effective Born radii is the Coulomb field approximation (CFA). In CFA, the effective Born radii are calculated as follows:

$$R_i^{-1} = \rho_i^{-1} - \frac{1}{4\pi}\int \theta(|\mathbf{r}| - \rho_i) r^{-4} \mathbf{dr} \qquad (3)$$

where the integral is over the solute volume surrounding atom (i). For a realistic molecule, the solute boundary (molecular surface) is anything but trivial, and further approximations are made to obtain a closed-form analytical expression for the above equation, e.g., the so-called pairwise de-screening approach of Hawkins, Cramer, and Truhlar, [27]. Onufriev, Bashford, and Case improved the method and a modified function to calculate R_i was proposed [28]:

$$R_i^{-1} = \tilde{\rho}_i^{-1} - \rho_i^{-1} \tanh(\alpha\Psi - \beta\Psi^2 + \gamma\Psi^3) \qquad (4)$$

where $\tilde{\rho} = \rho - offset$ and offset is equal to 0.09 Å in ref. [28]. In the equation, $\Psi = I\tilde{\rho}_i$

$$I = \frac{1}{4\pi}\int \theta(|\mathbf{r}| - \tilde{\rho}_i) r^{-4} \mathbf{dr} \qquad (5)$$

In equation (4), α, β, and γ are treated as adjustable dimensionless parameters to be optimized. Onufriev, Bashford, and Case proposed the following two types of parameter settings: α = 0.8, β = 0.0, and γ = 2.909125 and α = 1.0, β = 0.8, and γ = 4.85. Another improved version of the GB method (GBn), in which a simple analytic correction term is introduced for solvent-excluded volume of each pair of atoms, has been proposed [29]. In addition to the GB method, the analytical linearized PB (ALPB) method [30, 31] is a promising approach for continuous solvents. The ALPB equation approximates ΔG_{el} as

$$\Delta G_{el} = -\frac{1}{2}\left(\frac{1}{\varepsilon_{in}} - \frac{1}{\varepsilon_{ex}}\right)\frac{1}{1+\alpha\beta}\sum_{ij} q_i q_j \left(\frac{1}{f_{GB}} + \frac{\alpha\beta}{A}\right) \qquad (6)$$

where $\beta = \varepsilon_{in}/\varepsilon_{ex}$ is the ratio of the internal and external dielectrics, α = 0.571412, and A characterizes its overall dimensions and global shape, i.e., the so-called effective electrostatic size of the molecule.

2. MD SIMULATIONS OF Aβ PEPTIDES CONTAINING D-ASP RESIDUES

2.1. Aβ Peptides Containing D-Asp Residues

Computationally constructed $A\beta_{1\text{–}42}$ peptide structures containing 42 residues were used in this study. Two types of initial structures were prepared for the wild-type peptide; one was the experimentally obtained α-helical structure from the Protein Data Bank [32] (PDB ID: 1Z0Q) [11] and the other was the linear structure generated by the tleap module of AmberTools. These initial structures are shown in Figure 2. Based on the 3D structure of the wild-type peptide (named peptide **1**), the mutant $A\beta_{1\text{–}42}$ peptide containing D-Asp23 (peptide **2**), the mutant peptide containing D-Asp1 and D-Asp23 (peptide **3**), and the mutant peptide containing D-Asp7 and D-Asp23 (peptide **4**) were generated. These peptides were experimentally synthesized and the molecular properties were observed. The tendency for β-sheet formation is increased by stereoinversion of Asp23, and the aggregation properties of $A\beta_{1\text{–}42}$ peptides are also increased [7]. On the other hand, the experimental data suggest that stereoinversion of Asp7 plays only an auxiliary role in the aggregation, and stereoinversion of Asp1 seems to prevent the β-sheet formations caused by D-Asp23.

(a) α-helical structure

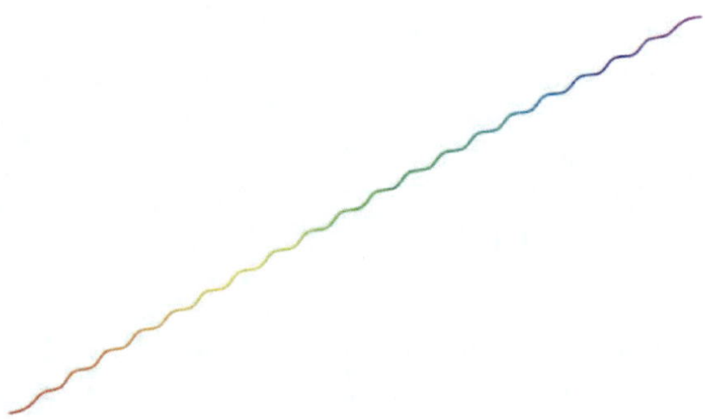

(b) Linear structure

Figure 2. Initial structures. The initial α-helical structure was constructed using the PDB entry 1Z0Q, and the linear initial was constructed using the tleap module of AmberTools. The helix is illustrated by a ribbon diagram.

Table 1. Aβ peptides used in this study

Peptides	D-Amino acid residues	Secondary structure[a]
1	none (wild type)	random
2	D-Asp23	β sheet
3	D-Asp1, D-Asp23	random
4	D-Asp7, D-Asp23	β sheet

[a]The secondary structure content was approximately 60% or greater in each peptide immediately after dissolution [4].

The wild type and mutants used in this study are shown in Table 1. Experimentally observed tendencies to form β-sheet structures, which were measured by CD spectra [7], are also described in the table.

Sakai-Kato et al. reported that 60% of peptide **1** formed a random coil structure, and 65% and 69% of peptides **2** and **4** formed β-sheet structures, respectively. In addition, they reported that the spectrum for peptide **3** was similar to that for peptide **1** [7]. To generate mutants, the side chain atoms, except for Cβ in Asp residues, were deleted and Cβ was interchanged with Hα (Figure 3). After the interchange, complements of the side chains were generated using tleap. In addition, hydrogen atoms not observed by experiments were added. The monomer for each generated peptide was used in our computations.

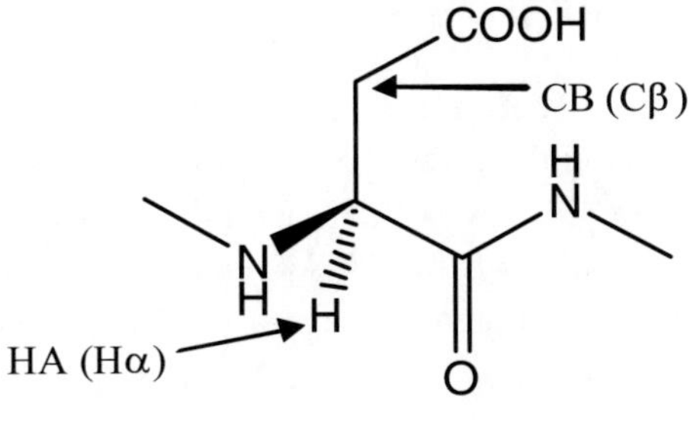

Figure 3. Cβ and Hα of the Asp residue. D-Asp was constructed from L-Asp by interchanging these two atoms.

2.2. Structural Analysis of MD Trajectories

MD results were analyzed using the ptraj module of AmberTools. The structures of $A\beta_{1-42}$ peptides were extracted every 0.5 ps (for normal MD) or 1.0 ps (for REMD), and secondary structures for all residues of the extracted peptide structures were assigned by the DSSP method [33]. The secondary structure contents of peptides were calculated using the DSSP results, and the tendencies of residues to form some particular secondary structures were also investigated.

2.3. MD Simulations Using the GB Method Developed by Hawkins, Cramer, and Truhlar

In the preparatory calculations before MD simulations, structural optimizations of computationally constructed peptides were performed. Optimizations of hydrogen atoms added by tleap were performed first and side-chain atoms were secondarily optimized, followed by whole structure optimizations in the last step of the preparatory calculations. The maximum number of optimization cycles was 500, 10000, and 30000 for hydrogen atoms, side-chain atoms, and whole structures, respectively. Optimizations were performed until the number of cycles reached the maximum or the energy gradients were less than 0.1 kcal/(mol Å). Hydrogen and side-chain optimizations were performed in vacuo, and whole peptide optimizations were performed in an aqueous environment by the continuous solvent model using the GB method developed by Hawkins, Cramer, and Truhlar (GB^{HCT}) [27]. The 3D structures of $A\beta_{1-42}$ peptides obtained by preparatory calculations were used for MD simulations to investigate the influence of D-Asp residues. MD simulations were termed "production MD." For production MD, 20-ns simulations were performed using a time step of 1 fs at a temperature of 300 K. The continuous solvent model using the GB method was conducted with water as the solvent. The concentration of counterions for the ion screening of interactions according to the Debye–Hückel limiting law was set at 0.1 M. Lengths of bonds containing hydrogen atoms were constrained by SHAKE [34]. The AMBER ff99SB force field [35, 36] was used for preparatory optimizations and production MD. The force field parameters of L-Asp were also used for D-Asp. The cutoff for nonbonded interactions was 999 Å, and the maximum distance between atom pairs considered while performing pairwise summation involved in calculating the effective Born radii was also set at 999 Å. Because all distances between the two atoms of $A\beta_{1-42}$ peptides did not exceed 999 Å during optimizations and simulations, the cutoff of 999 Å was equivalent to the infinite distance, i.e., no cutoff. Optimizations and MD simulations were performed using the sander module of AMBER9. The α-helical structure shown in Figure 2(a) was used as the initial structure.

The total number of residues contributing to a β-extended structure identified through MD simulations is shown in Table 2. More residues of mutant peptides **2**, **3**, and **4**, which contained D-Asp residues, contributed to β-extended structures than those of wild-type peptide **1**, which did not contain D-Asp. These results, which indicate that D-Asp residues affect the 3D structures of proteins and peptides, are consistent with the experimental results. In addition, it should be noted that these results were observed for monomer peptides simulations, and not for multimer simulations of $A\beta_{1-42}$ peptides. Thus, in contrast to peptide **1**, mutant peptides **2**, **3**, and **4** might form β-structures by themselves.

Table 2. Total number of residues contributing to the β-extended structure. The total number of residues assigned as "parallel beta-sheet" and "antiparallel beta-sheet" by DSSP are shown in the "parallel" and "antiparallel" columns, respectively

Peptide	Parallel	Antiparallel	Total
1	2148	1604	3752
2	35759	9107	44866
3	17733	13915	31648
4	49710	161	49871

This could be one of the reasons why peptides **2** and **4** experimentally formed β-sheet structures immediately after dissolution [7]. Because peptides **2** and **4** might form β structures by themselves (without aggregation of many peptides), they can form β-sheet structures much faster than peptide **1**, which was assumed to form a β structure after aggregation [37]. For peptide **3,** in which the stereochemistry of Asp1 was inverted, the total number of residues contributing to β structures was smaller than those for peptides **2** and **4**. This result is consistent with the experimental finding that D-Asp1 blocks fibril formation caused by stereoinversion of Asp23.

2.4. Longer MD Simulations Using Several Implicit Solvents

The results of the preceding section using MD simulations with GB^{HCT} reproduced the experimental data in which the tendency of D-Asp-containing Aβ peptides to form β-sheet structures was higher than that of the wild-type peptides. However, for more detailed investigations, longer and/or generalized ensemble simulations are desirable. On the other hand, because GB^{HCT} tends to underestimate the effective radii for buried atoms, it may be unsuitable for MD simulations of large proteins containing many buried atoms [38]. Thus, comparative tests of implicit solvent models other than GB^{HCT} were conducted. In addition, REMD, which is one of the generalized ensemble methods, was examined.

The structural features of the four types of $A\beta_{1-42}$ peptides shown in Table 1 were evaluated by MD simulations. The procedures used in this study are summarized in Table 3.

Table 3. MD methods used in the test calculations

	Implicit solvent[a]	Initial structure	MD type
I	GB^{OBC} (igb = 2)	α helix	normal MD
II	GB^{OBC} (igb = 5)	α helix	normal MD
III	GBn (igb = 7)	α helix	normal MD
IV	ALPB (alpb = 1, igb = 7)	α helix	normal MD
V	GB^{OBC} (igb = 5)	Linear	normal MD
i	GB^{OBC} (igb = 5)	Linear	REMD
ii	GBn (igb = 7)	Linear	REMD

[a]The values of flags for AMBER input files are given in parentheses.

As shown in Table 2, MD results calculated using seven types of procedures, which included four types of implicit solvent models, two types of initial structures, and one normal MD or REMD, were compared. Although MD simulations were performed in the same fashion as in section 2.3, 100 ns simulations were performed and implicit solvent models were used for these calculations. For implicit solvent models, the GB model proposed by Onufriev, Bashford, and Case (GB^{OBC}) [28], GBn [29], and ALPB [30,31] were used. For GB^{OBC}, two types of settings, i.e., $\alpha = 0.8$, $\beta = 0.0$, and $\gamma = 2.909125$ ("igb = 2" setting) and $\alpha = 1.0$, $\beta = 0.8$, and $\gamma = 4.85$ ("igb = 5" setting), were examined. The effective electrostatic size for ALPB was determined to be 20 Å from the result of the utility *elsize* calculation. The effective Born radii for ALPB were obtained from GBn. By comparing the results obtained by several implicit solvent models, influences of the models on MD simulations can be estimated.

REMD calculations were performed in addition to normal MD. Compared to normal MD, powerful structure sampling can be performed using REMD because high-temperature simulations are performed together with low-temperature simulations in REMD. In this study, REMD was used for simulations with a linear initial structure. Because the linear peptide structure appears to be different from the native structure, expanded ensemble methods such as REMD should be used. Multiple MD simulations at different temperatures are performed in parallel for REMD. In this study, the following 16 temperatures were used: 269.7 K, 284.4 K, 300.0 K, 316.4 K, 333.8 K, 352.0 K, 371.3 K, 391.7 K, 413.1 K, 435.7 K, 459.6 K, 484.8 K, 511.3 K, 539.3 K, 568.8 K, and 600.0 K. In preparation for REMD, 200-ps normal MD simulations with a 2-fs time step were performed to achieve equilibria at the given temperatures. Then, 50-ns REMD calculations with a 2-fs time step were performed. The number of MD steps between each exchange attempt was set at 500.

Because high-temperature MD calculations were included in REMD, artificial (undesirable) chiral inversions frequently occurred. Thus, structural restrictions using initial structures were adopted to avoid chiral inversions. The atomic radii of mbondi2 were used for GB^{OBC} and the bondi radii were used for GBn and ALPB.

To compare the results of MD simulations with the experimental data, the secondary structure contents were computed from the number of residues contributing to certain secondary structures and the number of all residues as follows:

$$P_{\text{beta}} = N_{\text{beta}}/N_{\text{all}}$$
$$P_{\text{random}} = N_{\text{random}}/N_{\text{all}}$$

where P_{beta} and P_{random} were the β-structure contents and random coil contents, respectively, and N_{beta}, N_{random}, and N_{all} were the number of β-structure residues, random coil residues, and all residues, respectively. For calculations of the secondary structure contents, the residues classified as "b" (parallel β sheet), "B" (antiparallel β sheet), and "T" (hydrogen bonded turn) were defined as β-structure residues. The residues classified as "0" (random coil) were defined as random coil residues. The definitions of secondary structures are same as those used in ref. [15] and reflect the fact that the experimental β-structure contents include β-turn contents [39].

In Figure 4, the secondary structure contents obtained by seven MD simulations are shown. The values illustrated in the figure are the averages throughout the MD trajectories. The experimentally observed helical, β, and random coil contents for wild-type peptide **1** were approximately 5%, 25%, and 70%, respectively [7,39]. Furthermore, the experimentally observed β content for peptide **2**, in which Asp23 was sterically inverted, was approximately 60% [7]. In the comparison between the calculated results of peptides **1** and **2**, the β contents obtained by methods **II**, **III**, **IV**, **i**, and **ii** for peptide **2** were larger than those for peptide **1** (Figure 4). On the other hand, for the random coil contents, methods **III**, **IV**, **i**, and **ii** provided qualitatively reasonable results in which the contents for peptide **1** were larger than those for peptide **2**. Because methods **III**, **IV**, **i**, and **ii** could qualitatively reproduce the β and random coil contents, these methods appear capable of reasonably calculating the effect of stereoinversion of Asp residues on 3D structures.

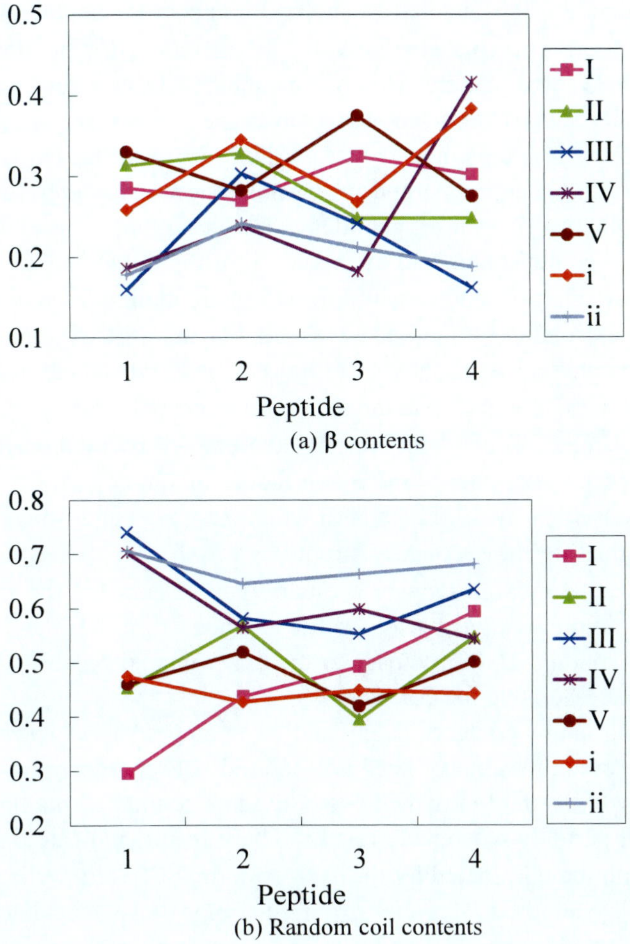

Figure 4. Comparisons of the secondary structure contents. Secondary structures were defined by the DSSP program included in the ptraj module of AmberTools. The vertical axes of the figure are the secondary structure contents. (a) β contents, (b) Random coil contents.

When considering peptides **3** and **4** in addition to peptides **1** and **2**, only two methods, **IV** and **i**, could qualitatively reproduce the experimental results in which the β contents of peptides **2** and **4** were greater than those of peptides **1** and **3,** and the random coil contents of peptides **1** and **3** were greater than those of peptides **2** and **4**. Although the computational results of method **ii** were consistent with the results of experiments for peptides **1**, **2**, and **3**, the results for the β content of peptide **4** were inconsistent.

Method **IV** could reproduce the experimental results; however, the random coil contents were overestimated by this method. For example, the calculated random coil content of peptide **2** by this method was 56.5%, although the experimental random coil content was less than approximately 40%. In this study, the setting "igb = 7" was used for method **IV**. The random coil contents obtained by other methods using the "igb = 7" setting, i.e., methods **III** and **ii**, were also greater than those obtained by method **i**; the random coil content of peptide **2** was 58.2% by method **III** and 64.6% by method **ii**, although that by method **i** was 42.8%. In addition to the random coil contents, the β contents calculated for the peptides by method **IV** were much smaller than those calculated by method **i,** except for peptide **4**, although method **IV** could reproduce the β contents qualitatively. In particular, for peptide **2**, the β content obtained by method **IV** was 23.9%, which was the smallest value obtained among the methods. These results suggest that secondary structure formation was restrained and that random coil structure was overestimated in the "igb = 7" setting. On the other hand, the random coil contents calculated by method **i** were not unacceptably high for peptides **2** and **4**; the experimental contents for these peptides were less than approximately 40%. The calculated random coil contents obtained by method **i** for these two peptides were the smallest among the methods used in this study. In addition to the random coil contents, although the β content of peptide **2** calculated by method **i** was smaller than the experimental value, it was closest to the experimental value, as shown in Figure 4. These results indicate that method **i** is the most appropriate method to calculate the structural features of $A\beta_{1-42}$ peptides.

For the β and random coil contents, the results obtained by method **I**, in which more β structures were observed in peptides **1** and **3** and more random coils were present in peptides **2** and **4**, were inconsistent with the experimental results, suggesting that method **I** is inappropriate for evaluating the secondary structures of $A\beta_{1-42}$ peptides. On the other hand, the β contents obtained by method **II** were comparatively reasonable for peptides **1**, **2,** and **3**. Although the GB^{OBC} implicit solvent was used in methods **I** and **II**, the secondary structure contents obtained by method **II** were closer to the experimental results. Thus, the "igb = 5" setting appeared to be reasonable for GB^{OBC}.

The secondary structure contents calculated by method **V,** in which the linear initial structure was used, were unreasonable because they did not reproduce the experimental data. As discussed previously, method **II,** which used the same settings as method **V** except for the initial structure, gave partially reasonable results. These results indicate that the effect of the initial structures cannot be eliminated by the 100-ns normal MD. In particular, because the β contents calculated by method **V** were different from those calculated by method **II**, hydrogen-bonding networks appeared to be affected by the initial structures. In contrast to normal MD, REMD with linear initial structures (method **i**) could reproduce the experimental results qualitatively, although the settings were the same as those for method **V**. These results suggest that REMD is useful even if highly artificial initial structures are used, and powerful sampling methods such as REMD should be applied for artificially generated initial structures. In addition, the results indicate that 50-ns REMD simulations can provide reasonable results, although the 100-ns normal MD is insufficient.

The occurrence ratios of helical and β structures for each residue of peptides **1** and **2** calculated from MD trajectories by method **i**, which could qualitatively reproduce experimental results, are shown in Figure 5.

The helical structures include α, 3_{10}, and π helices, while β structures include parallel and antiparallel sheets as well as hydrogen-bonded turns. As shown in the figure, the occurrence ratios of β structures in the C- and N-terminal regions of peptide **2** were larger than those in the terminal regions of peptide **1**.

Thus, large portions of the $A\beta_{1-42}$ monomer appeared to form β structures by stereoinversion of Asp23. The peak of the helical structures located in the C terminus of peptide **1** shifted from residue 34 to 31. This result indicates that the tendency of the C-terminal region of the $A\beta_{1-42}$ peptide to form helical structures is decreased because of the presence of D-Asp23. Because the C-terminal helix is known to play an important role in aggregation of the wild-type Aβ peptide [11,40], the difference in C-terminal secondary structures between peptides **1** and **2** may suggest that the mechanism of aggregation for D-Asp-containing Aβ peptides is different from that for wild-type peptides.

In Figure 6, the representative 3D structures of $A\beta_{1-42}$ peptides containing D-Asp23 (peptide **2**) obtained by the MD simulation using method **i** are shown. In the figure, snapshot structures at 22 and 38 ns, in which β structures were observed, are illustrated. The helical and arrowed ribbons represent the α helix and β sheet, respectively.

As shown in the figure, β structures were clearly observed in the C-terminal region for snapshot (a) and in the N-terminal region for snapshot (b).

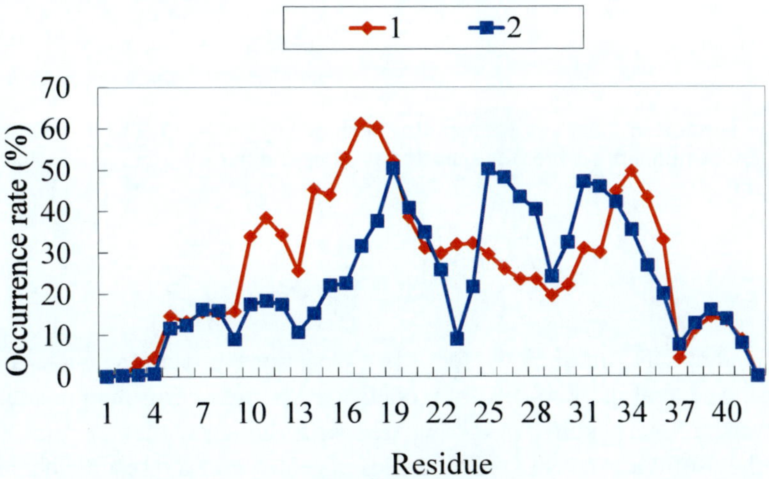

(a) Helical structure

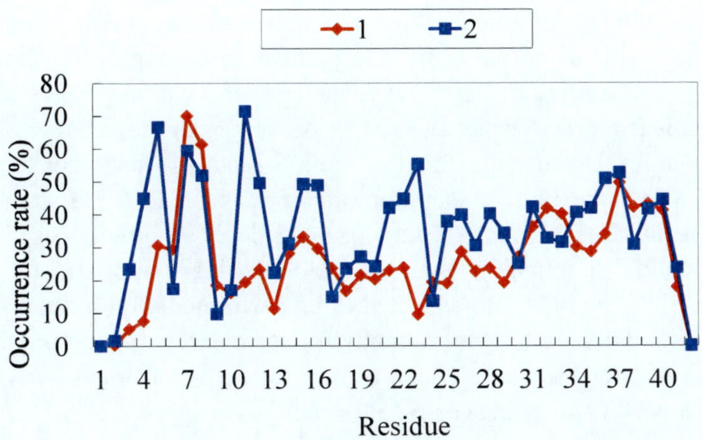

(b) β structure

Figure 5. Occurrence ratios of secondary structures for residues obtained by REMD. The ratios were calculated as N_{sec}/N_{all}, where N_{sec} is the number of snapshots in which the secondary structure appeared in the residue and n_{all} is the number of all snapshots used for analyses (in case of REMD, N_{all} = 30000). (a) Helical structure, (b) β structure.

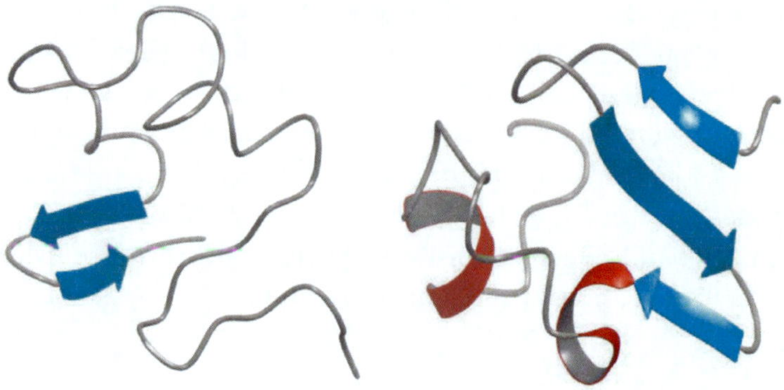

(a) Snapshot at 22 ns; (b) Snapshot at 38 ns

Figure 6. Two snapshots in trajectory for peptide **2** obtained by method **i**. The β structures were observed in the C-terminal region for the snapshot at 22 ns and in the N-terminal region for the snapshot at 38 ns.

CONCLUSION

In this study, MD simulations for Aβ_{1-42} peptides containing D-Asp residues were performed using several implicit solvent models. REMD was performed in addition to normal MD. The computational results were compared with the experimental data obtained by CD spectra, and the usefulness of MD procedures was evaluated. The results indicate that REMD with GB^{OBC} is useful for simulations of D-Asp-containing peptides. In addition, because force field parameters for L-Asp residues were used for D-Asp residues without modification in this study, the qualitatively reasonable results obtained by REMD suggest that the existing force fields can also be applied for calculations of D-amino acid-containing peptides. For normal MD calculations, ALPB using the GBn radii appears to be appropriate. On the other hand, results of the "igb = 2" setting of GB^{OBC} were inconsistent with the experimental data, and the "igb = 5" setting was more suitable for GB^{OBC}. As shown in the results of simulations using linear initial structures, because the effect of initial structures was larger for normal MD than for REMD, REMD appears to be appropriate if reasonable initial structures cannot be prepared. The results of this study could be considered as guidelines for simulations of proteins and peptides containing D-amino acid residues, and structural bioinformatics investigations of these peptides are expected to be performed with reference to the results of this study. MD simulations are useful not only for secondary structure predictions mentioned in this study but also for the investigation of other structural features such as evaluations of hydrophobicities, which are influenced by 3D structures.

The results of this study suggest that the monomer peptide **2** forms β-sheet structures by itself, and the results were consistent with the experimental data in which β structures were observed immediately after dissolution. Because peptide **1** is known to form β-sheet structures after aggregation [37], the computational results in which peptide **2** forms β-sheet structures without aggregation indicate that the mechanisms of β-sheet formation were different between peptides **1** and **2**.

The finding of this study is consistent with the experimental observations in which the β content of peptide **2** was higher than that of peptide **1** immediately after dissolution. On the other hand, the absolute value of the β content for peptide **2** was smaller than the experimental value (approximately 60%) even when method **i**, which can reproduce the experimental results qualitatively, was used. This result may have been obtained because of the limitations of the implicit solvent and/or the ff99SB force field. We intend to further investigate these aspects in a future study via simulations using the explicit solvent and/or other force fields.

REFERENCES

[1] Kang, J., Lemaire, H.G., Unterbeck, A., Salbaum, J.M., Masters, C.L., Grzeschik, K.-H., Multhaup, G., Beyreuther, K. and Mullaer-Hill, B. (1987). The precursor of Alzheimer's disease amyloid A4 protein resembles a cell-surface receptor. *Nature*, *325*, 733–736.

[2] Corder, E.H., Saunders, A.M., Strittmatter, W.J., Schmechel, D.E., Gaskell, P.C., Small, G.W., Roses, A.D., Haines, J.L. and Pericak-Vance, M.A. (1993). Gene dose of apolipoprotein E type 4 allele and the risk of Alzheimer's disease in late onset families. *Science*, *261*, 921–923.

[3] Wisniewski, T., Ghiso, J. and Frangione, B. (1991). Peptides homologous to the amyloid protein of Alzheimer's disease containing a glutamine for glutamic acid substitution have accelerated amyloid fibril formation. *Biochem. Biophys. Res. Commun.* *179*, 1247–1254.

[4] Saido, T.C., Iwatsubo, T., Mann, D.M., Shimada, H., Ihara, Y. and Kawashima, S. (1995). Dominant and differential deposition of distinct β-amyloid peptide species. $A\beta_{N3(pE)}$, in senile plaques. *Neuron*, *14*, 457–466.

[5] Roher, A.E., Lowenson, J.D., Clarke, S., Wolkow, C., Wang, R., Cotter, R.J., Reardon, I.M., Zürcher-Neely, H.A., Heinrikson, R.L., Ball, M.J. and Greenberg, B.D. (1993). Structural alterations in the peptide backbone of β-amyloid core protein may account for its deposition and stability in Alzheimer's disease. *J. Biol. Chem.*, *268*, 3072–3083.

[6] Tomiyama, T., Asano, S., Furiya, Y., Shirasawa, T., Endo, N. and Mori, H. (1994). Racemization of Asp23 residue affects the aggregation properties of Alzheimer amyloid β protein analogues. *J. Biol. Chem.*, *269*, 10205–10208.

[7] Sakai-Kato, K., Naito, M. and Utsunomiya-Tate, N. (2007). Racemization of the amyloidal β Asp^1 residue blocks the acceleration of fibril formation caused by racemization of the Asp^{23} residue. *Biochem. Biophys. Res. Commun.*, *364*, 464–469.

[8] Fujii, N. and Saito, T. (2004). Homochirality and life. *Chem. Rec.*, *4*, 267–278.

[9] Fujii, N., Satoh, K., Harada, K. and Ishibashi, Y. (1994). Simultaneous stereoinversion and isomerization at specific aspartic acid residues in αA-crystallin from human lens. *J. Biochem.*, *116*, 663–669.

[10] Fujii, N., Tajima, S., Tanaka, N., Fujimoto, N., Takata, T. and Shimo-Oka, T. (2002). The presence of D-β-aspartic acid-containing peptides in elastic fibers of sun-damaged skin: a potent marker for ultraviolet-induced skin aging. *Biochem. Biophys. Res. Commun.*, *294*, 1047–1051.

[11] Tomaselli, S., Esposito, V., Vangone, P., van Nuland, N.A.J., Bonvin, A.M.J.J., Guerrini, R., Tancredi, T., Temussi, P.A. and Picone, D. (2006). The α-to-β conformational transition of Alzheimer's Aβ-(1-42) peptide in aqueous media is reversible: a step by step conformational analysis suggests the location of β conformation seeding. *ChemBioChem*, *7*, 257–267.

[12] Lührs, T., Ritter, C., Adrian, M., Riek-Loher, D., Bohrmann, B., Döbeli, H., Schubert, D. and Riek, R. (2005). 3D structure of Alzheimer's amyloid-β(1-42) fibrils. *Proc. Natl. Acad. Sci. USA*, *102*, 17342–17347.

[13] Baumketner, A., Bernstein, S.L., Wyttenbach, T., Bitan, G., Teplow, D.B., Bowers, M.T. and Shea, J.-E. (2006). Amyloid β-protein monomer structure: a computational and experimental study. *Protein Sci.*, *15*, 420–428.

[14] Kent, A., Jha, A.K., Fitzgerald, J.E. and Freed, K.F. (2008). Benchmarking implicit solvent folding simulations of the amyloid β(10-35) fragment. *J. Phys. Chem.*, *112*, 6175–6186.

[15] Ikebe, J., Kamiya, N., Ito, J., Shindo, H. and Higo, J. (2007). Simulation study on the disordered state of an Alzheimer's β amyloid peptide Aβ(12-36) in water consisting of random-structural, β-structural, and helical clusters. *Protein Sci.*, *16*, 1596–1608.

[16] Simmerling, C., Strockbine, B. and Roitberg, A.E. (2002). All-atom structure prediction and folding simulations of a stable protein. *J. Am. Chem. Soc.*, *124*, 11258–11259.

[17] Oda, A., Kobayashi, K. and Takahashi, O. (2010). Computational study of the three-dimensional structure of, *N*-acetyltransferase 2–acetyl coenzyme A complex. *Biol. Pharm. Bull.*, *33*, 1639–1643.

[18] Senapati, S., Bui, J.M. and McCammon, J.A. (2005). Induced fit in mouse acetylcholinesterase upon binding a femtomolar inhibitor: A molecular dynamics study. *J. Med. Chem.*, *48*, 8155–8162.

[19] Oda, A., Yamaotsu, N. and Hirono, S. (2004). Studies of binding modes of (S)-mephenytoin to wild types and mutants of cytochrome P450 2C19 and 2C9 using homology modeling and computational docking. *Pharm. Res.*, *21*, 2270–2278.

[20] Mitsutake, A., Sugita, Y. and Okamoto, Y. (2001). Generalized-ensemble algorithms for molecular simulations of biopolymers. *Biopolymers*, *60*, 96–123.

[21] Jorgensen, W.L., Chandrasekhar, J., Madura, J. and Klein, M.L. (1983). Comparison of simple potential functions for simulating liquid water. *J. Chem. Phys.*, *79*, 926–935.

[22] Jorgensen, W.L. and Madura, J.D. (1985). Temperature and size dependence for Monte Carlo simulations of TIP4P water. *Mol. Phys.*, *56*, 1381–1392.

[23] Sharp, K.A. and Honig, B. (1990). Electrostatic interactions in macromolecules: Theory and experiment. *Annu. Rev. Biophys. Biophys. Chem.*, *19*, 301–332.

[24] Davis, M.E. and McCammon, J.A. (1990). Electrostatics in biomolecular structure and dynamics. *Chem. Rev.*, *90*, 509–521.

[25] Still, W.C., Tempczyk, A., Hawley, R.C. and Hendrickson, T. (1990). Semianalytical treatment of solvation for molecular mechanics and dynamics. *J. Am. Chem. Soc.*, *112*, 6127–6129.

[26] Case, D.A., Darden, T.A., Cheatham, T.E., Simmerling, C.L., Wang, J., Duke, R.E., Luo, R., Crowley, M., Walker, R.C., Zhang, W., Merz, K.M., Wang, B., Hayik, S., Roitberg, A., Seabra, G., Kolossváry, I., Wong, K.F., Paesani, F., Vanicek, J., Wu, X., Brozell, S.R., Steinbrecher, T., Gohlke, H., Yang, L., Tan, C., Mongan, J., Hornak, V.,

Cui, G., Mathews, D.H., Seetin, M.G., Sagui, C., Babin, V. and Kollman, P.A. (2008) *AMBER 10*, San Francisco, CA: University of California.

[27] Hawkins, G.D., Cramer, C.J. and Truhlar, D.G. (1996). Parametrized models of aqueous free energies of solvation based on pairwise descreening of solute atomic charges from a dielectric medium. *J. Phys. Chem.*, *100*, 19824–19839.

[28] Onufriev, A., Bashford, D. and Case, D.A. (2004). Exploring protein native states and large-scale conformational changes with a modified generalized Born model. *Proteins, struct. funct. genet.*, *55*, 383–394.

[29] Mongan, J., Simmerling, C., McCammon, J.A., Case, D.A. and Onufriev, A. (2006). Generalized Born with a simple, robust molecular volume correction. *J. Chem. Theory Comput.*, *3*, 156–169.

[30] Sigalov, G., Scheffel, P. and Onufriev, A. (2005). Incorporating variable environments into the generalized Born model. *J. Chem. Phys.*, *122*, 094511.

[31] Sigalov, G., Fenley, A. and Onufriev, A. (2006). Analytical electrostatics for biomolecules: Beyond the generalized Born approximation. *J. Chem. Phys.*, *124*, 124902.

[32] Berman, H.M., Westbrook, J., Feng, Z., Gilliland, G., Bhat, T.N., Weissig, H., Shindyalov, I.N. and Bourne, P.E. (2000). The protein data bank. *Nucleic Acids Research*, *28*, 235–242.

[33] Kabsch, W. and Sander, C. (1983). Dictionary of protein secondary structure: pattern recognition of hydrogen-bonded and geometrical features. *Biopolymers*, *22*, 2577–2637.

[34] Ryckaert, J.-P., Ciccotti, G. and Berendsen, H.J.C. (1977). Numerical integration of the cartesian equations of motion of a system with constraints: Molecular dynamics of *n*-alkanes. *J. Comput. Phys.*, *23*, 327–341.

[35] Wang, J., Cieplak, P. and Kollman, P.A. (2000). How well does a restrained electrostatic potential (RESP) model perform in calculating conformational energies of organic and biological molecules? *J. Comput. Chem.*, *21*, 1049–1074.

[36] Hornak, V., Abel, R., Okur, A., Strockbine, B., Roitberg, A. and Simmerling, C. (2006). Comparison of multiple Amber force fields and development of improved. *Proteins, struct. funct. bioinf.*, *65*, 712–725.

[37] Kirkitadze, M., Condron, M.M. and Teplow, D.B. (2001). Identification and characterization of key kinetic intermediates in amyloid β-protein fibrillogenesis. *J. Mol. Biol.*, *312*, 1103–1119.

[38] Onufriev, A., Bashford, D. and Case, D.A. (2000). Modification of the generalized Born model suitable for macromolecules. *J. Phys. Chem. B*, *104*, 3712–3720.

[39] Fezoui, Y. and Teplow, D.B. (2002). Kinetic studies of amyloid β-protein fibril assembly: differential effects of α-helix stabilization. *J. Biol. Chem.*, *277*, 36948–36954.

[40] Pike, C.J., Burdick, D., Walencewicz, A.J., Glabe, C.G. and Cotman, C.W. (1993). Neurodegeneration induced by β-amyloid peptides *in vitro*: the role of peptide assembly state. *J. Neurosci.*, *13*, 1676–1687.

In: Amyloids
Editors: I. P. Halcheck et al.

ISBN: 978-1-62100-538-4

Chapter 7

AMYLOID, ARISES ON THE MEMBRANES AND ACTS ON THE MEMBRANES

Katsuhiko Yanagisawa*
National Center for Geriatrics and Gerontology
35 Gengo, Morioka, Obu 474-8522, Japan

ABSTRACT

Assembly and deposition of physiologically generated, amyloidogenic proteins occur in various human diseases, referred to as amyloidoses, through misfolding that is either partial unfolding or partial folding of the proteins. A growing body of evidence suggests that misfolding of a given protein is driven by point mutation(s) or by the presence of environmental factors favorable for amyloidogenesis. One of the most extensively studied partial-folding-type amyloidogenic proteins is amyloid ß-protein (Aß), a proteinaceous component of senile plaques in Alzheimer disease. Although the soluble Aß monomer does not show any fixed secondary structure under physiological conditions, Aß adopts an α-helix-rich structure in the presence of gangliosides in particular membrane microdomains such as lipid rafts. Previous studies suggest that the ganglioside-induced α-helix-rich conformer acts as an endogenous seed for the consecutive assembly of Aß, resulting in the formation of ß-sheet-rich amyloid fibrils in the brain. Notably, it has also been suggested that the alterations of the lipid composition of membrane microdomains likely occur prior to the induction of the partially folded endogenous seeds and the formation of amyloid fibrils. Regarding the emergence of toxicities of protein aggregates, it is widely accepted that amyloid fibrils as well as smaller oligomers, which may be intermediates generated on the pathway to amyloid fibril formation or derivative fragments of amyloid fibrils, attack membranes targeting or not targeting specific proteins expressed on the surface of the membranes. In this review, the impact of the presence of biological membranes in terms of the initiation of amyloid fibril formation and causing amyloid toxicities will be discussed.

* TEL: +81-562-44-6594; FAX: +81-562-44-6594; E-mail: katuhiko@ncgg.go.jp

Introduction

Amyloidoses include various human neurodegenerative diseases such as Alzheimer disease (AD), Parkinson disease, prion diseases and peripheral amyloidoses such as type II diabetes mellitus and immunoglobulin light-chain amyloidosis. One of the fundamental questions regarding the pathogenesis of amyloidoses is how physiologically generated proteins transform into pathologically aggregated amyloid fibrils. In general, amyloidoses can emerge by either the partial unfolding of proteins with ordered structures or the partial folding of natively unfolded, unordered proteins. Lysozyme is an example of unfolding-type amyloidogenic proteins [1, 2] whereas partial-folding-type proteins include amyloid ß-protein (Aß), α-, ß- and γ-synucleins, islet amyloid polypeptide (IAPP), tau, ABri and ADan which are causative proteins for familial British and Danish dementia, respectively (see ref. [3] for review) The molecular mechanisms underlying amyloid fibril formation from two types of amyloidogenic protein, namely, ordered proteins and natively unordered proteins, may be mutually different; however, it is generally accepted that abnormal misfolding of proteins is requisite for the subsequent intermolecular binding through electrostatic, hydrogen bonding and hydrophobic interactions, leading to the formation of large amyloid fibrils [4]. Additionally, it has been generally accepted that misfolded intermediates of a given amyloidogenic protein invariably possess a certain α-helix structure that favorably acts as a seed for the subsequent amyloidogenetic process [5-7]. Mechanisms underlying the initial conformational transition of amyloidogenic proteins to an assembly-prone, α-helix-containing structure are largely unclarified; however, it was previously suggested that conformational changes may be caused by point mutation(s), which lead to the alteration of their amino-acid sequences or, as in most of the cases, by the presence of environmental factors favorable for amyloidogenesis. In this context, it should be stressed that the environment can affect the secondary structures of proteins, which are primarily determined by the amino-acid sequence [8]. With this line of evidence, it is possible to explain why a single organ and even specific regions of an organ are selectively affected in certain amyloidoses. Among the environmental factors, membrane lipids should be one of the most potent factors that affect the structures of proteins.

1. Membrane-Guided Amyloid Fibril Formation: Particular Commitment of Gangliosides

It has been widely accepted that amyloidogenic proteins, particularly natively unfolded proteins, start to assemble into amyloid fibrils upon interaction with cellular membranes. Mounting evidence suggests that the interaction of proteins with gangliosides on the cellular surface appears to be one of the driving forces that convert their native forms to assembly-prone forms in various cases of amyloidogenesis as indicated below. Aß amyloidogenesis initiated by a ganglioside-induced conformational change of Aß has been extensively studied (see refs. [9-11] for review).

1) Amyloid ß-Protein

a) Discovery of Ganglioside-Bound Aß (GAß) in Human Brains

Aß assembly into fibrils and Aß deposition in the brain are critical steps in the pathogenesis of AD; however, it remains to be clarified how monomeric, nontoxic Aß is converted to assembled, toxic fibrils in the brain. Previously, human brains were carefully examined to explore the mechanism underlying Aß assembly and deposition in the AD brain, resulting in a final discovery of a novel Aß species, which was characterized by its tight binding to GM1 ganglioside (GM1). On the basis of the molecular characteristics of the ganglioside-bound Aß (GAß), including its high potential to induce the formation of large Aß assemblies and its altered immunoreactivity, it was hypothesized that Aß adopts an altered conformation through its binding to GM1, and then, starts to assemble into amyloid fibrils by acting as an endogenous seed in the brain [12, 13]. To date, various *in vitro* and *in vivo* studies on GAß have been performed, providing evidence that supports the GAß hypothesis and information on GAß as follows.

b) Binding Specificity of Aß to Gangliosides

The specific binding of Aß to gangliosides was validated by previous careful examinations. For example, Choo-Smith et al. investigated the binding behavior of Aß on artificial membranes composed of various lipids, including gangliosides [14]. In their experiment, they found that Aß selectively binds to gangliosides at physiological ionic strength. Nonspecific Aß binding to acidic phospholipids can be detected only at a very low, nonphysiological ionic strength. In addition, Kakio et al. investigated the ganglioside species specificity in the potency to induce conformational change of Aß using GM1, GD1a, GD1b and GT1b, and concluded that GM1-Aß binding is the strongest [15].

c) Favorable Membrane Microenvironment for GAß Generation

GM1 is a major lipid constituent of membrane microdomains such as lipid rafts, which are also rich in cholesterol and sphingomyelin, in plasma membranes. Thus, it was reasonably assumed that GAß generation favorably occurs in these membrane microdomains. Matsuzaki's group previously reported that GAß generation *in vitro* is highly sensitive to the lipid composition of membranes [15, 16]. They found that GAß is readily generated in the sphingomyelin/cholesterol/GM1 matrix mimicking lipid rafts in contrast to a phosphatidylcholine environment. In terms of the requirement of cholesterol for inducing a favorable microenvironment for GAß generation, it was previously reported that cholesterol enhances ganglioside clustering, in which GAß formation is facilitated [17].

Interestingly, it has recently been shown that cholesterol has a dual function, either facilitating or inhibiting Aß-ganglioside interactions through the fine tuning of ganglioside conformation [18]. Significance of lipid rafts in the induction of Aß assembly has recently been reviewed [19].

d) Structural Analyses of Aß Bound to Gangliosides

Conformational rearrangements of Aß upon binding to gangliosides have been indicated by several previous studies by circular dichroism (CD) and nuclear magnetic resonance (NMR) spectroscopies. McLaurin and Chakrabartty first reported that the Aß mixtures incubated in the presence of gangliosides showed significant changes in CD spectra [20]. Choo-Smith and Surewicz also reported the conformational transition of Aß upon binding to gangliosides [21]. They stressed that this interaction appears to be ganglioside-specific as no change in Aß conformation was found in the presence of other lipids, including phospholipids and sphingomyelin. Furthermore, it was of note in the study of Choo-Smith and Surewicz that the presence of only isolated oligosaccharides of gangliosides is insufficient to induce a conformational change of Aß. Matsuzaki and Horikiri investigated the molecular interaction between Aß and gangliosides in more detail by CD spectroscopy and fourier transform infrared-polarized attenuated total reflection spectroscopy and found that Aß specifically binds to ganglioside-containing membranes and adopts an altered structure distinct from that of solubilized Aß [15]. Conformational transition upon Aß binding to gangliosides has also been studied in etail by NMR spectroscopy. Williamson et al. have attempted to characterize the interaction of Aß with GM1 and micelles and found that chemical shift changes, which imply the conformational transition from a random coil to an α-helical structure, occur primarily in the N-terminal region of Aß [22]. Utsumi et al. have employed ultrahigh-field NMR techniques to analyze the Aß structure associated with ganglioside micelles, and found that Aß forms discontinuous α-helices at particular segments of Aß upon binding to the ganglioside micelles, leaving the remaining regions unstructured [23]. Furthermore, they also found that the sugar-lipid interface is required for the Aß binding, demonstrating that Aß lies on the hydrophobic/hydrophilic interface of a ganglioside cluster exhibiting an up-and-down topological mode [23]. Putative mechanisms underlying GAß generation and subsequent formation of amyloid fibrils and oligomers are shown in Figure 1.

e) Evidence for GAß Generation and GAß Involvement in Aß Assembly in Brain

To validate whether GAß is indeed generated and acts as a seed for amyloid formation in the brain, careful immunohistochemical examination of human brains, including AD brains, was previously performed using a monoclonal antibody specific to GAß. The immunohistochemistry revealed GAß being stainied selectively in intraneuronal organelles but not senile plaques [13]. Results of a subsequent study of aged monkey brains suggest that neuronal GAß-containing organelles are likely endosomes [13]. To obtain direct evidence that GAß acts as an endogenous seed for Aß assembly and deposition in the brain, Fab fragments of an antibody specific to GAß were administered into transgenic mice expressing a mutant *amyloid precursor protein (APP)*, following the conjugation of the Fab fragments with a protein transduction domain (tat) to assist with their delivery into the brain through the blood-brain barrier [24].

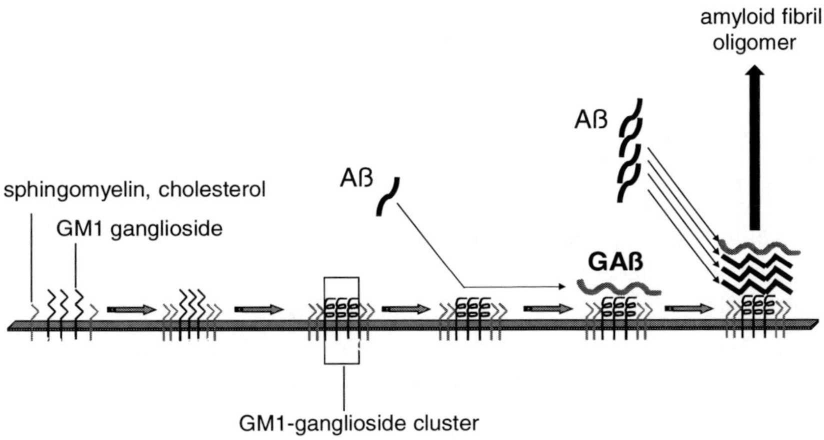

Figure 1. Proposed scheme for the formation of amyloid fibrils and oligomers from soluble monomeric Aß through GAß generation. Note that GAß is favorably generated on the surfaces of clustered GM1, that likely provide unique microenvironment induced by the alteration in the lateral arrangement of the oligosaccharide chain of the GM1 molecule. It should also be noted that GAß and the Aß at the end of the growing fibrils share the same α-helix-rich structure, that is required for acting as a seed for subsequent amyloidogenesis process.

In this experiment, the levels of deposited Aß in the Fab-administered mouse brains were significantly lower than those in control mouse brains. Furthermore, a recent study using ganglioside-synthase deficient mice indicates that gangliosides determine the Aß assembly and deposition in brain [25]. This finding strongly suggests that GAß is a *bona fide* product and facilitates Aß assembly in the brain by acting as an endogenous seed.

f) Endocytic Pathway Abnormality as Putative Background for Gaß Generation

Although lipid-raft-like membrane microdomains are a favorable platform for GAß generation, it remains unclarified how the formation of such membrane microdomains can be enhanced in the neurons of brains that are prone to develop AD. The previous findings of Cataldo and Nixon may provide a key to the clarification of this issue [26]. They showed that early endosomes are abnormally enlarged in the neurons of the brains obtained from sporadic AD patients and individuals with Down syndrome [26]. Importantly, abnormalities of the endocytic pathway seemed to precede extraneuronal Aß deposition [26]. On the basis of these findings, it was speculated that the perturbation of the endocytic pathway in neurons may be responsible for the pathological GM1 accumulation inside or on the surface of neurons, prior to extraneuronal senile plaque formation. This possibility has been examined in a series of cell biological experiments and it has been shown that blocking the endocytic pathway by knocking down a small GTPase protein, Rab7, induces GM1 accumulation on the cell surface and the appearance of enlarged early endosomes, and, furthermore, facilitated amyloid fibril formation from exogenously added Aß on the cell surface in a GAß-dependent manner [27-29]. These results suggest that impairment of the late but not early endocytic pathway contributes to amyloid fibril formation on the surface of neurons through the facilitation of GM1 accumulation, leading to GAß generation. Supporting this possibility, GAß has been immunohistochemically detected in the endosomal systems in the neurons of aged nonhuman primate brains [30].

2) Amyloidogenic Proteins Other Than Aß

a) α-Synuclein

α-Synuclein is a proteinaceous component of Lewy bodies and Lewy neurites, which are pathological structures in brains of patients with Parkinson disease and Lewy body disease.

Similar to Aß, α-synuclein is a natively unfolded protein and possesses no ordered structure as a soluble monomer under physiological conditions [31]. α-Synuclein also has a lipid-binding domain similarly to Aß and recent studies indicate that its binding to charged lipids such as acidic phospholipids and gangliosides leads to a significant conformational change [31-33]. Again, similarly to the case of Aß, the interaction of monomeric α-synuclein with negatively charged lipid membranes causes a significant conformational change from a random coil to an α-structure [31, 33-35].

Fantini and Yahi have recently investigated the interaction between α-synuclein and glycosphingolipids in detail and found that the binding selectively occurrs with the binding to monosilaoganglioside GM3 being the most active and that to disialoganglioside GD3 being less active [36]. Furthermore, they suggest that subtle amino acid variations in the consensus ganglioside-binding motif of both α-synuclein and Aß confer distinct ganglioside-binding properties [36].

b) Human Islet Amyloid Polypeptide (hIAPP)

The human islet amyloid polypeptide (hIAPP), also referred to as amylin, is a proteinaceous component of amyloid fibrils deposited in the pancreas of patients with type 2 diabetes mellitus. It has been shown that, similarly to Aß and α-synuclein, hIAPP favorably interacts with negatively charged membranes, and then, starts to assemble into amyloid fibrils [37, 38]. This finding suggests that weakly, not strongly, stable α-helical conformation of hIAPP is an intermediate, which is required for the initiation of consecutive assembly of the protein, leading to the formation of amyloid fibrils with ß-sheet conformation. Furthermore, it was also shown that the membrane-bound α-helical form of hIAPP retains a unique region, which is significantly important for the consequent amyloidogenic process of hIAPP [39]. The transient adoption of α-helical conformation by hIAPP on the negatively charged lipid surface has been validated as a crucial step to the subsequent hIAPP assembly process by infrared reflection absorption spectroscopy by Lopes et al. [40]. Interestingly, Lopes et al. stressed that hIAPP readily adsorbs at hydrophobic as well as anioinic hydrophilic surfaces, leading to the formation of amyloid fibrils, whereas at the purely hydrophobic interface, only small amorphous aggregates are formed. In this context, it may be noteworthy that gangliosides in membranes provide a unique surface microenvioronment, containing both hydrophobic forces and negative electrostatic potentials produced by the oligosaccharaide chains of gangliosides [41]. Supporting these possibilities, it has been shown that, first, the amyloid-forming fragment of hIAPP works as glycolipid-binding domain with aromatic residues in the fragment that is particularly crucial for the binding process [42], and second, amyloid fibril formation from hIAPP is accelerated in the presence of gangliosides, furthermore, this event specifically occurs in membrane microdomains such as lipid rafts [43].

c) Serum Amyloid A

Serum amyloid A (SAA) is an acute-phase reactant protein, which is mainly generated by hepatocytes under regulation by various cytokines. The amino terminus of SAA (AA protein) forms into amyloid fibrils (AA amyloids) *in vivo* under various pathological conditions such as rheumatoid arthritis. Gellermann et al. previously reported that the formation of AA amyloid fibrils readily occurs in GM1-rich lipid rafts on the surface of cultured human monocytes, and the destruction of lipid rafts following the addition of a cholesterol-reducing agent inhibits the formation of AA amyloid fibrils [44]. These findings strongly support a possibility that gangliosides in lipid rafts on the cell surface are intimately involved in the promotion of the assembly of the AA proteins, which is very similar to the mechanism underlying assembly of other amyloidogenic proteins. On the basis of these findings, it can be assumed that an endogenous seed for the consecutive assembly of the AA protein is likely generated on the surface of monocytes. Indeed, Sponarova et al. showed that monocytes isolated from AA amyloidosis animal models possess amyloid seeds, that potently enhance AA amyloidosis in recipient mice [45].

2. Action of Amyloid Fibrils and/or Oligomers against Cellular Membranes

1) Amyloid Fibrils

In terms of the toxicities of Aß amyloid fibrils, it should be noted that their structure is significantly affected by the mechanical stress during the preparation of fibrils [46]. Matsuzaki's group previously prepared Aß fibrils in the presence of GM1/sphingomyelin/cholesterol membranes (Mem-fibrils) and in solution (Sol-fibrils) and compared their morphologies, surface hydrophobicities and cytotoxicities [47, 48]. Interestingly, transmission electron microscopy (TEM) revealed that Mem-fibrils are typical nonbranched amyloid fibrils (12.0 ± 0.7 nm, width), whereas Sol-fibrils are thinner fibrils or protofilaments (6.7 ± 1.3 nm, width) [48]. Importantly, it was also noted that Mem-fibrils are significantly cytotoxic against PC12 cells differentiated with nerve growth factor, whereas Sol-fibrils are much less toxic [48]. Furthermore, it was observed that Mem-fibrils exhibit a significantly stronger binding to the cell membranes than Sol-fibrils, as revealed by Congo red assay [48]. Notably, they stressed in their report that this difference can be explained by the fact that the surface hydrophobicity of Mem-fibrils, as estimated from the binding of 1-anilinonaphthalene-8-sulfonate, is higher than that of Sol-fibrils [48].

2) Oligomers

Currently, it is considered that oligomers, not amyloid fibrils, are the intimate toxic aggregate forms of any amyloidogenic proteins. Although the mechanism underlying oligomer generation has been largely unclarified, it should be noted that oligomers can be generated on the pathway to amyloid fibril formation or through the degradation of fragmented *parent* amyloid fibrils (Figure 2). Indeed, amyloid fibrils likely become more toxic upon fragmentation [49]. In the case of Aß oligomers, besides their toxicities being through ubiquitous mechanisms such as calcium dysregularion and membrane disruption [50], recent studies have suggested that they bind to several particular proteins that are expressed on the cell surface, leading to the emergence of specific neurotoxicities as follows. First, it has been shown that Aß oligomers specifically bind to the cellular prion protein (PrP^c); interestingly, the interaction of Aß oligomers with PrP^c does not require prior conformational alteration to its pathogenic form, PrP^{sc} [51]. In this study, it has been shown that the knockdown of PrP^c rescues the long-term potentiation inhibited by Aß oligomers. Second, *N*-methyl-D-aspartate receptors (NMDARs) have also been reported to potently function as a receptor of Aß oligomers [52]. The knockdown of NMDARs readily blocks the neuronal oxidative stress induced by Aß oligomers. Notably, NMDAR knockdown markedly abolishes the Aß oligomer binding to dendrites of cultured neurons.

It has also been reported that an early detrimental action of Aß oligomers on synapses is likely linked to the activation of NR-2B-containing NMDARs [53].

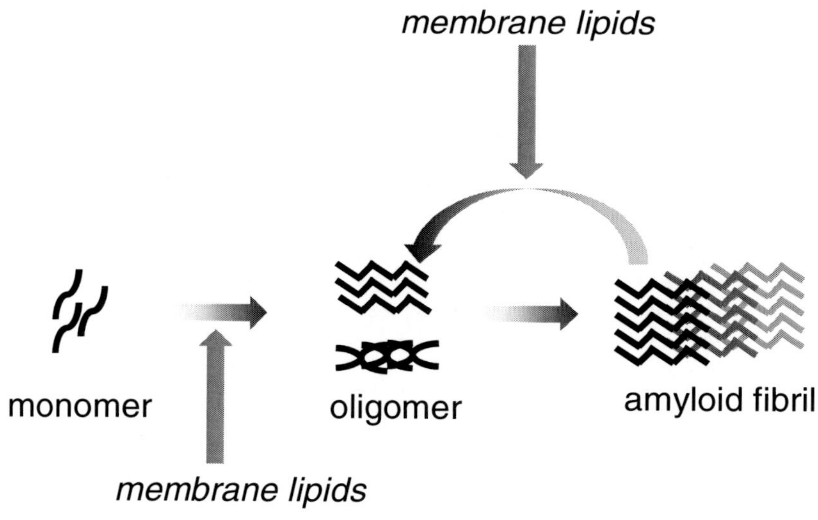

Figure 2. Membrane-guided assembly of soluble monomeric Aß to oligomers and amyloid fibrils. At least, a part of oligomers is likely generated through the degradation of amyloid fibrils as shown in a recent study [55].

In addition, ionotropic glutamate receptors, which include not only NMDARs but also AMPA receptors, can be overactivated by Aß oligomers, leading to neuronal death through mitochondrial dysfunction associated with Ca^{2+} overload [54].

Conclusion

As is reviewed in this article, it is suggested that cellular membranes are deeply involved in the initiation of pathologic assembly of amyloidogenic proteins. Evidence supporting this possibility is accumulating, particularly in the case of amyloidogenesis of natively unfolded proteins. It remains to be clarified how a pathologic interaction between a given amyloidogenic protein and cellular membranes occurs selectively in disease-development-prone organs or tissues.

However, as is taken into account that amyloidoses favorably develop with age, age-dependent alteration in the lipid composition of particular membranes may be crucial for the pathogenesis. Currently, it is assumed that amyloid fibrils act against bulk plasma membranes, whereas oligomers target rather specific, functional proteins on the cell surface (Figure 3).

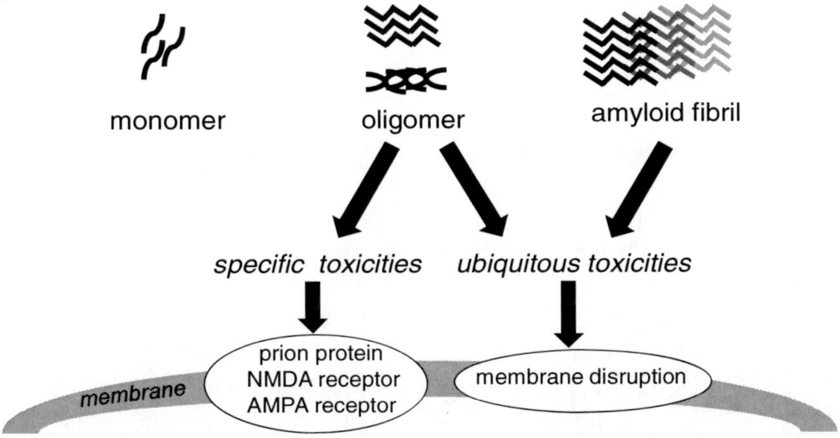

Figure 3. Putative mechanism of the toxicities induced by the assembled Aßs, including oligomers and amyloid fibrils. Amyloid fibrils may induce ubiquitous toxicities such as membrane disruption, whereas oligomers likely affect the function of particular proteins on the surface of the cells in addition to their ubiquitous toxicities.

It will be challenging in the future to clarify the mechanisms underlying cellular dysfunction and/or cell death caused by pathologic protein assemblies, including amyloid fibrils and oligomers in various human amyloidoses.

References

[1] Booth, DR ; Sunde, M ; Bellotti, V ; Robinson, CV ; Hutchinson, WL ; Fraser, PE ; Hawkins, PN ; Dobson, CM ; Radford, SE ; Blake, CC ; Pepys, MB. Instability, unfolding and aggregation of human lysozyme variants underlying amyloid fibrillogenesis. (1997) *Nature,* 385, 787-793.

[2] Xu, M ; Shashilov, VA ; Ermolenkov, VV ; Fredriksen, L ; Zagorevski, D ; Lednev, IK. The first step of hen egg white lysozyme fibrillation, irreversible partial unfolding, is a two-state transition. (2007) *Protein Sci*, 16, 815-832.

[3] Uversky, VN. Amyloidogenesis of natively unfolded proteins. (2008) *Curr. Alzheimer Res.*, 5,260-287.

[4] Dobson, CM. Protein folding and misfolding. (2003) *Nature*, 426, 884-890.

[5] Fezoui, Y ; Teplow, DB. Kinetic studies of amyloid ß-protein fibril assembly. Differential effects of α-helix stabilization. (2002) *J. Biol. Chem*, 277, 36948-36954.

[6] Apostolidou, M ; Jayasinghe, SA ; Langen, R. Structure of α-helical membrane-bound human islet amyloid polypeptide and its implications for membrane-mediated misfolding. (2008) *J. Biol. Chem.*, 283,17205-17210.

[7] Liu, G ; Prabhakar, A ; Aucoin, D ; Simon, M ; Sparks, S ; Robbins, KJ ; Sheen, A ; Petty, SA ; Lazo, ND. Mechanistic studies of peptide self-assembly: transient α-helices to stable ß-sheets. (2010) *J. Am. Chem. Soc.,* 132, 18223-18232.

[8] Zhong, L ; Johnson, WC, Jr. Environment affects amino acid preference for secondary structure. (1992) *Proc. Natl. Acad. Sci. USA*, 89,4462-4465.

[9] Yanagisawa, K. Role of gangliosides in Alzheimer's disease. (2007) *Biochim. Biophys. Acta*, 1768, 1943-1951.

[10] Ariga, T ; McDonald, MP ; Yu, RK. Role of ganglioside metabolism in the pathogenesis of Alzheimer's disease--a review. (2008) *J. Lipid Res.,* 49, 1157-1175.

[11] Matsuzaki, K ; Kato, K ; Yanagisawa, K. Aß polymerization through interaction with membrane gangliosides. (2010) *Biochim. Biophys Acta*, 1801, 868-877.

[12] Yanagisawa, K ; Odaka, A ; Suzuki, N ; Ihara, Y. GM1 ganglioside-bound amyloid ß-protein (Aß): a possible form of preamyloid in Alzheimer's disease. (1995) *Nat. Med*, 1, 1062-1066.

[13] Hayashi, H ; Kimura, N ; Yamaguchi, H ; Hasegawa, K ; Yokoseki, T ; Shibata, M ; Yamamoto, N ; Michikawa, M ; Yoshikawa, Y ; Terao, K ; Matsuzaki, K ; Lemere, CA ; Selkoe, DJ ; Naiki, H ; Yanagisawa, K. A seed for Alzheimer amyloid in the brain. (2004) *J. Neurosci.*, 24, 4894-4902.

[14] Choo-Smith, LP; Garzon-Rodriguez, W ; Glabe, CG ; Surewicz, WK. Acceleration of amyloid fibril formation by specific binding of Aß-(1-40) peptide to ganglioside-containing membrane vesicles. (1997) *J. Biol. Chem.*, 272, 22987-22990.

[15] Kakio, A ; Nishimoto, S ; Yanagisawa, K ; Kozutsumi, Y ; Matsuzaki, K. Interactions of amyloid ß-protein with various gangliosides in raft-like membranes: importance of GM1 ganglioside-bound form as an endogenous seed for Alzheimer amyloid. (2002) *Biochemistry,* 41, 7385-7390.

[16] Matsuzaki, K ;Horikiri, C. Interactions of amyloid ß-peptide (1-40) with ganglioside-containing membranes. (1999) *Biochemistry*, 38, 4137-4142.

[17] Kakio, A ; Nishimoto, SI ; Yanagisawa, K ; Kozutsumi, Y ; Matsuzaki, K. Cholesterol-dependent formation of GM1 ganglioside-bound amyloid ß-protein, an endogenous seed for Alzheimer amyloid. (2001) *J. Biol. Chem.*, 276, 24985-24990.

[18] Yahi, N ; Aulas, A ; Fantini, J. How cholesterol constrains glycolipid conformation for optimal recognition of Alzheimer's ß amyloid peptide (Aß1-40). (2010) *PLoS One,* 5, e9079.

[19] Rushworth, JV ;Hooper, NM. Lipid Rafts: Linking Alzheimer's Amyloid-ß Production, Aggregation, and Toxicity at Neuronal Membranes. (2010) I*nt. J. Alzheimers Dis.*, 2011, 603052.

[20] McLaurin, J ; Chakrabartty, A. Membrane disruption by Alzheimer ß -amyloid peptides mediated through specific binding to either phospholipids or gangliosides. Implications for neurotoxicity. (1996) *J. Biol. Chem.*, 271, 26482-26489.

[21] Choo-Smith, LP ; Surewicz, WK. The interaction between Alzheimer amyloid ß(1-40) peptide and ganglioside GM1-containing membranes. (1997) *FEBS Lett*, 402, 95-98.

[22] Williamson, MP ; Suzuki, Y ; Bourne, NT ; Asakura, T. Binding of amyloid ß-peptide to ganglioside micelles is dependent on histidine-13. (2006) *Biochem. J.*, 397, 483-490.

[23] Utsumi, M ; Yamaguchi, Y ; Sasakawa, H ; Yamamoto, N ; Yanagisawa, K ; Kato, K. Up-and-down topological mode of amyloid ß-peptide lying on hydrophilic/hydrophobic interface of ganglioside clusters. (2009) *Glycoconj J.,* 26, 99-1006.

[24] Yamamoto, N ; Yokoseki, T ; Shibata, M ; Yamaguchi, H ; Yanagisawa, K. Suppression of Aß deposition in brain by peripheral administration of Fab fragments of anti-seed antibody. (2005) *Biochem. Biophys. Res. Commun.*, 335, 45-47.

[25] Oikawa, N ; Yamaguchi, H ; Ogino, K ; Taki, T ; Yuyama, K ; Yamamoto, N ; Shin, RW ; Furukawa, K ; Yanagisawa, K. Gangliosides determine the amyloid pathology of Alzheimer's disease. (2009) *Neuroreport*, 20, 1043-1046.

[26] Cataldo, AM ; Peterhoff, CM ; Troncoso, JC ; Gomez-Isla, T ; Hyman, BT ; Nixon, RA. Endocytic pathway abnormalities precede amyloid ß deposition in sporadic Alzheimer's disease and Down syndrome: differential effects of APOE genotype and presenilin mutations. (2000) *Am. J. Pathol*, 157,277-286.

[27] Yuyama, K ; Yamamoto, N ; Yanagisawa, K. Chloroquine-induced endocytic pathway abnormalities: Cellular model of GM1 ganglioside-induced Aß fibrillogenesis in Alzheimer's disease. (2006) *FEBS Lett,* 580, 6972-6976.

[28] Yuyama, K ; Yamamoto, N ; Yanagisawa, K. Accelerated release of exosome-associated GM1 ganglioside (GM1) by endocytic pathway abnormality: another putative pathway for GM1-induced amyloid fibril formation. (2008) *J. Neurochem.,* 105, 217-224.

[29] Yuyama, K ; Yanagisawa, K. Late endocytic dysfunction as a putative cause of amyloid fibril formation in Alzheimer's disease. (2009) *J. Neurochem*, 109, 1250-1260.

[30] Kimura, N ; Yanagisawa, K. Endosomal accumulation of GM1 ganglioside-bound amyloid ß-protein in neurons of aged monkey brains. (2007) *Neuroreport*, 18, 1669-1673.

[31] Davidson, WS ; Jonas, A ; Clayton, DF ; George, JM. Stabilization of α-synuclein secondary structure upon binding to synthetic membranes. (1998) J. *Biol.Chem.*, 273, 9443-9449.

[32] Perrin, RJ ; Woods, WS ; Clayton, DF ; George, JM. Interaction of human α-Synuclein and Parkinson's disease variants with phospholipids. Structural analysis using site-directed mutagenesis. (2000) *J. Biol. Chem.*, 275, 34393-34398.

[33] Eliezer, D ; Kutluay, E ; Bussell, R, Jr. ; Browne, G. Conformational properties of α-synuclein in its free and lipid-associated states. (2001) *J. Mol. Biol.,* 307, 1061-1073.

[34] Jao, CC ; Der-Sarkissian, A ; Chen, J ; Langen, R. Structure of membrane-bound α-synuclein studied by site-directed spin labeling. (2004) *Proc. Natl. Acad. Sci. USA*, 101, 8331-8336.

[35] Martinez, Z ; Zhu, M ; Han, S ; Fink, AL. GM1 specifically interacts with α-synuclein and inhibits fibrillation. (2007) Biochemistry, 46, 1868-1877.

[36] Fantini, J ; Yahi, N. Molecular basis for the glycosphingolipid-binding specificity of α-synuclein: key role of tyrosine 39 in membrane insertion. (2011) *J. Mol. Biol.,* 408, 654-669.

[37] Knight, JD ; Miranker, AD. Phospholipid catalysis of diabetic amyloid assembly. (2004) *J. Mol. Biol*, 341, 1175-1187.

[38] Jayasinghe, SA ; Langen, R. Lipid membranes modulate the structure of islet amyloid polypeptide. (2005) *Biochemistry,* 44, 12113-12119.

[39] Apostolidou, M ; Jayasinghe, SA ; Langen, R. Structure of α-helical membrane-bound human islet amyloid polypeptide and its implications for membrane-mediated misfolding. (2008) *J. Biol. Chem.,* 283, 17205-17210.

[40] Lopes, DH ; Meister, A ; Gohlke, A ; Hauser, A ; Blume, A ; Winter, R. Mechanism of islet amyloid polypeptide fibrillation at lipid interfaces studied by infrared reflection absorption spectroscopy. (2007) *Biophys. J.,* 93, 3132-3141.

[41] Kuroda, Y ; Maeda, Y ; Sawa, S ; Shibata, K ; Miyamoto, K ; Nakagawa, T. Effects of detergents on the secondary structures of prion protein peptides as studied by CD spectroscopy. (2003) *J. Pept Sci.,* 9, 212-220.

[42] Levy, M ; Garmy, N ; Gazit, E ; Fantini, J. The minimal amyloid-forming fragment of the islet amyloid polypeptide is a glycolipid-binding domain. (2006) *FEBS J.,* 273, 5724-5735.

[43] Wakabayashi, M; Matsuzaki, K. Ganglioside-induced amyloid formation by human islet amyloid polypeptide in lipid rafts. (2009) *FEBS Lett*, 583, 2854-2858.

[44] Gellermann, GP ; Appel, TR ; Tannert, A ; Radestock, A ; Hortschansky, P; Schroeckh, V ; Leisner, C ; Lutkepohl, T ; Shtrasburg, S ; Rocken, C ; Pras, M ; Linke, RP ; Diekmann, S ; Fandrich, M. Raft lipids as common components of human extracellular amyloid fibrils. (2005) *Proc. Natl. Acad. Sci. USA*, 102, 6297-6302.

[45] Sponarova, J; Nystrom, SN; Westermark, GT. AA-amyloidosis can be transferred by peripheral blood monocytes. (2008) *PLoS One*, 3, e3308.

[46] Petkova, AT ; Leapman, RD ; Guo, Z ; Yau, WM ; Mattson, MP ; Tycko, R. Self-propagating, molecular-level polymorphism in Alzheimer's ß-amyloid fibrils. (2005) *Science,* 307, 262-265.

[47] Okada, T ; Wakabayashi, M ; Ikeda, K ; Matsuzaki, K. Formation of toxic fibrils of Alzheimer's amyloid ß-protein-(1-40) by monosialoganglioside GM1, a neuronal membrane component. (2007) *J. Mol. Biol.,* 371, 481-489.

[48] Okada, T; Ikeda, K ; Wakabayashi, M ; Ogawa, M ; Matsuzaki, K. Formation of toxic Aß(1-40) fibrils on GM1 ganglioside-containing membranes mimicking lipid rafts: polymorphisms in Aß(1-40) fibrils. (2008) *J. Mol.Biol.*, 382, 1066-1074.

[49] Xue, WF ; Hellewell, AL ; Gosal, WS ; Homans, SW ; Hewitt, EW ; Radford, SE. Fibril fragmentation enhances amyloid cytotoxicity. (2009) *J. Biol. Chem.*, 284, 34272-34282.

[50] Demuro, A ; Mina, E ; Kayed, R ; Milton, SC ; Parker, I ; Glabe, CG. Calcium dysregulation and membrane disruption as a ubiquitous neurotoxic mechanism of soluble amyloid oligomers. (2005) *J. Biol. Chem.,* 280, 17294-17300.

[51] Lauren, J ; Gimbel, DA ; Nygaard, HBGilbert, JW ; Strittmatter, SM. Cellular prion protein mediates impairment of synaptic plasticity by amyloid-ß oligomers. (2009) *Nature*, 457, 1128-1132.

[52] Decker, H ; Jurgensen, S ; Adrover, MF ; Brito-Moreira, J ;Bomfim, TR ; Klein, WL ; Epstein, AL ; De Felice, FG ; Jerusalinsky, D ; Ferreira, ST. N-methyl-D-aspartate receptors are required for synaptic targeting of Alzheimer's toxic amyloid-ß peptide oligomers. (2010) *J. Neurochem.*, 115, 1520-1529.

[53] Ronicke, R ; Mikhaylova, M ; Ronicke, S ; Meinhardt, J ; Schroder, UH; Fandrich, M ; Reiser, G ; Kreutz, MR ; Reymann, KG. Early neuronal dysfunction by amyloid ß oligomers depends on activation of NR2B-containing NMDA receptors. (2010) *Neurobiol Aging* (in press).

[54] Alberdi, E ; Sanchez-Gomez, MV ; Cavaliere, F ; Perez-Samartin, A ; Zugaza, JL ; Trullas, R ; Domercq, M ; Matute, C. Amyloid ß oligomers induce Ca2+ dysregulation and neuronal death through activation of ionotropic glutamate receptors. (2010) *Cell Calcium*, 47, 264-272.

[55] Martins, IC ; Kuperstein, I ; Wilkinson, H ; Maes, E ; Vanbrabant, M ; Jonckheere, W ; Van Gelder, P ; Hartmann, D ; D'Hooge, R ; De Strooper, B ; Schymkowitz, J ; Rousseau, F. Lipids revert inert Aß amyloid fibrils to neurotoxic protofibrils that affect learning in mice. (2008) *EMBO J.,* 27, 224-233.

In: Amyloids
Editors: I. P. Halcheck et al.
ISBN: 978-1-62100-538-4

Chapter 8

INSIGHTS INTO THE AGGREGATION PATHWAY OF α-SYNUCLEIN

Jillian Madine*
Institute of Integrative Biology, University of Liverpool,
Crown Street, Liverpool L69 7ZB,
United Kingdom

ABSTRACT

The aggregation of α-synuclein to form insoluble fibrillar inclusions is the probable key event in the onset of many neurodegenerative diseases including Parkinson's disease and the Lewy body variant of Alzheimer's disease. Previous studies and aggregation prediction programmes have identified key regions within the α-synuclein sequence that are implicated in pathogenic aggregation of the protein. Here, I have examined the self-association properties of short synthetic peptides corresponding to regions of the full-length protein. The peptides examined were selected from within the fibril core based on secondary structure previously identified by NMR. A range of biophysical techniques have been used to try and elucidate the properties of the amyloid formation pathway. One of the major unknown properties of amyloid proteins is the nature of the toxic species. It is commonly thought to be the early oligomeric or protofibrillar species that are the culprit for toxic effects seen in neuronal death rather than insoluble fibrils. The data presented here provides a useful insight into amyloid formation and will be used to guide the synthesis of peptide inhibitors of α-synuclein aggregation and provide potential new therapies for neurodegenerative diseases.

Keywords: α-synuclein, aggregation, amyloid, neurodegenerative diseases.

---

* E-mail: j.madine@liv.ac.uk, tel: +44 151 7954507.

Abbreviations

CD, circular dichroism;
DLB, dementia with Lewy bodies;
DLS, dynamic light scattering;
NMR, nuclear magnetic resonance;
NAC, non-amyloid component of Alzheimer's disease plaques;
PD, Parkinson's disease;
TEM, transmission electron microscopy;
ThT, thioflavin T.

Introduction

Dementia affects 7% of the population over 65 years of age and 30% of those over 80 years of age [1]. The two most common forms of dementia after Alzheimer's disease are Dementia with Lewy bodies (DLB), accounting for 10-15% of cases [2] and Parkinson's disease (PD) affecting 1-2% of the population over 65 years of age [3]. These diseases are associated with misfolded protein aggregation and deposition in areas of the brain. The protein aggregates can take the form of Lewy bodies, with a dense core of insoluble filamentous proteins and lipids, surrounded by a peripheral halo. The major protein component of these inclusions is α-synuclein [4], a natively unfolded protein capable of misfolding into a state with a high propensity to aggregate into amyloid fibrils [5].

At present there are no methods available to prevent or cure these dementias, with treatments concentrating on improving symptoms [6]. The major treatment for PD is levodopa, the precursor of the neurotransmitter dopamine. It acts to replace the dopamine deficiency caused by death of dopamine producing cells. Other treatments for PD include dopamine agonists which directly stimulate remaining dopamine receptors. Cholinesterase inhibitors are currently used to treat DLB patients [7]. These also act to boost neurotransmission from remaining nerve termini, but they often have gastrointestinal side effects.

α-Synuclein is a single polypeptide of 140 amino acids (Figure 1a). Previous studies have identified key regions within the α-synuclein sequence as necessary for the aggregation of the protein. Many of these segments lie within a central hydrophobic region (residues 61-95) identified as the non-amyloid component (NAC) of Alzheimer's disease plaques [8]. El-Agnaf and co-workers synthesised 7-mer peptides spanning the whole sequence and identified the N-terminal half of the NAC region (residues 64-86) as responsible for α-synuclein self-association [9]. Other sequences within the NAC region including residues 66-74 [10], 68-78 [11], 71-82 [12], 69-79 [13] and 77-82 [14] have also been proposed to be specifically involved in aggregation (Figure 1b). Many studies have been carried out involving the mutation of key residues within α-synuclein in order to determine critical regions that may be involved in amyloid formation. Mutations introduced within the fibril core region of 37-89, had the greatest effect on fibril formation rates [16]. Specifically, polar or charge substitutions within the region 63-74 caused a significant reduction in fibrillisation. Also, an increase in hydrophobicity within this region showed enhanced fibril formation

(T72A), consistent with hydrophobicity being a critical factor in stabilising the fibril core [17]. Several studies have predicted that the fibril core of α-synuclein may be split into several shorter peptide regions (66-74, 77-82 and 86-92) [18, 19].

Nuclear magnetic resonance (NMR) data studying the secondary structure of full-length α-synuclein fibrils revealed that the sequence encompassing residues 38-95 forms the core fibrillar region [15]. Within this region residues 69-82 are found in a continuous β-strand flanked by two turn regions (residues 66-68 and 83-86) (Figure 1a). Small chain alanine and glycine residues, known to have little aggregation propensity make up the turn sequences [20]. In addition, a slightly larger structural core spanning residues 34-101 has been identified by site-directed spin labelling [21].

At present it is unclear as to the nature of the toxic species involved in disease, but it is known that α-synuclein and α-synuclein peptide fragments show an increase in neurotoxicity upon aggregation [22]. This implies that prevention of aggregation will reduce production of the toxic species. Inhibition of aggregation of amyloidogenic peptides is already being investigated as a potential therapeutic strategy for the treatment of neurodegenerative diseases.

(a) MDVFKMGLSKAKEGVVAAAEKTKQGVAEAAGKTKEGV**LYVGS**KTKEGVVH
10 20 30 40 50

G**VATVAE**KTK**EQVTN**VGG**AVVT**G**VTAVAQKTV**EGAG**SIAAAT**G**FV**KKDQL
60 70 80 90 100

GKNEEGAPQEGILEDMPVDPDNEAYEMPSEEGYQDYEPEA
110 120 130 140

(b)
^{71}**VT**G**VTAVAQKTV**^{82}
^{77}**VAQKTV**^{82}
^{68}G**AVVT**G**VTAVA**^{78}
^{66}VGG**AVVT**G**V**^{74}
^{69}**AVVT**^{72}

(c)
^{61}**EQVTN**VGG**AVVT**G**VTAVAQKTV**EGAG**SIAAAT**G**FV**^{95}
^{74}**VTAVAQKTV**EGAG**SIAAAT**G**FV**^{95}
^{51}G**VATVAE**KTK**EQVTN**^{65}
^{61}**EQVTN**VGG**AVVT**^{72}

Figure 1. a) Sequence of α-synuclein showing identified β-sheet (bold) and turn (underlined) regions from published NMR data [15]. b) Regions previously identified as critical for aggregation. c) Peptides presented here encompassing larger sections of the core β-strand region.

The development of any drug requires a lead compound that can bind to the target. For inhibition of amyloid formation, the lead compounds are the key amyloid regions. The amyloidogenic peptide corresponding to key regions may then be modified to retain its binding and specificity, but to prevent further peptide association. There are several methods that have already been pioneered to modify peptides for use as aggregation inhibitors. One modification technique that has been shown to successfully inhibit amyloidogenic peptide aggregation is N-methylation [23].

This has been applied to the α-synuclein peptide fragments; 68-78 [24] and 71-82 [14]. A related method of peptide modification involving introduction of alternative non-native amino acids that disrupt the β-sheet packing, has also been introduced into α-synuclein peptide-based inhibitors [25]. Further investigation of the detailed study using overlapping 7-mer peptides proposed modified peptides based on residues 69-72 as potential inhibitors capable of blocking α-synuclein oligomerisation [9].

In addition to these methods there are many other known modification procedures; for example, incorporation of a proline residue within a β-sheet peptide will introduce a bend in the β-strand, preventing associations required for β-sheet propagation [26]. Alternatively, a large bulky group at the terminus of a peptide (e.g. a steroid) will hinder the association of neighbouring peptide molecules [27].

Coupling recognition sequences with disrupting elements (e.g. polylysine) is also effective [28]. While peptides normally make poor drugs, modification gives them many beneficial properties; including high proteolytic resistance, increased membrane solubility, low immunogenicity and blood-brain-barrier permeability [29]. This has been demonstrated by detection of modified peptides in the brain following uptake, with no indication of peptide degradation [30].

The work presented in this chapter set out to identify key regions of α-synuclein that are responsible for the aggregation of the protein into insoluble filamentous deposits found within disease-associated inclusions. This has been achieved by studying the aggregation characteristics of peptides of varying lengths which correspond to the core β-strand region, in turn determining which parts of the peptides can form the critical amyloid core. This information can be used to aid the design of potential inhibitors of aggregation.

Analysis of Aggregation

Preparation of Samples

Synthetic peptides were purchased from Peptide Protein Research Ltd (U.K.). The peptides were subjected to three dissolution-evaporation cycles with hexafluoroisopropanol to break up any initial aggregates. Aqueous solutions of 200, 400 and 800 μM 51-65, 61-72, 74-95 and 200 μM 61-95 were suspended in 10 mM phosphate, pH7 and incubated with continuous agitation, at 37°C. Samples were taken for analysis immediately after preparation of the solution and again at specified time points throughout the analysis up to 7 days.

Analysis of Fibril Growth and Morphology

Amyloid formation was monitored with the fluorescent dye Thioflavin T (ThT). Peptide sample (50 μl) was added to 950 μl of 10 μM ThT in 10 mM phosphate, pH7, vortexed and transferred to a 1 cm path length cuvette. Spectra were recorded on a Cary Eclipse Varian fluorescence spectrophotometer with excitation at 450 nm and emission at 482 nm, with band pass 5 nm. A new excitation band at 450 nm and enhanced emission at 482 nm are observed when amyloid is present.

The amount of insoluble material formed by aggregation of the α-synuclein peptides was quantified by sedimentation. Following incubation for one week at 37°C with agitation, insoluble aggregates were removed by centrifugation at 13,000 rpm for 30 mins. The amount of peptide remaining in the supernatant was analysed after alkaline hydrolysis using the 1° amine-reactive agent ninhydrin [31].

Turbidity measurements were also monitored by measuring the absorbance at 400 nm on a Cary Eclipse Varian spectrophotometer.

Aggregation was monitored by dynamic light scattering (DLS). Peptide (50 μl in 10 mM phosphate, pH 7) was added to a 3 mm cell. A series of 20 measurements of 10 seconds each were taken at 30°C on a Zetasizer Nano DLS machine from Malvern Instruments, and averaged profiles were produced.

Changes in secondary structure were monitored by far-UV circular dichroism (CD). Spectra were recorded on a Jasco J-810 spectropolarimeter at 25°C, over the 250 to 190 nm range at a scan rate of 10 nm/min with step size of 0.1 nm. Spectra were recorded as the average of 4 scans. 40 μM peptide in 10 mM phosphate, pH7 was added to a 0.1 cm cuvette. After subtracting control spectra, spectral amplitudes were converted to molar elipticity values (θ) using the equation:

$$\theta = \frac{\mathrm{CD}}{c \times 10 \times l \times n}$$

where CD is spectral amplitude (mdeg), c is peptide concentration (M), l is path length (cm), and n is number of amino acids.

Morphologies of the insoluble aggregates formed following incubation of the peptides were analysed by transmission electron microscopy (TEM) using negative staining methods (4 % uranyl acetate). Peptide suspensions (10 μl) were loaded onto carbon coated copper grids and visualised on a Tecnai 10 electron microscope at 100 kV.

Cell Viability Assay

Aliquots of peptide solutions (20 μl) were taken at specified time points throughout the incubation process and frozen until required. SH-SY5Y neuronal cells (ECACC) were plated on 96 well plates at 30,000 cells/well (80 μl) and grown for 24 h. Peptide samples (20 μl, preincubated at 200 μM) were added in aggregation buffer and incubated for 48 h prior to addition of 10 μl Cell Counting Kit-8 (CCK-8) (NBS Biologicals) dye, and further incubation for 2 h. Absorbance was read at 450 nm. Percentage cell viability was calculated based on the absorbance measured relative to that of control cells exposed to buffer alone.

Results and Discussion

The peptides studied in this chapter were chosen based on previous NMR data which identified β-sheet and turn regions within α-synuclein fibrils (Figure 1a) [15]. Previous work to identify a core region critical for α-synuclein aggregation in neurodegenerative disease has focussed mainly on regions encompassing one β-strand (^{69}AVVTGVTAVAQKTV^{82}) and accompanying turn (^{66}VGG^{68}) (Figure 1b). Based on the fibril architecture proposed from the NMR data (Figure 1a), peptides were chosen to expand the search for a core region to include surrounding turn and β-strand sequences (Figure 1c). This work aims to elucidate what role these sequences may play in the initiation of disease-associated fibrillisation by studying the aggregation and amyloidogenic properties, morphology and cytotoxicity of the peptides.

The peptides were incubated at a range of concentrations for up to 7 days and a variety of techniques were used to follow the aggregation profile and study the final species produced.

ThT is a dye commonly used to distinguish amyloid fibril formation from amorphous aggregation. ThT is able to bind to peptides 61-95 and 61-72, indicating the presence of amyloid following incubation for 7 days (Figure 2). In contrast, peptides 51-65 and 74-95 show little or no ThT fluorescence suggesting an absence of amyloid structure.

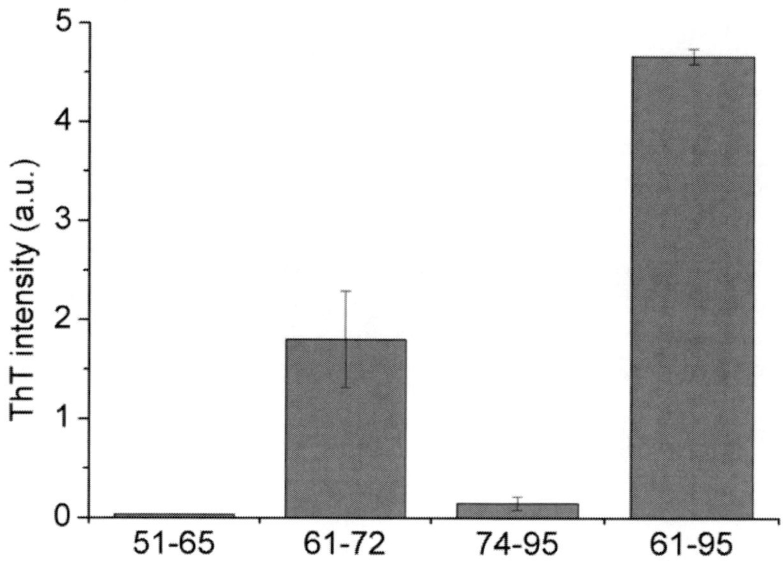

Figure 2. ThT fluorescence intensity of peptides incubated at 200 μM in 10 mM phosphate, pH7, at 37°C, for 7 days. The mean and standard deviation are shown (n=3).

In agreement with the ThT results shown in Figure 2, incubation of peptides 61-95 and 61-72 show an increased amount of aggregated peptide present after 7 days (Figure 3a). However, in contrast with the ThT results, peptide 74-95 also shows an increase in amount of aggregated peptide. This suggests peptide 74-95 forms aggregates that do not have the characteristic cross β-sheet structure of amyloid that is required to bind ThT. The peptide 51-65 consistently shows a lack of aggregation even at an increased concentration of 800 μM (Figure 3a). Absorbance values indicative of turbidity show a large increase for peptide 61-95 over the 7 day time-course (Figure 3b). Turbidity measurements for 74-95 also show a significant increase over 7 days. Taken together with previous data (Figure 2 and 3a), it appears that peptide 74-95 is undergoing amorphous aggregation. The peptide 61-72 shows low levels of turbidity at all time-points studied (Figure 3b). This technique measures levels of light passing through the sample, suggesting that the peptide 61-72 forms amyloidogenic species that are too small to be detected by this method. As predicted from previous data, 51-65 maintains low absorbance values consistent with a lack of aggregation (Figure 3b).

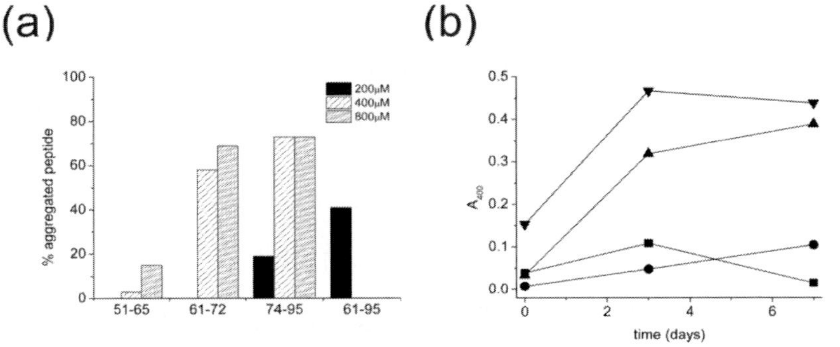

Figure 3. a) Percentage aggregated peptide measured by concentration of peptide remaining in the supernatant following centrifugation of insoluble aggregates after incubation at 37°C in 10 mM phosphate, pH7, for 7 days, at 200 μM (solid), 400 μM (sparse) and 800 μM (dense). 61-95 was tested at 200 μM only due to peptide availability. b) Turbidity measurements for peptides incubated at 200 μM. Absorbance values at 400 nm are shown for 51-65 (■), 61-72 (●), 74-95 (▲) and 61-95 (▼).

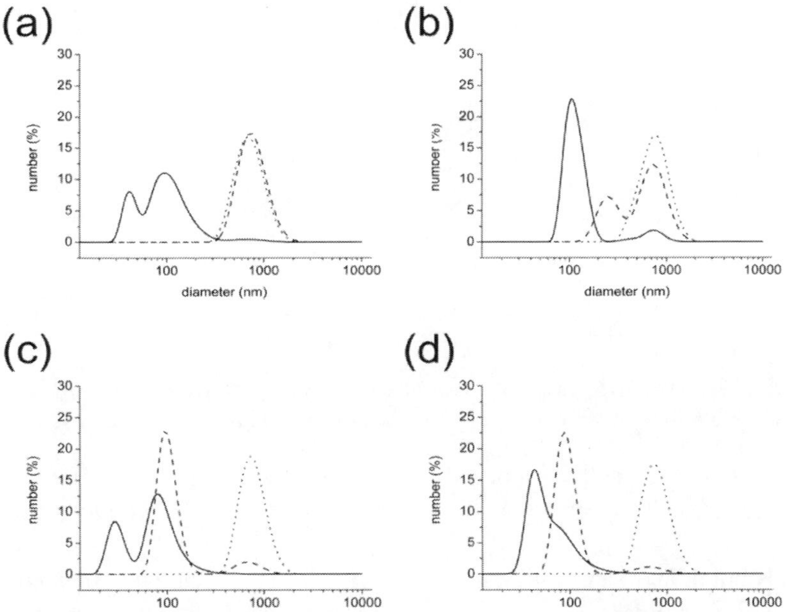

Figure 4. DLS data of peptides 51-65 (a), 61-72 (b), 74-95 (c) and 61-95 (d), freshly prepared (—), incubated at 37°C for 3 days (- - -) and 7 days (···), at 200 μM in 10 mM phosphate, pH7.

DLS data shows that all 4 peptides increase in size over the 7 day time-period. It is worth noting that the peptides forming non-amyloid aggregates (Figure 4, a and c; corresponding to 51-65 and 74-95) have a more disperse size range at the initial stages of incubation, than the amyloidogenic peptides (Figure 4, b and d; corresponding to 61-72 and 61-95). Interestingly, peptide 51-65 shows an increase in size that is not detected by turbidity measurements. This observation could provide information relating to the pathway of amorphous versus amyloidogenic aggregation, however this would require further study involving a larger selection of aggregation prone peptides.

All of the peptides studied have a random coil conformation at the initial stages of observation, as assessed by CD (Figure 5). The peptide 51-65 remains random coil following incubation for 7 days, however the amount of peptide remaining in solution is reduced (Figure 5a). The other 3 peptides (61-72, 74-95 and 61-95) show a lack of detectable CD signal by day 7, indicating peptide falling out of solution over the 7 day incubation (Figure 5, b-d).

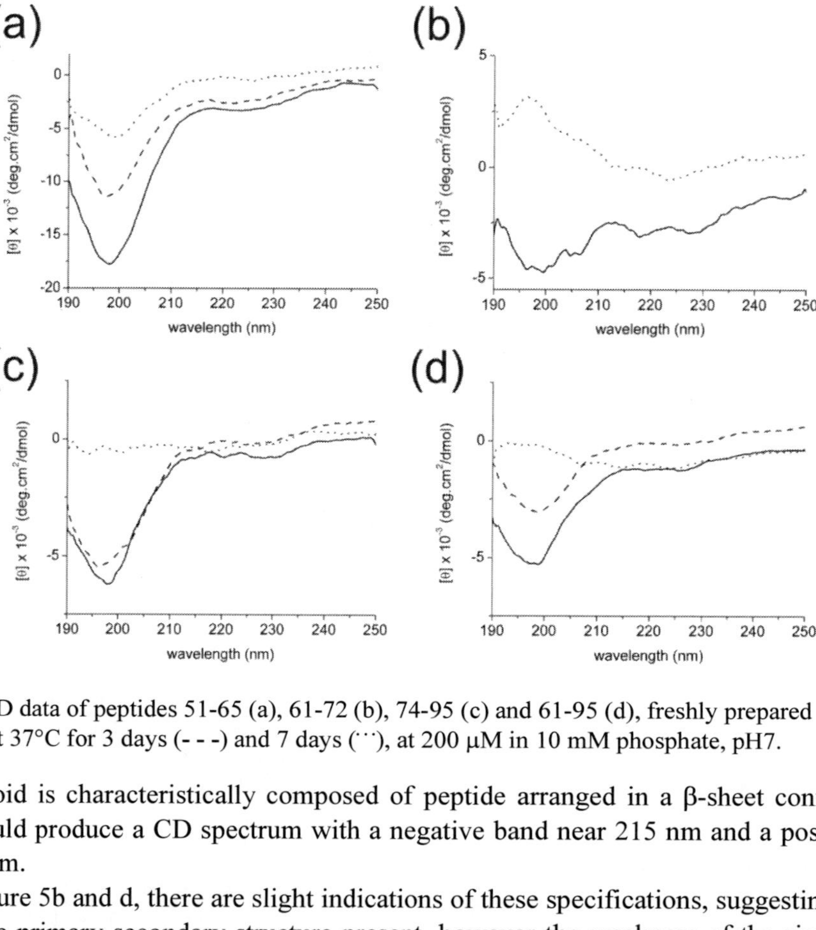

Figure 5. CD data of peptides 51-65 (a), 61-72 (b), 74-95 (c) and 61-95 (d), freshly prepared (—), incubated at 37°C for 3 days (- - -) and 7 days (···), at 200 μM in 10 mM phosphate, pH7.

Amyloid is characteristically composed of peptide arranged in a β-sheet conformation, which would produce a CD spectrum with a negative band near 215 nm and a positive band near 198 nm.

In Figure 5b and d, there are slight indications of these specifications, suggesting that this may be the primary secondary structure present, however the weakness of the signal due to lack of soluble peptide makes this diagnosis difficultThe morphology of aggregates produced following incubation for 7 days were visualised using negative stain TEM.

As expected peptides 61-95 and 61-72 show fibrillar species characteristic of amyloid (Figure 6, b and d), whereas peptides 51-65 and 61-72 show amorphous species (Figure 6, a and c).

The techniques presented so far study the aggregation process over time and the morphology of the species produced. It has been established that amyloid aggregation is involved in the progression of neurodegenerative disease, however it is currently unclear which species along the aggregation pathway is the toxic agent responsible for disease-associated cell death. In order to elucidate the toxic properties of the peptides studied here, cell viability of neuronal cells in culture was examined upon addition of peptides at varying stages throughout the aggregation process.

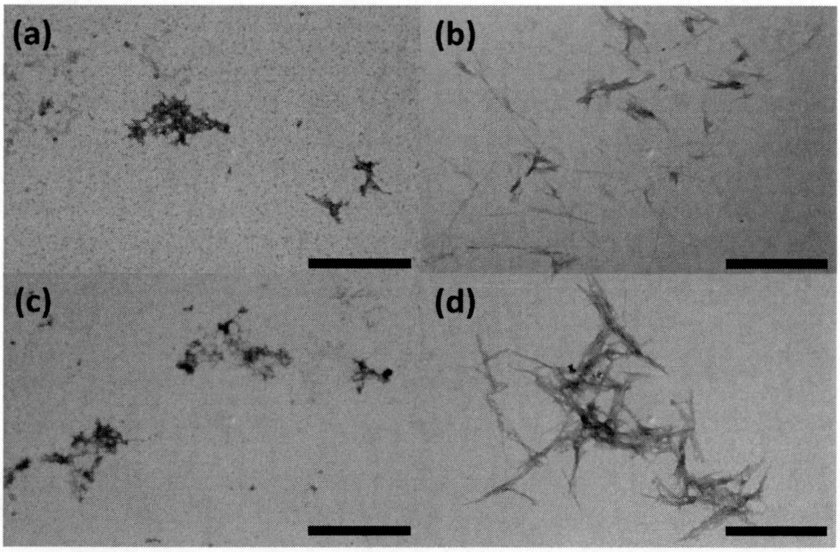

Figure 6. Negative stain TEM of peptides 51-65 (a), 61-72 (b), 74-95 (c) and 61-95 (d), harvested after incubation of 200 μM peptide in 10 mM phosphate, pH7, at 37°C for 7 days. Scale bar is 500 nm.

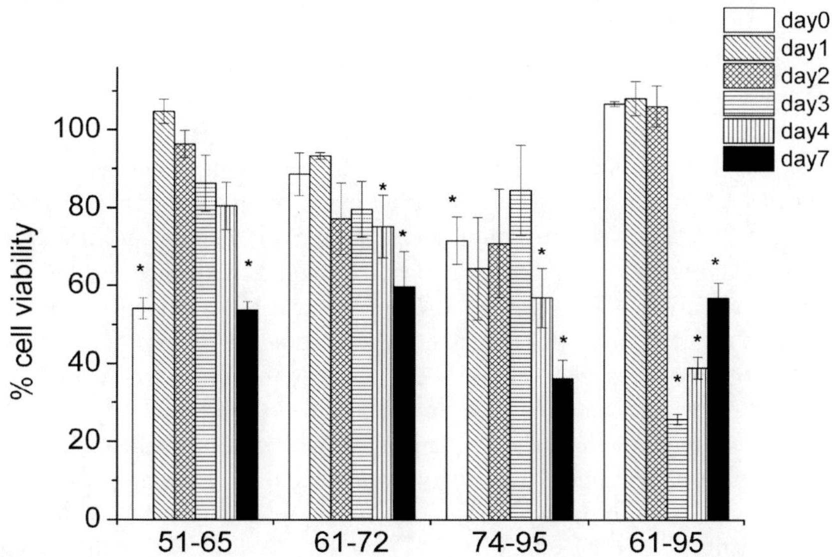

Figure 7. Cell viability of SH-SY5Y cells after addition of the peptides, preincubated at 200 μM in 10 mM phosphate, pH7, 37°C. The mean and standard error are shown (n=4). Statistical comparisons were carried out between toxicity in the presence of the peptides and buffer alone, using Student's *t*-test, with significant results ($P<0.05$) indicated by *.

Following incubation for 7 days all peptides show significant toxicity to cells in contrast to buffer alone (Figure 7). This indicates that both fibrillar and amorphous aggregates affect cell viability of SH-SY5Y cells. The two amyloid forming peptides (61-72 and 61-95) show no toxicity towards SH-SY5Y cells in the early timepoints (days 0-2) with species becoming more cytotoxic as incubation continues. There is also some indication that for the peptide 61-95 there is a recovery of cell viability towards the later stages of peptide incubation.

This suggests that higher aggregates may be less toxic than early stage species. Overall these results are consistent with amyloid aggregation being the primary cause of cell death in neurodegenerative disease. Suprisingly, the peptides which are hypothesised to form amorphous aggregates (74-95 and 51-95), show 30-50 % cell death upon addition of freshly prepared peptide (day 0). However, incubation of these peptides for 1 to 3 days appears to give rise to peptide species that do not significantly affect cell viability. This suggests that there are changes happening within these samples during incubation which affect the toxicity of the peptide towards neuronal cells.

CONCLUSION

It is not suprising that the peptide 61-95 encompassing the entire NAC region is able to aggregate to form amyloid fibrils over a 7 day period. In addition, the peptide 61-72 is also able to fibrillise over the same time period. These two peptides contain a common region of residues (^{66}VGGAVVT^{72}) which are not present in any of the other peptides (Figure 1c), suggesting that this may be the critical sequence required for the formation of amyloid within the NAC region of α-synuclein. This identified sequence contains residues ^{69}AVVT^{72} which have been previously identified to show a significant contribution to aggregation of α-synuclein (Figure 1b) [32].

Aggregation prediction programmes are commonly employed to predict the propensity of peptides and proteins to aggregate compared with a similar size peptide with random sequence. Full-length α-synuclein is predicted to have a low propensity to aggregate at pH7 using the Zygreggator prediction method (Z_{agg} = -1.68) [33]. This is not suprising as it is a natively unfolded protein that requires an initiation event or specific conditions to promote amyloid formation. It is known that amino acids of high aggregation propensity are often found within buried regions of native proteins. When the sequence of α-synuclein is shortened to consider only the NAC region (61-95) the aggregation propensity increases (Z_{agg} = 1.22) (Table 1). Predicted values for the other peptides studied in this chapter fit well with experimental data that peptides 61-95 and 61-72 aggregate to form amyloid fibrils, whereas peptides 51-65 and 74-95 form amorphous aggregates at pH7 (Table 1).

The entire region encompassing the sequences of the peptides studied (51-95) was run through the prediction programme to identify which residues have intrinsic aggregation propensity; this may assist with predicting the core region responsible for amyloid formation (Figure 8). The region ^{65}NVGGAVVTGVTA^{76} shows the highest intrinsic aggregation propensity out of the identified core β-sheet sequence (residues 51-65). This sequence spans the identified region ^{66}VGGAVVT^{72} that is proposed to be required for fibril formation from data presented here. This also supports data suggesting that modified peptides based on residues [69]AVVT[72] can act as potential inhibitors capable of blocking α-synuclein oligomerisation [9].

Table 1. Amyloid aggregation propensities of the peptides studied in this chapter

Peptide sequence	Z_{agg}
51-65	0.66
61-72	1.08
74-95	0.83
61-95	1.22

Peptides with Z_{agg} scores above the threshold ($Z_{agg} > 1$) are highlighted in bold

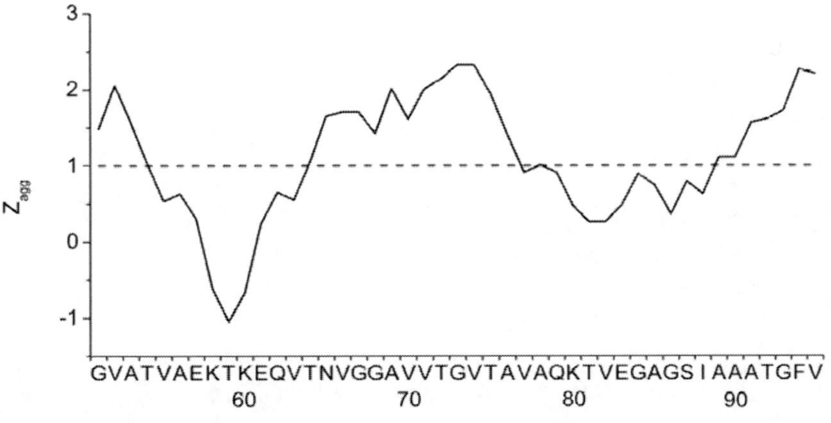

Figure 8. Predicted aggregation propensity of the region of α-synuclein studied in this work, residues 51-95. The horizontal line indicates the threshold ($Z_{agg} > 1$) used to define areas of high intrinsic aggregation propensity using the Zygreggator method.

In addition, the prediction programme also highlights the region [90]AATGFV[95] as having high intrinsic aggregation propensity (Figure 8). Previous studies have shown that F94 is required for α-synuclein oligomerisation, but may be redundant in fibril elongation [34], this conclusion was a result of work studying peptides 76-96 and 76-94, which showed fibrillisation over a longer timescale of 450 hours [34]. Other studies have shown that π-π stacking may be important in amyloid formation which could explain the role of F94 in initiating α-synuclein oligomerisation [19]. Under conditions tested here, peptide 74-95 which contains the residue F94, shows increased turbidity and aggregated peptide, however these aggregates appear amorphous. Increased incubation of 74-95 for up to 21 days did not show any further aggregation (data identical to that at 7 days, therefore not shown). This indicates that incubation conditions may be a critical factor in determining the specific aggregation pathway of peptide fragments. The peptide 51-65 does not show turbidity or aggregated peptide, however cell viability and DLS data indicate changes are occurring within the sample during the incubation period suggesting that species formed are small and soluble. Aggregation prediction shows negative areas within this region (Figure 8), which could explain the lack of amyloid aggregation observed for this peptide.

The work shown in this chapter fits with the theory that amyloidogenic aggregation of α-synuclein is initiated within the central core region, specifically residues ^{66}VGGAVVT^{72} or ^{69}AVVT^{72} which then propagates to the remainder of the protein.

Acknowledgments

This work was supported financially with a fellowship from Alzheimer's Research UK.

References

[1] McKeith, IG; Mintzer, J; Aarsland, D; Burn, D; Chiu, H; Cohen-Mansfield, J; Dickson, D; Dubois, B; Duda, JE; Feldman, H; Gauthier, S; Halliday, G; Lawlor, B; Lippa, C; Lopez, OL; Machado, JC; O'Brien, J; Playfer, J and Reid, W, Dementia with Lewy bodies. *Neurology*, 2004 3, 19-28.

[2] McKeith, IG; Galasko, D; Kosaka, K; Perry, EK; Dickson, DW; Hansen, LA; Salmon, DP; Lowe, J; Mirra, SS; Byrne, EJ; Lennox, G; Quinn, NP; Edwardson, JA; Ince, PG; Bergeron, C; Burns, A; Miller, BL; Lovestone, S; Collerton, D; Jansen, ENH; Ballard, C; de Vos, RAI; Wilcock, GK; Jellinger, KA and Perry, RH, Consensus guidelines for the clinical and pathologic diagnosis of dementia with Lewy bodies (DLB): report of the consortium on DLB international workshop. *Neurology*, 1996 47, 1113-1124.

[3] Goedert, M, α-Synuclein and neurodegenerative diseases. *Nature Reviews Neuroscience*, 2001 2, 492-501.

[4] Spillantini, LC; Schmidt, ML; Lee, VMY; Trojanowski, JQ; Jakes, R and Goedert, M, a-Synuclein in Lewy bodies. *Nature*, 1997 388, 839-840.

[5] Conway, KA; Harper, JD and Lansbury, PTJ, Fibrils formed in vitro from α-synuclein and two mutant forms linked to Parkinson's disease are typical amyloid. *Biochemistry*, 2000 39, 2552-2563.

[6] Stocchi, F, Prevention and treatment of motor fluctuations. *Parkinsonism and Related Disorders*, 2003 9, S73-S81.

[7] McKeith, I; Del Ser, T; Spano, P; Emre, M; Wesnes, K; Anand, R; Cicin-Sain, A; Ferrara, R and Spiegel, R, Efficacy of rivastigmine in dementia with Lewy bodies: a randomised, double-blind, placebo-controlled international study. *Lancet*, 2000 356, 2031-2036.

[8] Han, H; Weinreb, PH and Lansbury, PTJ, The core Alzheimer's peptide NAC forms amyloid fibrils which seed and are seeded by β-amyloid: is NAC a common trigger or target in neurodegenerative disease? *Chemistry and Biology*, 1995 2, 163-169.

[9] El-Agnaf, OMA; Paleologou, KE; Greer, B; Abogrein, AM; King, JE; Salem, SA; Fullwood, NJ; Benson, FE; Hewitt, R; Ford, KJ; Martin, FL; Harriott, P; Cookson, MR and Allsop, D, A strategy for designing inhibitors of α-synuclein aggregation and toxicity as a novel treatment for Parkinson's disease and related disorders. *FASEB Journal*, 2004 18, 1315-1349.

[10] Du, H; Tang, L; Luo, X; Li, H; Hu, J; Zhou, J and Hu, H, A peptide motif consisting of glycine, alanine, and valine is required for the fibrillisation and cytoxicity of human α-synuclein. *Biochemistry*, 2003 42, 8870-8878.

[11] Bodles, AM; Guthrie, DJS; Greer, B and Irvine, GB, Identification of the region of non-Abeta component (NAC) of Alzheimer's disease amyloid responsible for its aggregation and toxicity. *Journal of Neurochemistry*, 2001 78, 384-395.

[12] Giasson, BI; Murray, IVJ; Trojanowski, JQ and Lee, VMY, A hydrophobic stretch of 12 amino acid residues in the middle of α-synuclein is essential for filament assembly. *Journal of Biological Chemistry*, 2001 276, 2380-2386.

[13] El-Agnaf, OMA and Irvine, GB, Aggregation and neurotoxicity of a-synuclein and related peptides. *Biochemical Society Transactions*, 2002 30, 559-565.

[14] Madine, J; Doig, AJ and Middleton, DA, Design of an N-methylated peptide inhibitor of α-synuclein aggregation guided by solid-state NMR. *Journal of American Chemical Society*, 2008 130, 7873-7881.

[15] Heise, H; Hoyer, W; Becker, S; Andronesi, OC; Riedel, D and Baldus, M, Molecular-level secondary structure, polymorphism, and dynamics of full-length α-synuclein fibrils studied by solid-state NMR. *Proceedings of National Academy of Science, USA*, 2005 102, 15871-15876.

[16] Koo, HJ; Choi, MY and Im, H, Aggregation-defective alpha-synuclein mutants inhibit the fibrillation of Parkinson's disease-linked alpha-synuclein variants. *Biochemical and Biophysical Research Communications*, 2009 386, 165-169.

[17] Koo, HJ; Lee, HJ and Im, H, Sequence determinants regulating fibrillation of human alpha-synuclein. *Biochemical and Biophysical Research Communications*, 2008 368, 772-778.

[18] Sawaya, MR; Sambashivan, S; Nelson, R; Ivanova, MI; Sievers, SA; Apostol, MI; Thompson, MJ; Balbirnie, M; Wiltzius, JJW; McFarlane, HT; Madsen, AO; Riekel, C and Eisenberg, D, Atomic structures of amyloid cross-β spines reveal varied steric zippers. *Nature*, 2007 447, 453-457.

[19] Tzotzos, S and Doig, AJ, Amyloidogenic sequences in native protein structures. *Protein Science*, 2010 19, 327-348.

[20] Parrini, C; Taddei, N; Ramazzotti, M; Degl'Innocenti, D; Ramponi, G; Dobson, CM and Chiti, F, Glycine residues appear to be evolutionarily conserved for their ability to inhibit aggregation. *Structure*, 2005 13, 1143-1151.

[21] Der-Sarkissian, A; Jao, CC; Chen, J and Langen, R, Structural organization of α-synuclein fibrils studied by site-directed spin labelling. *Journal of Biological Chemistry*, 2003 278, 37530-37535.

[22] El Agnaf, OM; Jakes, R; Curran, MD; Middleton, D; Ingenito, R; Bianchi, E; Pessi, A; Neill, D and Wallace, A, Aggregates from mutant and wild-type α-synuclein proteins and NAC peptide induce apoptotic cell death in human neuroblastome cells by formation of β-sheet and amyloid-like filaments. *FEBS Letters*, 1998 440, 71-75.

[23] Hughes, E; Burke, RM and Doig, AJ, Inhibition of toxicity in the β-amyloid peptide fragment Aβ(25-35) using N-methylated derivatives. *Journal of Biological Chemistry*, 2000 275, 25109-25115.

[24] Bodles, AM; El Agnaf, OMA; Greer, B; Guthrie, DJS and Irvine, GB, Inhibition of fibril formation and toxicity of a fragment of α-synuclein by an N-methylated peptide analogue. *Neuroscience Letters*, 2004 359, 89-93.

[25] Madine, J; Wang, X; Brown, DR and Middleton, DA, Evaluation of beta-Alanine- and GABA-Substituted Peptides as Inhibitors of Disease-Linked Protein Aggregation. *ChemBioChem*, 2009 10, 1982-1987.

[26] Soto, C; Kindy, MS; Baumann, M and Frangione, B, Inhibition of amyloidosis by peptides that prevent β-sheet conformation. *Biochemical and Biophysical Research Commumications*, 1996 226, 672-680.

[27] Findeis, MA; Musso, GM; Arico-Muendel, CC; Benjamin, HW; Hundal, AM; Lee, J-J; Chin, J; Kelley, M; Wakefield, J; Hayward, NJ and Molineaux, SM, Modified-peptide inhibitors of amyloid β-peptide polymerization. *Biochemistry*, 1999 38, 6791-6800.

[28] Pallitto, MM; Ghanta, J; Heinzelman, P; Kiessling, LL and Murphy, RM, Recognition sequence design for peptidyl modulators of b-amyloid aggregation and toxicity. *Biochemistry*, 1999 38, 3570-3578.

[29] Adessi, C; Frossard, M-J; Boissard, C; Fraga, S; Bieler, S; Ruckle, T; Vilbois, F; Robinson, SM; Mutter, M; Banks, WA and Soto, C, Pharmacological profiles of peptide drug candidates for the treatment of Alzheimer's disease. *Journal of Biological Chemistry*, 2003 278, 13905-13911.

[30] Permanne, B; Adessi, C; Saborio, GP; Fraga, S; Frossard, M-J; Dorpe, JV; Dewachter, I; Banks, WA; Leuven, FV and Soto, C, Reduction of amyloid load and cerebral damage in a transgenic mouse model of Alzheimer's disease by treatment with a β-sheet breaker peptide. *The FASEB Journal*, 2002.

[31] Rosen, H, A modified ninhydrin colorimetric analysis of amino acids. *Archives Biochemistry and Biophysics*, 1957 67, 1-15.

[32] Rivers, RC; Kumita, JR; Tartaglia, GG; Dedmon, MM; Pawar, A; Vendruscolo, M; Dobson, CM and Christodoulou, J, Molecular determinants of the aggregation behavior of alpha- and beta-synuclein. *Protein Science*, 2008 17, 887-898.

[33] Pawar, AP; DuBay, KF; Zurdo, J; Chiti, F; Vendruscolo, M and Dobson, CM, Prediction of "aggregation-prone" and "aggregation-susceptible" regions in proteins associated with neurodegenerative diseases. *Journal of Molecular Biology*, 2005 350, 379-392.

[34] Yagi, H; Takeuchi, H; Ogawa, S; Ito, N; Sakane, I; Hongo, K; Mizobata, T; Goto, Y and Kawata, Y, Isolation of short peptide fragments from alpha-synuclein fibril core identifies a residue important for fibril nucleation: A possible implication for diagnostic applications. *Biochimica Et Biophysica Acta-Proteins and Proteomics*, 2010 1804, 2077-2087.

In: Amyloids
Editors: I. P. Halcheck et al.
ISBN: 978-1-62100-538-4

Chapter 9

A CRITICAL STUDY OF THE ISOSTERES OF 10TH CYSTEINE RESIDUE OF TRANSTHYRETIN INDUCED WITHCYSTEINE-*S*-SULFONATE BY ULTRA-HIGH RESOLUTION MASS SPECTROMETRY

Toyofumi Nakanishi*[*1], Noriyuki Iwasaki[2], Kazunori Saito[2], Shinnichi Wada[3,4], Masanori Yoshioka[1], Takashi Nirasawa[2] and Takayuki Takubo[1]

[1]Department of Clinical and Laboratory Medicine, Osaka Medical College
[2]Technical Support Center, Japan Bruker Daltonics
[3]Central Clinical Laboratory, Osaka Medical College Hospital
[4]Department of Functional Diagnostic Science,
Osaka University Graduate School of Medicine

ABSTRACT

The formation of unusual modifications on the sulfhydryl residue of 10th cysteine of transthyretin (TTR) caused to induce the amyloidogenicity. Based on the mass spectrometric analysis of patients' plasma with TTR related amyloidosis and other similar diseases and the profiles of TTR itself and TTR-related octapeptide under the alkaline condition at 37 degrees, we illustrated the schema of reaction steps for the formation of TTR-related amyloid fibril but there are still unknown events to well-understand the acquisition of the amyloidogenicity.

We investigated intermediate reactants *via* an unusual radical reaction following *S*-sulfonation to clarify precursors of proto-amyloidogenic TTR by ultra-high resolution time-of-flight MS (UHR-TOFMS). These reactants were not only mixed disulfide TTRs but also three abnormal structural products, such as dehydroalanine and glycine

* Corresponding author: Toyofumi Nakanishi, 2-7 Daigaku-cho, Takatsuki-city, 569-8686, Osaka, Japan. e-mail : nakanisi@poh.osaka-med.ac.jp

degradated from 10th cysteine residue of TTR with the reaction of β-elimination and threonine/ homoserine was also degraded on 13th methionine using the TTR modified with cysteine-*S*-sulfonate *in vitro*. Especially the dehydroalanine, α-acryl amino acid, having a high-reactivity, could be randomly polymerized between TTR itself and co-existence other molecules with a covalent bridge.

In conclusion, we revealed the induction of the dehydroalanine moiety from 10th cysteine residue of TTR irreversibly and progressively must be an essential molecule to form TTR-related amyloid fibril.

Keywords: radical reaction, transthyretin, dehydrated amino acid, polymerization, amyloidogenicity.

1. Introduction

Mechanisms of amyloid fibril formations have been investigated in various types of amyloidosis, such as Alzheimer's disease, Purion disease, Parkinson's disease and senile systemic amyloidosis (SSA) [1-7]. Transthyretin (TTR), one of the known amyloidogenic proteins has been shown to be deposited in the heart, lung and aorta of patients with SSA and also pathogenic in its mutated form in familial amyloid polyneuropathy (FAP). Recently, it has been shown that a key step in amyloid fibril formation is the dissociation of tetrameric TTR into monomers and that mutated forms of TTR lead to destabilization of the tetramer *in vitro* [8,9] . Furthermore, partial un-folding and reassembly of the monomer structure is necessary for fibril aggregation [10-14]. However, not only the initiating and universal step of the fibril deposition of TTR as well as the other amyloidogenic proteins but also the abnormal structure with induction of random polymerization is still undetermined both *in vivo* and *in vitro*.

Over the past decade, we have reported that the TTR molecule is altered at the thiol residue on the 10th cysteine within its amino acid sequence not only in SSA and FAP [15-18] but also in other conditions such as molybdenum cofactor deficiency (MoCD) [19-20] and sulfite oxidase deficiency (SOD) [21] and demonstrated the same structural modifications of the TTR molecule using an octapeptide (SK^{10}CPLMVK) derived from TTR incubated in alkaline conditions at 37 degrees by electrospray ionization and matrix-assisted laser desorption mass spectrometry [22]. Moreover, we previously demonstrated that the *S*-sulfonation of TTR was an important trigger step to form TTR-related amyloid fibrils using modified TTR with a cysteine-*S*-sulfonate *in vitro* and also illustrated the scheme of possible major steps in modification of the thiol group at the 10th cysteinyl residue of TTR [23], but there would be still unknown events to well-understand the mechanism of the formation of TTR-related amyloid fibril and support our hypothetical schema of radical reaction for the induction of amyloidogenicity of TTR molecule.

Until now, there are many reports which correlated between cross-linked amino acids and denaturation of proteins. For example, it is well known that wool, its main component was a keratin, is chemically converted to be the unusual structure of the composited cysteine residue of the keratin by the treatment of alkaline pH and heating [24]. Moreover, foods and antimicrobial proteins spontaneously formed cross-linked amino acids, such as lanthionine with heating under an alkali condition, and the presence of the residue along a protein chain decreases digestibility in rodents [25]. The cross-linked amino acid of the wool, a lanthionine, was formed by analogous nucleophilic addition reaction of the hydrosulfinyl residue of cysteine to dehydroalanine. Tubulin subunits were chemically formed a non-disulfide crosslinking, the formation of lysinoalanine, with manners of pH and temperature dependence [26]. Collagen in rat skin was also produced an oxidative cross-linkage to form a dehydro hydrolysinonorleucine in food deprivation [27]. However, there are a few reports correlation between these cross-linking and amyloidogenicity *in vivo* and *in vitro*.

In this present study, we identified ultra-high accuracy molecular masses of precursor products of dehydroalanine moiety *via* a radical reaction, which we had previously illustrated the scheme of major steps in modification of the thiol group at the 10th cysteinyl residue of TTR, such as dehydroalanine and glycine, and threonine/ homoserine replaced on 13th methionine using the modified TTR *in vitro* by the UHR-TOFMS. Our results concluded that the formation of an α-amino acrylic acid, such as a dehydroalanine moiety, must be a trigger and critical product for the inductions of not only TTR-related amyloid fibril but also the other proteins *in vitro*.

2. Material and Methods

2.1. Materials

Acetonitrile, formic acid and ammonium bicarbonate were purchased from Nakalai tesque (Kyoto, Japan). Trifluoroacetic acid and Congo red were from Sigma-Aldrich Japan (Tokyo, Japan). TTR standard was purchased from LeeBio Co.Ltd. (Lot No.0P0-4, Tokyo, Japan). We synthesized a cysteine-*S*-sulfonate by previously described methods [28].

2.2. Test Samples of Modified TTR Solution Treated with the Cysteine-*S*-Sulfonate

Fifty μL of standard TTR solution (2 mg/mL) was mixed with 100 mM ammonium bicarbonate (pH=8.3 and 4.0) and then 50 μL of cysteine -*S*-sulfonate (CSS) solution at a concentration of 13.3 mg/mL. The control was standard alkaline TTR without sulfonate. The mixtures were incubated at 37 degrees for one week. The reference was a freshed TTR standard.

2.3. Ultra High-Resolution Time-of-Flight Mass Spectrometry (UHR-TOFMS)

UHR-TOFMS analyses were performed on a maXis (Bruker Daltonics, Bremen, Germany) equipped with electrospray source (positive mode) and connected with an UltiMate 3000 (Dionex Co.Ltd, Tokyo, Japan) equipped a Proteonavi column (2.0 x 150mm; Shiseido Co.Ltd; Tokyo, Japan). The elution of the samples was performed with a linear gradient from 20 to 90 %B for 44 min. The solvent A was 0.1 % formic acid (FA) and solvent B was acetonitrile added 0.1 % FA. The measurement conditions of UHR-TOFMS were as follows; the scan range was m/z 50-3000, capillary voltage was 4.5 kV and nebulizer gas pressure was 2 bars, dry gas flow was 8.0 L/min and dry temperature was 200 degrees. Mass calibration was carried out using peptide mixture standard I.

In this study, we selected eight kinds of deconvoluted molecular masses of the modified TTRs to monitor the radical reactions of β-eliminations with Maximum Entropy method (Bruker Daltonics).

3. RESULTS

Figure 1 shows the mass chromatogram of heptavalent ions derived from six modified and an unknown TTRs incubated with a CSS at alkaline condition for a week by the UHR-TOFMS analysis. As shown in Figure 2, the profile revealed free (13760.94969), *S*-sulfinylated (+48 Da), *S*-thionylated (+32 Da), glycine degraded with cysteine residue (-46 Da) and threonine or homoserine (-30 Da) degraded on 13th methionine residue of TTR. Also, we identified a cysteinylated (+119 Da), cysteinyl-glycylated (+176Da), and glutathionylated (+307Da) without the CSS (data not shown).

Moreover, we identified dehydroalanine (-34 Da), an α-acryl amino acid, degraded with 10th cysteine residue using the modified TTR sample under the alkaline at 37 degrees for a week with the CSS. The measured molecular mass of the modified TTR degraded cysteine to dehydroalanine was a 13725.90120 Da and the accuracy was only 1.4 ppm differed from the theoretical molecular mass. The profiles of the TTR modified with/without cysteine-S-sulfonate *in vitro* were free, conjugations with cysteine, cysteinyl-glycine and glutathione, and *S*-sulfonated, *S*-sulfinylated, *S*-thionylated as same as those of TTR standard itself.

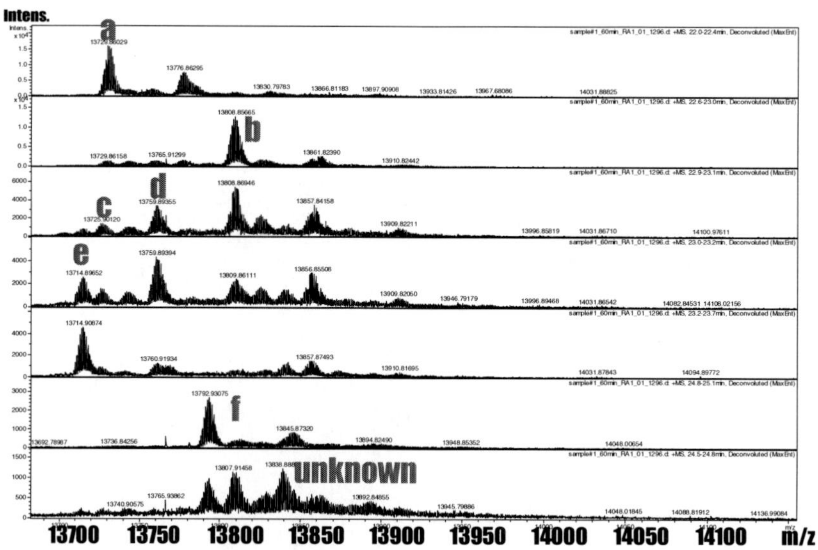

Figure 1. The base peak chromatogram and heptavalent ion mass spectra of modified TTR standard incubated with a cysteine-*S*-sulfonate at alkaline condition for a week.

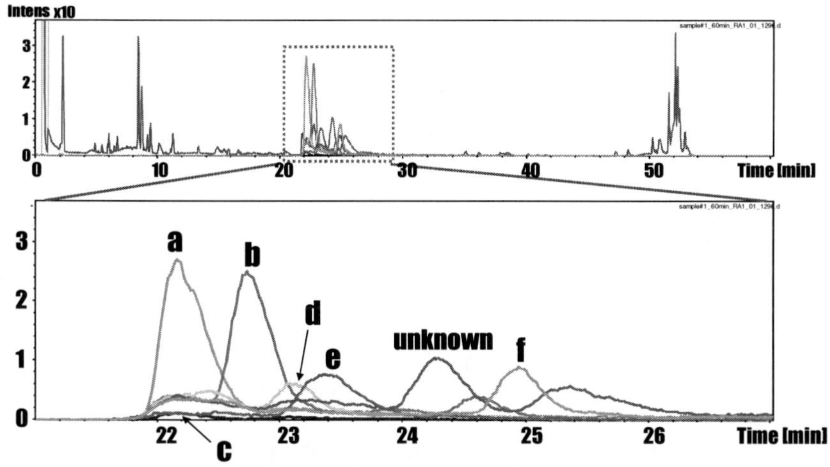

Figure 2. The deconvoluted mass spectra of the 6 modified TTRs and an unmodified TTR standard after incubation at alkaline condition with cysteine-*S*-sulfonate.

The transformed mass of an unknown peak, 13838.888, corresponded to an *S*-sulfonate TTR. Also, we identified a threonine/homoserine (-30 Da) was degraded on 13th methionine residue of TTR. The both analytes were positive with the Congo-red staining (the data not shown).

Table I showed the results of the measured molecular masses of TTRs by the UHR-TOFMS analysis and the corresponding to their theoretical molecular masses of the modified TTRs.

Table I. The list of the TTR isosteres induced with CSS under the alkaline pH

	TTR isosteres	Heptavalent ion		Δ (ppm)	Simulated Molecular Formula
		Measured	Theoretical		
a	Met->Thr/homoSer	1962.57054	1952.56696	1.8	$C_{616}H_{948}N_{160}O_{194}S$
b	*S*-O3	1973.71140	1973.70656	2.5	$C_{617}H_{950}N_{160}O_{193}S_2(O_3)$
c	Dehydroalanine	1961.99896	1961.99626	1.4	$C_{617}H_{948}N_{160}O_{193}S$
d	Free	1966.85970	1966.85160	4.1	$C_{617}H_{950}N_{160}O_{193}S_2$
e	Cys->Gly	1960.29041	1960.28198	4.3	$C_{616}H_{948}N_{160}O_{193}S$
f	*S*-SH	1971.43767	1971.41899	9.5	$C_{617}H_{950}N_{160}O_{193}S_2(S)$

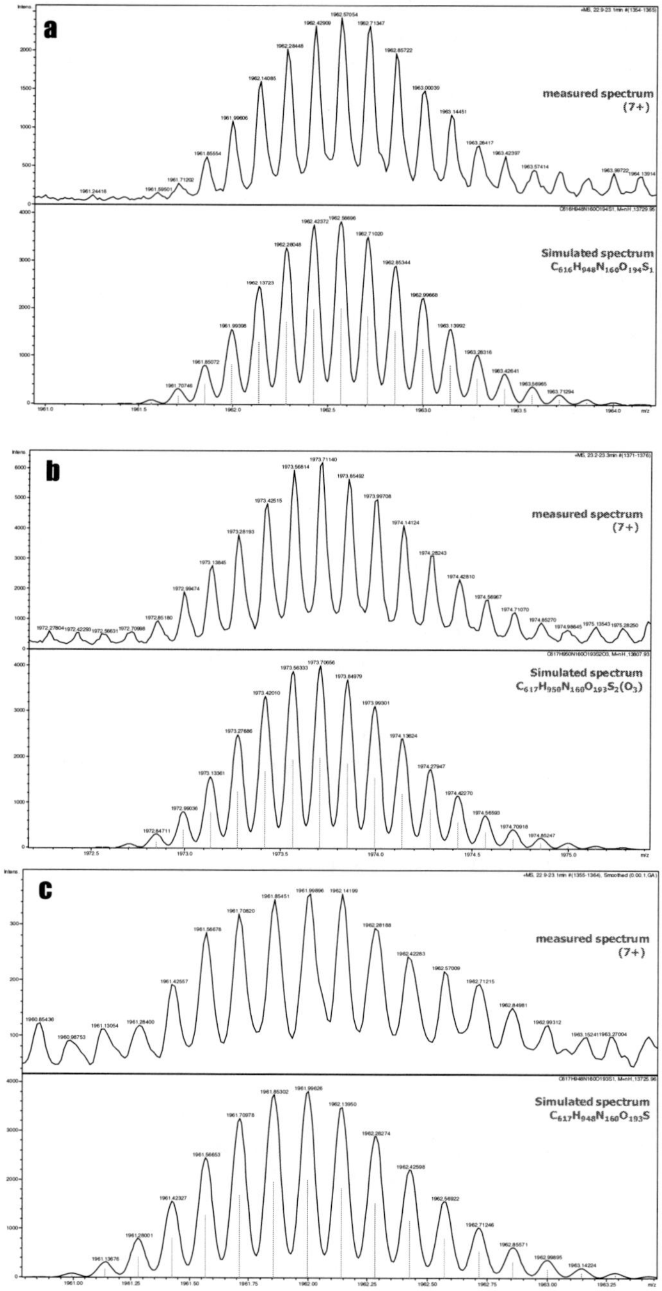

Figure continued on next page.

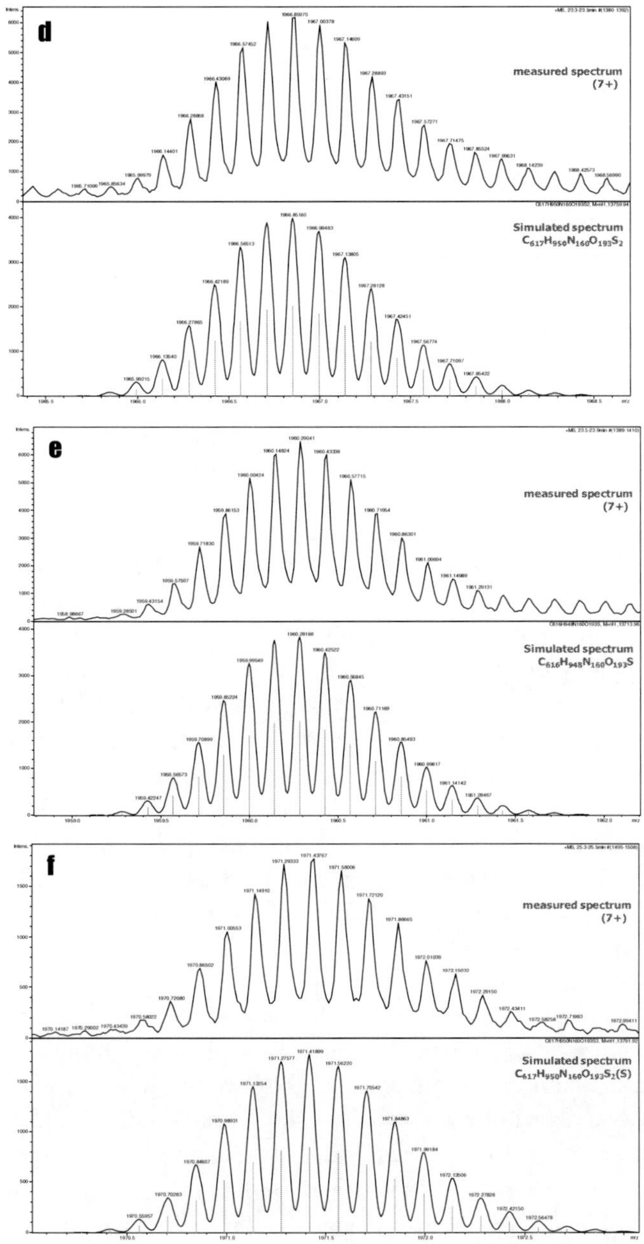

Figure 3. The heptavalent ion mass spectra of 5 identified modified TTRs, unmodified TTR and an unknown TTR.

4. Discussion

In this present study, we identified ultra accurate molecular masses of *S*-sulfonation reactants to form the proto-dehydroalanine moiety *via* the radical reaction, which we had previously presented the scheme of major steps in an unusual modification of the sulfhydryl residue at the 10th cysteinyl residue to induce the TTR-related amyloid fibril, such as dehydroalanine and glycine, and threonine/ homoserine degraded on 13th methionine using the modified TTR *in vitro* by the UHR-TOFMS. Although we previously demonstrated that this *S*-sulfonation of TTR was an important trigger step to form TTR-related amyloid fibril using modified TTR with a cysteine-*S*-sulfonate *in vitro*, but there are still unknown events to well-understand the formation of TTR-related amyloid fibril.

Generally, the sulfhydryl residue of cysteine is in turn reversibly converted to a sulfenyl residue which progressively and irreversibly changes to a sulfinyl and sulfonyl under oxidative conditions *in vivo* and *in vitro*. Also, the hydroxyl residues of serine, threonine and tyrosine are chemically modified to form the same α-acryl amino acid. Moreover the β-elimination following the irreversible *S*-sulfonation induced the dehydroalanine residue on the 10th cysteine of the TTR, which randomly could be polymerized with neighboring co-existence molecules. The dehydroalanine residue, with a very high reactivity, could be bound protein monomer with a carbon-carbon bond and so acquire resistance to proteinase. We previously confirmed that the *S*-modified TTR *in vivo* and *in vitro* formed dimer, tetramer and oligomers even though it was under a high reduction and a high denature conditions using an electrophoresis/western blot analysis. In the present study, we identified not only free, cysteinylated, cysteinyl-glycylated, glutathionylated, *S*-sulfonated, *S*-sulfinylated, *S*-thionylated, but also dehydroalanine and glycine degraded with 10th cysteine residue of TTR. The present data additionally supported our hypothesis that the dehydroalanine moiety of modified TTR monomer could be polymerized TTR itself or TTR-protein oligomers with covalent bridge *in vivo* and *in vitro* due to the formation of an α-amino acrylic acid *via* the radical reaction.

Lindley H. et.al.[24] reported that wool, its main component was a keratin, is chemically converted to be the unusual structure of the composited cysteine residue of the keratin by the treatment of alkaline pH and boiling. The cross-linked amino acid of the wool, a lanthionine, was easily formed by analogous nucleophilic addition reaction of the hydrosulfinyl residue of cysteine to dehydroalanine.

Kuipers O.P. et.al.[29] presented that antimicrobial proteins, such as nisin and subtilin spontaneously were formed cross-linked amino acids, such as lanthionine and lysinoalanine with heating under an alkali condition, and the presence of the abnormal amino acid along a protein chain decreases digestibility in rodents. The formation of the cross-linking essentially needed the presence of sulfhydryl residue of cysteine, N-acetyl-cysteine, glutathione, dephosphorylation of O-phosphoryl esters, and acylation of ε-amino group of lysine side chains. The antimicrobial activities closely related to their own structures.

Correia J.J. et.al [26] reported that tubulin subunits were chemically formed a non-disulfide crosslinking lysinoalanine with manners of pH and temperature dependence and the denaturation process partially involves an irreversible aggregation. The aggregated tubulin resisted to β-mercaptoethanol. The lysinoalanine formation is a two-step process; dehydroalanine formed by β-elimination at serine and then ε-amino residue of lysine were bound to be the crosslinking.

Deyl Z. et.al.[27] indicated that Collagen in rat skin was also produced an oxidative cross-linkage to form a dehydro hydrolysinonorleucine in food deprivation for a long period and increased in γ-chain polymerization. The increase of the polymerized skin collagen in rat caused to induce this unusual amino acid from reutilization of soluble type under the oxidative stress such as food restriction.

Rangachari V. et al. [30] recently reported that two or three dehydroalanine moieties containing penta-peptide fragment corresponding to residue 16-20 of Aβ inhibit to form Aβ fibril *in vitro*. The inhibition of the Aβ aggregation with the penta-peptides was due to bind an identical segment of Aβ peptide with a non-covalent bridge according to a stereo-specific manner. The dehydroalanine-isoleucine-dehydroalanine motif in the penta-peptides would disrupt the nucleation of Aβ aggregation by imposing a linear extended conformation to a segment containing residues of Aβ peptides. The α, β-dehydro-amino acid, such as dehydroalanine, have been observed to form β-sheet and acquire the protease resistance so that they were just characteristic of amyloid fibril. Our present studies demonstrated that the dehydroalanine moiety was generated from the 10th cysteine residue of *S*-sulfonated TTR *via* the radical reaction following the *S*-sulfonation with the cysteine-*S*- sulfonate as an accelerator and then the dehydroalanine moiety randomly polymerized with another TTR or co-existed molecules.

In conclusion, we identified ultra-high accuracy molecular masses of intermediate TTR products with a radical reaction *via* major steps in modification of the thiol group at the 10th cysteinyl residue of TTR, such as dehydroalanine and glycine, and threonine/ homoserine replaced on 13th methionine using the modified TTR *in vitro* by the UHR-TOFMS. It is very important to identify a trigger molecule, such as the dehydroalanine, for understanding of the etiology of amyloidogenicity and the structure of proto-drug of anti-amyloidosis.

ACKNOWLEDGMENTS

We thank Teruo.Ueno (Center of Research Laboratory, Osaka Medical College) for technical assistance with fluorescence microscopy. We are also grateful to late Emeritus Prof. Akira.Shimizu for helpful discussions and George Bouras for assistance with scientific communication. This work was supported by a Grant-in-Aid for Challenging Exploratory Research of the Japan Society for the Promotion of Science (No. 20659093).

REFERENCES

[1] P. Westermark, Amyloid: toward terminology clarification. Report from the Nomenclature Committee of the International Society of Amyloidosis, *Amyloid* 12 (2005) 1-4.

[2] D.J. Selkoe, Cell biology of the amyloid beta-protein precursor and the mechanism of Alzheimer's disease, *Annul. Rev. Cell. Biol.* 10 (1994) 373-403.

[3] T. Iwatsubo, A. Odaka, N. Suzuki, H. Mizusawa, N. Nukina, Y. Ihara, Visualization of A beta 42(43) and A beta 40 in senile plaques with end-specific A beta monoclonals: evidence that an initially deposited species is A beta 42(43), *Neuron* 13 (1994) 45-53.

[4] M.H. Polymeropoulos, C. Laveran, E. Lerov, S.E. Ide, A. Dehejia, A. Dutra, B. Pike, H. Root, J. Rubenstein, R. Bover, E.S. Stenroos, S. Chandrasekharappa, A. Athanassiadou, T. Papapetropoulos, W.G. Johnson, A.M. Lazzarini, R.C. Duvoisin, G. DiIorio, L.I.

Golbe, R.L, Nussbaum, Mutation in the alpha-synuclein gene identified in families with Parkinson's disease, *Science* 276 (1997) 2045-2047.

[5] P.D. Gorevic, F.C. Prelli, J. Wright, M. Pras, B. Frangione, Systemic senile amyloidosis. Identification of a new prealbumin (transthyretin) variant in cardiac tissue: immunologic and biochemical similarity to one form of familial amyloidotic polyneuropathy, *J. Clin. Invest.* 83 (1989) 836-843.

[6] F. Rahimi, A. Shanmugam, G. Bitan, Structure-function relationships of pre-fibrillar protein assemblies in Alzheimer's disease and related disorders, *Curr. Alzheimer Res.* 5 (2008) 319-341.

[7] K.P. Wu, S. Kim, D.A. Fela, J. Baum, Characterization of conformational and dynamic properties of natively unfolded human and mouse alpha-synuclein ensembles by NMR: implication for aggregation, *J. Mol. Biol.* 378 (2008) 1104-1115.

[8] J.S. Kingsburg, T.M. Laue, E.S. Klimtchuk, R. Theberge, C.E. Costello, L.H. Connors, The modulation of transthyretin tetramer stability by cysteine 10 adducts and the drug diflunisal. Direct analysis by fluorescence-detected analytical ultracentrifugation, *J. Biol. Chem.* 283 (2008) 11887-11896.

[9] P. Westermark, J. Bergstrom, A. Solomon, C. Murphy, K. Sletten, Transthyretin-derived senile systemic amyloidosis: clinicopathologic and structural considerations, *Amyloid* 10 Suppl 1 (2003) 48-54.

[10] C.M. Dobson, Protein folding and misfolding, Nature 426 (2003) 884-90.

[11] T. Sekijima, R.L. Wiseman, J. Matteson, P. Hammarsrom, S.R. Miller, A.R. Sawkar, W.E. Balch, J.W. Kelly, The biological and chemical basis for tissue-selective amyloid disease, *Cell* 121 (2005) 73-85.

[12] Z. Lai, W. Colon, J.W. Kelly, The acid-mediated denaturation pathway of transthyretin yields a conformational intermediate that can self-assemble into amyloid, *Biochemistry* 35 (1996) 6470-6482.

[13] P. Hammarstrom, Y. Sekijima, J.T. White, R.L. Wiseman, A. Lim, C.E. Costello, K. Altland, F. Gaezuly, H. Budka, J.W. Kelly, D18G transthyretin is monomeric, aggregation prone, and not detectable in plasma and cerebrospinal fluid: a prescription for central nervous system amyloidosis? *Biochemistry* 42 (2003) 6656-6663.

[14] G.J. Miroy, Z. Lai, H.A. Lashuel, S.A. Peterson, C. Strang, J.W. Kelly, Inhibiting transthyretin amyloid fibril formation via protein stabilization, *Proc. Natl.. Acad .Sci. USA*. 93 (1996) 15051-15056.

[15] M. Kishikawa, T.Nakanishi, A. Miyazaki, A. Shimizu, M. Nakazato, K. Kangawa, H. Matsuo, Simple detection of abnormal serum transthyretin from patients with familial amyloidotic polyneuropathy by high-performance liquid chromatography/electrospray ionization mass spectrometry using material precipitated with specific antiserum, *J. Mass Spectrom.* 31 (1996) 112-114.

[16] M.Kishikawa, T. Nakanishi, A. Miyazaki, M. Hatanaka, A. Shimizu, S. Tamoto,N. Ohsawa, H. Hayashi, M.Kanai, A new nonamyloid transthyretin variant, G101S, detected by electrospray ionization/mass spectrometry, Mutations in brief no.201. Online. *Hum. Mutat.* 12 (1998) 363.

[17] M. Kishikawa, T. Nakanishi, A. Miyazaki, A. Shimizu, A simple and reliable method of detecting variant transthyretins by multidimensional liquid chromatography coupled to electrospray ionization mass spectrometry, *Amyloid* 6 (1999) 48-53.

[18] M.Kishikawa, T. Nakanishi, A. Miyazaki, A. Shimizu, Enhanced amyloidogenesity of sulfonated transthyretin in vitro, a hypothetical etiology of senile amyloidosis, *Amyloid* 6 (1999) 183-186.

[19] M. Kishikawa, T. Nakanishi, A. Shimizu, M. Yoshino M, Detection by mass spectrometry of highly increased amount of S-sulfonated transthyretin in serum from a patient with molybdenum cofactor deficiency, *Pediatr. Res.* 47 (2000) 492-494.

[20] M. Kishikawa, J.O. Sass, N. Sakura N, T. Nakanishi, Shimizu A, Yoshioka, The peak height ratio of S-sulfonated transthyretin and other oxidized isoforms as a marker for molybdenum cofactor deficiency, measured by electrospray ionization mass spectrometry, *Biochim. Biophys. Acta,* 1588 (2002) 135-138.

[21] J.O. Saas, T. Nakanishi, T. Sato, W. Sperl, A. Shimizu, S-homocysteinylation of transthyretin is detected in plasma and serum of humans with different types of hyperhomocysteinemia, *Biochem. Biophys. Res. Commun.* 310 (2003) 242-246.

[22] T.Nakanishi, T. Sato, S. Sakoda, M.Yoshioka, A. Shimizu, Modification of cysteine residue in transthyretin and a synthetic peptide: analyses by electrospray ionization mass spectrometry, *Biochim. Biophys. Acta* 1698 (2004) 45-53.

[23] T.Nakanishi, T. Sato, S. Sakoda, M.Yoshioka, A. Shimizu, S-sulfonation of transthyretin is an important trigger step in the formation of transthyretin-related amyloid fibril. *Biochim. Biophys. Acta* 1804 (2010) 1449-56.

[24] H.Lindley, H.Phillips, The action of alkalis on wool. *Biochem.* 39 (1944) 17-23.

[25] R.G. Walsh, A.S. Nashef, R.E. Feeney, Intramolecular cross-linking of proteins by formation of lysinoalanine or lanthionine. Modification of disulfides in ovomucoids. *Int. J. Pept. Protein Res.* 14 (1979) 290-9.

[26] J.J. Correia, L.D. Lipscomb, S. Lobert, Nondisulfide crosslinking and chemical cleavage of tubulin subunits: pH and temperature dependence. *Arch. Biochem. Biophys.*300(1993)105-14.

[27] Z. Deyl, K. Macek, The role of oxidative cross-link stabilization in increased collagen polymerization in food deprived rats. *Physiol. Bohemoslov.* 31(1982)537-42.

[28] I.H. Segel, M.J, Johnson, Synthesis and characterization of sodium cysteine-S-sulfonate monohydrate. *Anal. Biochem.* 5 (1963) 330-337.

[29] O.P. Kuipers, H.S. Rollema, W.M.G.J. Yap, H.J. Boot, R.J. Siezen, W.M. de Vos, Engineering dehydrated amino acid residues in the antimicrobial peptide nisin. *J. Biol.Chem.* 267 (1992)24340-46.

[30] V.Rangachari, Z.S.Davey, B.Healy, B.D.Moore, L.K.Sonoda, B.Cusack, G.M.Maharvi, A.H.Fauq, T.L.Rosenberry, *Biopolymers* 91(2009)456-65.

In: Amyloids
Editor: I. P. Halcheck et al.

ISBN 978-1-62100-538-4

Chapter 10

ATOMIC-RESOLUTION STRUCTURES OF PRION AGAAAAGA AMYLOID FIBRILS

Jiapu Zhang*
School of Sciences, Information Technology and Engineering,
University of Ballarat, Mount Helen,
Ballarat, Victoria 3353, Australia

Abstract

To the best of the authors knowledge, there is little structural data available on the AGAAAAGA palindrome in the hydrophobic region (113-120) of prion proteins due to the unstable, noncrystalline and insoluble nature of the amyloid fibril, although many experimental studies have shown that this region has amyloid fibril forming properties and plays an important role in prion diseases. In view of this, the present study is devoted to address this problem from computational approaches such as local optimization steepest descent, conjugate gradient, discrete gradient and Newton methods, global optimization simulated annealing and genetic algorithms, canonical dual optimization theory, and structural bioinformatics. The optimal atomic-resolution structures of prion AGAAAAGA amyloid fibils reported in this Chapter have a value to the scientific community in its drive to find treatments for prion diseases or at least be useful for the goals of medicinal chemistry.

Keywords: Amyloid Fibrils; Prion AGAAAAGA Palindrome; Atomic-resolution Structures.

1. Introduction

Prion diseases are invariably fatal and highly infectious neurodegenerative diseases that affect humans and animals. Prion diseases are amyloid fibril diseases. Prion amyloid fibrils are believed rich in β-sheet structure and contain a cross-β core. Many experimental works

*E-mail addresses: jiapu_zhang@hotmail.com, j.zhang@ballarat.edu.au. Phone: 61-423487360, 61-3-5327 9809.

[1, 2, 3, 6, 7, 8, 9, 11, 13] show that the hydrophobic region AGAAAAGA of prion proteins (113-120) plays an important role in the conversion of amyloid fibrils. PrP lacking / deleting the palindrome (PrP 113-120) neither converted to PrP^{Sc}(infectious prion) nor generated proteinase K-resistant PrP [2, 6, 11, 13]. Brown et al. [1, 3] pointed out that the AGAAAAGA peptide was found to be necessary (though not sufficient) for blocking the toxicity and amyloidogenicity of PrP 106-126. The peptide AGAA did not form fibrils but the peptide AGAAAAGA formed fibrils in both water and PBS [1]. Thus, the minimum sequence necessary for fibril formation should be AGAAA, GAAAA, AAAAG, AAAGA, AGAAAA, GAAAAG, AAAAGA, AGAAAAG, GAAAAGA or AGAAAAGA. According to Brown [1], AGAAAAGA is important for fibril formation and is an inhibitor of PrP^{Sc} neurotoxicity.

In theory, for the sake of clarity, we use a program in Zhang et al. [16] to confirm that prion AGAAAAGA (113-120) segment has an amyloid fibril forming property. The theoretical computation results are shown in Fig. 1, from which we can see that prion AGAAAAGA (113-120) region is clearly identified as the amyloid fibril formation region because the energy is less than the amyloid fibril formation threshold energy -26 [16].

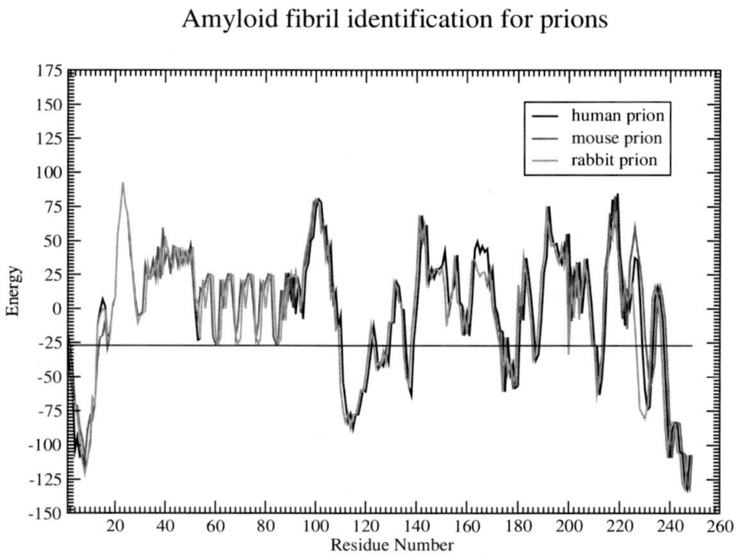

Figure 1. Prion AGAAAAGA (113-120) is clearly identified as fibril formation segment.

However, due to the unstable, noncrystalline and insoluble nature of the amyloid fibril, to date structural information on AGAAAAGA region (113-120) has been very limited. This region falls just within the N-terminal unstructured region PrP (1-123) of prion proteins. Traditional X-ray crystallography and nuclear magnetic resonance (NMR) spectroscopy experimental methods cannot be used to get its structural information. Under this background, computational approaches or introducing novel mathematical formulations and physical concepts into molecular biology can significantly stimulate the development of biological and medical science. The author has introduced novel mathematical global and local optimization computational approaches to produce the atomic-resolution structures of prion (113-120) AGAAAAGA amyloid fibrils, which are in Section 2. Section 3 summarizes this article.

2. Computational Optimization Approaches and Their AGAAAAGA Structural Results

In 2007, Sawaya et al. got a breakthrough finding: the atomic structures of all amyloid fibrils revealed steric zippers, with strong vdw interactions (LJ) between β-sheets and hydrogen bonds (HBs) to maintain the β-strands [12]. In this section, we will use suitable templates 2OMP.pdb (the LYQLEN peptide derived from human insulin residues 1318), 1YJP.pdb (the GNNQQNY peptide from the yeast prion protein Sup35), 3FVA.pdb (NNQNTF 173-178 segment from elk prion protein), 3NHC.pdb (GYMLGS segment 127-132 from human prion with M129), 3NVE.pdb (MMHFGN segment 138-143 from Syrian Hamster prion), 3NVF.pdb (IIHFGS segment 138-143 from human prion), 3NVG.pdb (MIHFGN segment 137-142 from mouse prion) and 3NVH.pdb (MIHFGND segment 137-143 from mouse prion) from the Protein Data Bank (http://www.rcsb.org/) to construct some amyloid fibril models for the prion AGAAAAGA region (113-120).

2.1. 2OMP, 1YJP Models

The author used the unmerge, mutate, and merge modules of Insight II (http://accelrys.com/) for 2OMP.pdb, 1YJP.pdb to build the 12-chain AGAAAA 2OMP-Model (MODEL01), 10-chain AGAAAAG 1YJP-Model (MODEL02), and 10-chain GAAAAGA 1YJP-Model (MODEL03) [14]. Then, using AMBER 10 [4], the MODEL01-03 were refined by the hybrid of steepest descent (SD) and conjugate gradient (CG) – simulated annealing (SA) – SDCG methods, where SA phase made the MODEL01-03 reach sufficient equilibration and stability. MODEL1 (Fig. 2) belongs to models of Class 7 (β-strand antiparallel, face=back, upup) of [12], and MODEL02-03 (Fig. 2) belong to models of Class 1 (β-strand parallel, face-to-face, upup) of [12].

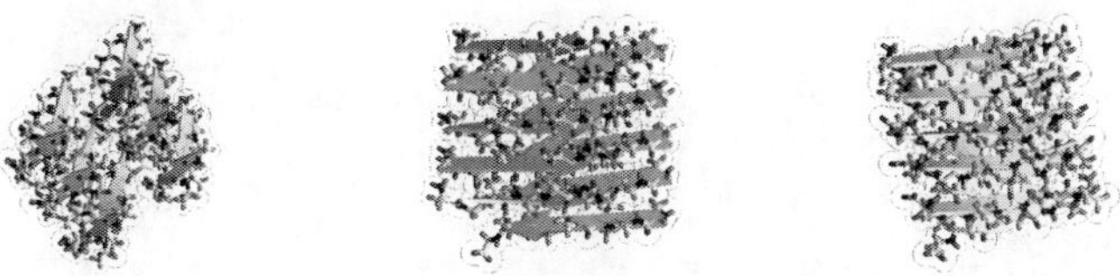

Figure 2. MODEL01 - MODEL03 for prion (113-120) AGAAAAGA amyloid fibrils.

Sawaya et al. [12] proposed eight classes of steric zipper structures for peptide segments of fibril forming proteins. For each Class, Zhang [14] constructed the molecular structures for the hydrophobic region AGAAAAGA palindrome of prion proteins (113120). Besides the above successful MODEL01-03, the unsuccessful molecular modeling experiences are: (1) For Class 1, based on the NNQQNY peptide from yeast prion protein Sup35 (1YJO.pdb), a hexamer model with six AAAAGA chains can be constructed by SDCG but cannot pass SA; (2) For Class 2 (i.e. β-strand parallel, face-to-back, up-up), a tetramer model with four AGAAAA chains can be constructed basing on the SNQNNF peptide of human prion 170175 (2OL9.pdb) by SDCG but cannot pass SA; (3) For other Classes (yeast Sup35 GNNQQNY form 2, 2OMM.pdb; yeast Sup35 NNQQ form 1, 2ONX.pdb; yeast Sup35 NNQQ form 2, 2OLX.pdb; human insulin VEALYL, 2OMQ.pdb; human Tau protein VQIVYK, 2ON9.pdb; human Aβ GGVVIA, 2ONV.pdb; human Aβ MVGGVV form 1, 2ONA.pdb; human Aβ MVGGVV form 2, 2OKZ.pdb; bovine RNase SSTSAA, 2ONW.pdb) in Sawaya et al. [12], the β-sheet structure of prion AGAAAAGA palindrome cannot pass the SDCG phase.

2.2. 3FVA Models

Instead of using Insight II, Zhang et al. used the mutate module of the free package Swiss-PdbViewer (SPDBV Version 4.01) (http://spdbv.vital-it.ch) and the hybrid discrete gradient (DG) simulated annealing method (i.e. DGSA) to build the prion AGAAAAGA amyloid fibril models [15]. The models were built based on the template 3FVA.pdb of NNQNTF segment from elk prion protein (173-178). After the models were built, the refinements were done completely same as [14]. A six chains AGAAAA model could not successfully pass SA of Amber 10. However, two prion AGAAAAGA palindrome amyloid fibril models (Fig. 3) - a six chains GAAAAG model (MODEL04) and a six chains AAAAGA model (MODEL05) - were successfully obtained. Compared with [14], the variations of RMSD, PRESS, and VOLUME (DENSITY) of the 4,400 ps' equilibration at 100 K are larger [15]; this might imply that DGSA is a little worse than Insight II.

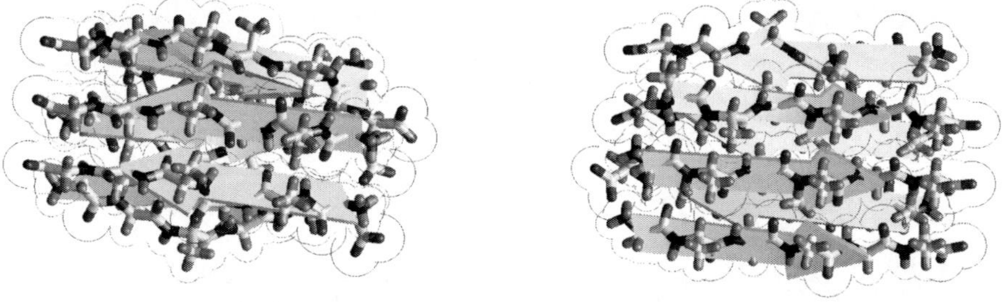

Figure 3. MODEL04 - MODEL05 for prion (113-120) AGAAAAGA amyloid fibrils.

2.3. 3NHC, 3NVE/F/G/H Models

Replacing the DGSA of Subsection 2.2, in this Subsection we will use any optimization solver, which can accurately solve an optimization problem with 3 or 6 variables, to build the models; thus the methods in this subsection are very simple and general for any problems in molecular modeling research area. The model building templates are: 3NHC.pdb (GYMLGS segment 127-132 from human prion with M129), 3NVE.pdb (MMHFGN segment 138-143 from Syrian Hamster prion), 3NVF.pdb (IIHFGS segment 138-143 from human prion), 3NVG.pdb (MIHFGN segment 137-142 from mouse prion), and 3NVH.pdb

(MIHFGND segment 137-143 from mouse prion). The mathematical theory is described as follows.

The atomic structures of all amyloid fibrils revealed steric zippers, with strong vdw interactions (LJ) between β-sheets and hydrogen bonds (HBs) to maintain the β-strands [12]. In mathematics, the potential energy mathematical formula for the vdw interactions between β-sheets is

$$V_{LJ}(r) = \frac{A}{r^{12}} - \frac{B}{r^6}, \tag{1}$$

and the potential energy mathematical formula for the HBs between the β-strands is

$$V_{HB}(r) = \frac{C}{r^{12}} - \frac{D}{r^{10}}, \tag{2}$$

where A, B, C, D are constants given. When V_{LJ} and V_{HB} are reaching their minimal values, the amyloid fibril structures should be in a most stable state. This is a molecular distance geometry problem (MDGP) [5], which arises in the interpretation of NMR data and in the determination of protein structure [as an example to understand MDGP, the problem of locating sensors in telecommunication networks is a DGP. In such a case, the positions of some sensors are known (which are called anchors) and some of the distances between sensors (which can be anchors or not) are known: the DGP is to locate the positions of all the sensors. Here we look sensors as atoms and their telecommunication network as a molecule]. The three dimensional structure of a molecule with n atoms can be described by specifying the 3-Dimensional coordinate positions $x_1, x_2, \ldots, x_n \in \mathbb{R}^3$ of all its atoms. Given bond lengths d_{ij} between a subset S of the atom pairs, the determination of the molecular structure is

$$(\mathcal{P}_0) \quad to \quad find \qquad x_1, x_2, \ldots, x_n \quad s.t. \quad ||x_i - x_j|| = d_{ij}, (i,j) \in S, \tag{3}$$

where $||\cdot||$ denotes a norm in a real vector space and it is calculated as the Euclidean distance 2-norm in this paper. (3) can be reformulated as a mathematical global optimization problem (GOP)

$$(\mathcal{P}) \qquad \min P(X) = \textstyle\sum_{(i,j)\in S} w_{ij}(||x_i - x_j||^2 - d_{ij}^2)^2 \tag{4}$$

in the terms of finding the global minimum of the function $P(X)$, where $w_{ij}, (i,j) \in S$ are positive weights, $X = (x_1, x_2, \ldots, x_n)^T \in \mathbb{R}^{n\times 3}$ [10] and usually S has many fewer than $n^2/2$ elements due to the error in the theoretical or experimental data [17, 5]. There may even not exist any solution $x_1, x_2, \ldots, x_n$ to satisfy the distance constraints in (3), for example when data for atoms $i, j, k \in S$ violate the triangle inequality; in this case, we may add a perturbation term $-\epsilon^T X$ to $P(X)$:

$$(\mathcal{P}_\epsilon) \qquad \min P_\epsilon(X) = \textstyle\sum_{(i,j)\in S} w_{ij}(||x_i - x_j||^2 - d_{ij}^2)^2 - \epsilon^T X, \tag{5}$$

where $\epsilon \geq 0$. So, the molecular model building problem is a problem to get an global minimal solution of (5). Specially, for the amyloid fibril molecular modeling problem, we find after mutations the hydrogen bonds between β-strands are still maintained but the vdw contacts become very far. Thus, the d_{ij} in (5) should be the sum of vdw radii of atoms i and j.

After mutations of 3NHC.pdb, 3NVE/F/G/H.pdb by Swiss-PdbViewer, we find the following least vdw contacts should be maintained respectively for the 3NHC, 3NVE/F/G/H models. 3NHC: A.ALA3.CB-G.ALA4.CB, B.ALA4.CB-H.ALA3.CB (A.ALA3.CB and B.ALA4.CB (two anchors) and G.ALA4.CB and H.ALA3.CB (two sensors), MODEL06-08). 3NVE: A.ALA4.CB-G.ALA3.CB, B.ALA4.CB-G.ALA3.CB (MODEL09), or A.ALA4.CB-G.ALA3.CB, A.ALA2.CB-G.ALA3.CB (MODEL10-11). 3NVF: A.GLY2.CA-H.GLY2.CA, A.ALA4.CB-H.GLY2.CA (MODEL12), A.ALA4.CB-H.ALA2.CB, A.ALA2.CB-H.ALA2.CB, A.ALA2.CB-H.ALA4.CB (MODEL13-14). 3NVG: A.GLY2.CA-H.GLY2.CA, A.GLY2.CA-H.ALA4.CB, A.ALA4.CB-H.GLY2.CA (MODEL15), A.ALA2.CB-H.ALA2.CB, A.ALA2.CB-H.ALA4.CB, A.ALA4.CB-H.ALA2.CB (MODEL16-17). 3NVH: A.ALA4.CB-H.ALA4.CB (MODEL18-19). We look at the former as anchor(s) and its partner as sensor(s). Thus in (5) the variables are 3 or 6 and its dual variables are 1 or 2 or 3. Any optimization solver should be able to accurately solve the prion AGAAAAGA amyloid fibril model building problem. Then the MODEL06-19 will be refined by the SDCG optimization program of Amber 10 and the perfect 3NHC, 3NVE/F/G/H prion AGAAAAGA amyloid fibril 3NHC, 3NVE/F/G/H models were got (Fig.s 4, 5, 6, 7, 8).

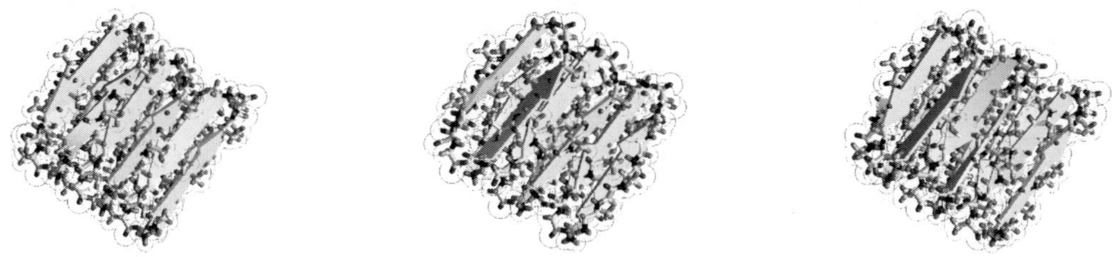

Figure 4. MODEL06 - MODEL08 for prion (113-120) AGAAAAGA amyloid fibrils.

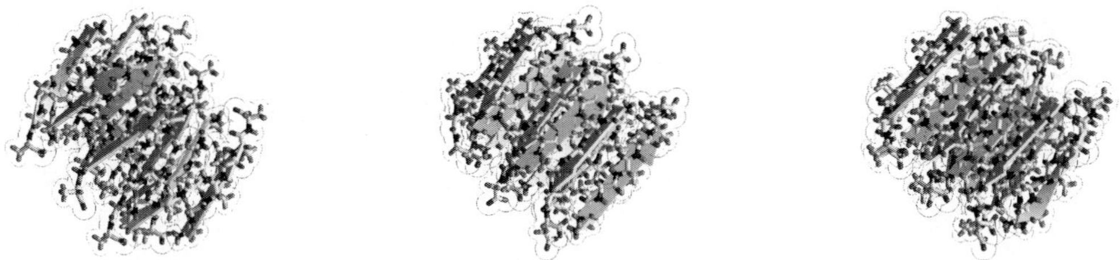

Figure 5. MODEL09 - MODEL11 for prion (113-120) AGAAAAGA amyloid fibrils.

3. Conclusion

Whenever traditional X-ray crystallography and NMR spectroscopy experimental methods cannot be used to get any structural information of proteins, computational approaches or

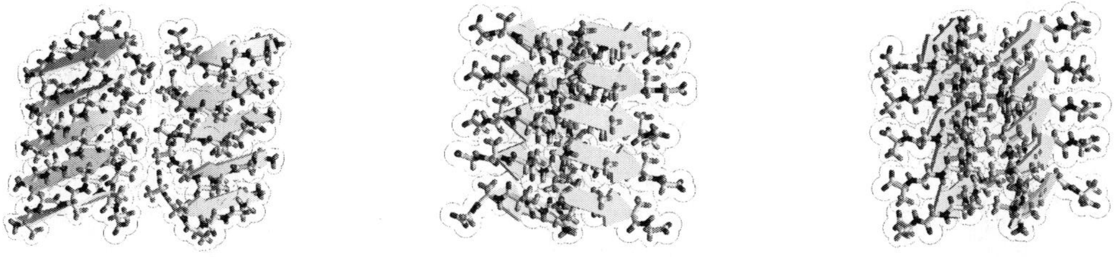

Figure 6. MODEL12 - MODEL14 for prion (113-120) AGAAAAGA amyloid fibrils.

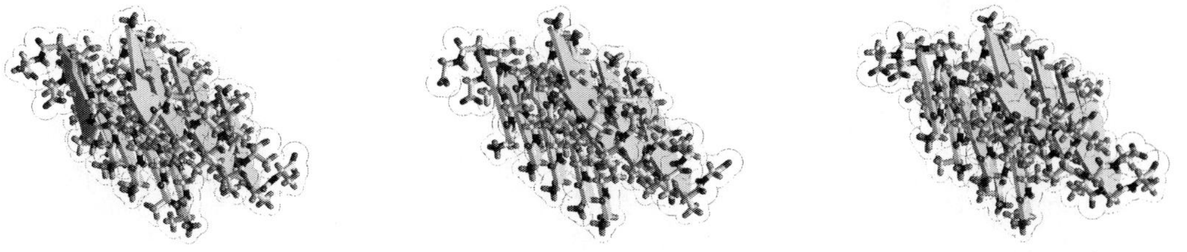

Figure 7. MODEL15 - MODEL17 for prion (113-120) AGAAAAGA amyloid fibrils.

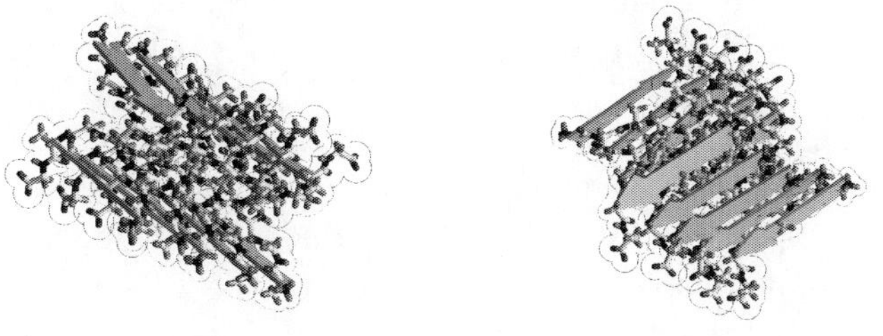

Figure 8. MODEL18 - MODEL19 for prion (113-120) AGAAAAGA amyloid fibrils.

introducing novel mathematical formulations and physical concepts into molecular biology can significantly stimulate the development of biological and medical science. The numerous optimal atomic-resolution structures of prion AGAAAAGA amyloid fibils reported in Fig.s 2-8 have a value to the scientific community in its drive to find treatments for prion diseases or at least be useful for the goals of medicinal chemistry.

Acknowledgments

This research was supported by a Victorian Life Sciences Computation Initiative (http://www.vlsci.org.au) grant number VR0063 on its Peak Computing Facility at the University of Melbourne, an initiative of the Victorian Government.

References

[1] Brown, D.R. (2000). Prion protein peptides: optimal toxicity and peptide blockade of toxicity. *Mol Cell Neurosci,* **15**, 66-78.

[2] Brown, D.R. (2001). Microglia and prion disease. *Microsc Res Tech,* **54**, 71–80.

[3] Brown, D.R., Herms, J., & Kretzschmar, H.A. (1994). Mouse cortical cells lacking cellular PrP survive in culture with a neurotoxic PrP fragment. *Neuroreport,* **5,** 2057-2060.

[4] Case, D.A., Darden, T.A., Cheatham III, T.E., Simmerling, C.L., Wang, J., Duke, R.E., Luo, R., Crowley, M., Walker, R.C., Zhang, W., Merz, K.M., Wang, B., Hayik, S., Roitberg, A., Seabra, G., Kolossvry, I., Wong, K.F., Paesani, F., Vanicek, J., Wu, X., Brozell, S.R., Steinbrecher, T., Gohlke, H., Yang, L., Tan, C., Mongan, J., Hornak, V., Cui, G., Mathews, D.H., Seetin, M.G., Sagui, C., Babin, V., & Kollman, P.A. (2008). *AMBER* **10**, University of California, San Francisco (Amber tutorials: http://ambermd.org/tutorials).

[5] Grosso, A., Locatelli, M., & Schoen, F. (2009). Solving molecular distance geometry problems by global optimization algorithms. *Comput. Optim. Appl.* **43**, 23–37.

[6] Holscher, C., Delius, H., & Burkle, A. (1998). Overexpression of nonconvertible PrPc delta114-121 in scrapie-infected mouse neuroblastoma cells leads to trans-dominant inhibition of wild-type PrPSc accumulation. *J Virol,* **72**, 1153-1159.

[7] Jobling, M.F., Huang, X., Stewart, L.R., Barnham, K.J., Curtain, C., Volitakis, I., Perugini, M., White, A.R., Cherny, R.A., Masters, C.L., Barrow, C.J., Collins, S.J., Bush, A.I., & Cappai, R. (2001). Copper and zinc binding modulates the aggregation and neurotoxic properties of the prion peptide PrP 106-126. *Biochemistry*, **40**, 8073-8084.

[8] Jobling, M.F., Stewart, L.R., White, A.R., McLean, C., Friedhuber, A., Maher, F., Beyreuther, K., Masters, C.L., Barrow, C.J., Collins, S.J., & Cappai, R. (1999). The hydrophobic core sequence modulates the neurotoxic and secondary structure properties of the prion peptide 106-126. *J Neurochem,* **73**, 1557-1565.

[9] Kuwata, K., Matumoto, T., Cheng, H., Nagayama, K., James, T.L., & Roder, H. (2003). NMR-detected hydrogen exchange and molecular dynamics simulations provide structural insight into fibril formation of prion protein fragment 106-126. *Proc Natl Acad Sci USA,* **100**, 14790-14795.

[10] More, J.J., & Wu, Z.J. (1997). Global continuation for distance geometry problems. *SIAM J. Optim.,* **7**, 814-836.

[11] Norstrom, E.M., & Mastrianni, J.A. (2005). The AGAAAAGA palindrome in PrP is required to generate a productive PrPSc-PrPC complex that leads to prion propagation. *J Biol Chem,* **280**, 27236-27243.

[12] Sawaya, M.R., Sambashivan, S., Nelson, R., Ivanova, M.I., Sievers, S.A., Apostol, M.I., Thompson, M.J., Balbirnie, M., Wiltzius, J.J., McFarlane, H.T., Madsen, A., Riekel, C., & Eisenberg, D. (2007). Atomic structures of amyloid cross-beta spines reveal varied steric zippers. *Nature*, **447**, 453–457.

[13] Wegner, C., Romer, A., Schmalzbauer, R., Lorenz, H., Windl, O., & Kretzschmar, H.A. (2002). Mutant prion protein acquires resistance to protease in mouse neuroblastoma cells. *J Gen Virol,* **83**, 1237-1245.

[14] Zhang, J.P., (2011). Optimal molecular structures of prion AGAAAAGA amyloid fibrils formatted by simulated annealing. *J. Mol. Model.,* **17**, 173-179 (Crystallography Times Newsletter 3 (1), January 2011, page 2, VerticalNews 2011 FEB 1/February 4th, 2011).

[15] Zhang, J.P., Sun, J., & Wu, C.Z. (2011). Optimal atomic-resolution structures of prion AGAAAAGA amyloid fibrils. *J. Theor. Biol.,* **279**, 17–28 (Nuclear Energy Research Today Volume 7 Issue 4, April 2011).

[16] Zhang, Z.Q., Chen., H., & Lai, L.H. (2007). Identification of amyloid fibril-forming segments based on structure and residue-based statistical potential. *Bioinformatics*, **23**, 2218-2225.

[17] Zou, Z.H., Bird, R.H., & Schnabel, R.B. (1997). A stochastic/perturbation global optimization algorithm for distance geometry problems. *J. Glob. Optim.*, **11**, 91-105.

In: Amyloids
Editors: I. P. Halcheck et al.
ISBN: 978-1-62100-538-4

Chapter 11

AMYLOIDOSIS: ETIOLOGY, PATHOGENESIS AND DIAGNOSIS

***Laura Díez*[1], *Antonio G. Prado*[2] *and Isabel G. Guijarro*[3]**
Department of Otolaryngology, El Bierzo Hospital[1],
POVISA[2], CHOU[3], Spain

ABSTRACT

Amyloidoses are a heterogeneous group of diseases characterized by the deposition of fibrillar protein in extracellular tissue space. Amyloid deposits consist of a stable, insoluble aggregate of precursor proteins that have adopted an antiparallel beta pleated sheet configuration, having an amorphous and eosinophilic appearance. Although the reason for the accumulation of fibrils is poorly understood, several factors may be involved seeing as there are certain characteristics in precursor molecules that make them amyloidogenic, thus predisposing them to amyloid fibril formation. Different types of amyloidosis exist depending on the fibrillar protein precursor. The most common are: AL, which is associated with plasma cell dyscrasia; AA, associated with chronic inflammation; and Aβ, which is found in patients on long-term haemodialysis for renal failure. Amyloid deposits may affect many organs (systemic amyloidoses), thus decreasing the patient's survival if vital organs are affected, or they may occur in a single tissue (localized amyloidoses) with a better prognosis. Biopsy is necessary to confirm the amyloid deposit, which acquires an apple green birefringence under polarized light after staining with Congo red. Abdominal fat aspirate is a non-invasive technique that presents high sensibility, 70-90%. Prompt diagnosis and evaluation for systemic disease is essential prior to treatment.

INTRODUCTION

Amyloidosis is a heterogeneous group of diseases caused by extracellular deposition of abnormal insoluble fibrils resulting from the aggregation of misfolded plasma proteins. The term amyloidosis, or "starch-like material" was coined in 1854 by Virchow to describe the reaction that these pathological accumulations suffered with iodine staining, taking on a blue colouring as occurs with starch. This term, though inappropriate given the proteinaceous nature of the amyloid, is still used today [1].

The amyloid substance is formed by a variable component, *fibrillar protein*, which is the basis for the pathogenic classification of amyloidosis. More than 25 biochemically distinct forms have been reported depending on the precursor protein, although they all maintain the same antiparallel configuration responsible for their insoluble nature and resistance to proteolytic digestion. They are linear unbranched fibrillar aggregates of 7.5-10 nm in diameter and variable length, in which each fibril is composed of polypeptide chains of ß-pleated sheets.

Table 1. Amyloid protein and their precursors

Amyloid Protein	Precursor	Distribution	Type	Syndrome or involved tissues
AL	Immunoglobulin light chain	Systemic Localized	Acquired	Primary Myeloma-associated
AA	Serum amyloid A	Systemic	Acquired	Secondary Reactive
$Aß_2$ M	$ß_2$-microglobulin	Systemic	Acquired	Chronic haemodialysis
ATTR	Transthyretin	Systemic	Hereditary Acquired	Familial Senility
Aß	Aß precursor	Localized	Acquired Hereditary	Alzheimer's disease Aging
APrP	Prion protein	Localized	Acquired	Sporadic CJD Familial CJD, GSSD,FFI
ABri	ABri precursor	Localized or Systemic?	Hereditary	British familial dementia
ACys	Cystatin C	Systemic	Hereditary	Icelandic hereditary cerebral amyloid angiopathy
AApoAI	Apolipoprotein A-I	Systemic	Hereditary	Liver, kidney, heart
AApoAII	Apolipoprotein A-II	Systemic	Hereditary	Kidney, heart
AGel	Gelsolin	Systemic	Hereditary	Finnish hereditary amyloidosis
ALys	Lysozyme	Systemic	Hereditary	Kidney, liver, spleen
AFib	Fibrinogen Aα chain	Systemic	Hereditary	Kidney

CJD: Creutzfeldt-Jakob disease. GSSD: Gestmann-Sträussler-Scheinker disease. FFI: Fatal familial insomnia.

These characteristics are what determine the progressive replacement and destruction of the parenchyma of the organs affected by amyloid deposition and associated clinical manifestations [2]. The constant element is the *P component*, a glycoprotein similar in all types of amyloid that acts as a basic skeleton on which the fibrillar proteins are deposited, thus contributing to the stabilization of the fibrillar protein and preventing its degradation and phagocytosis. It derives from the serum precursor (SAP) of hepatic synthesis, present in all individuals but only deposited in those with amyloidosis due to its affinity for amyloid fibrils, constituting a valuable diagnostic method in systemic amyloidoses. Other amyloid components consist of *glycosaminoglycans* (heparan sulphate and dermatan sulphate) and *apolipoproteins* (E and J), covalently linked to the fibrils, promoting fibril assembly and deposition of amyloid material in the tissues [3].

Amyloid deposition may involve multiple organs, decreasing patient survival if vital organs are affected (systemic amyloidosis) or it may affect a single tissue (localized amyloidosis), showing a better prognosis. It may be classified as *acquired*, being present in diseases that produce abnormal proteins or large amounts of normal proteins, or *hereditary*, being caused by genetic mutations that encode proteins with amyloidogenic properties. Another form of classification of amyloidosis and one that corresponds to the current nomenclature is based on the nature of the protein precursors. Each type of amyloidosis is called AX, where the first letter A corresponds to the term amyloidosis and the second is an abbreviation of the fibrillar component [4].

Etiology and Pathogenesis

The detection of serum protein precursors has been an important finding for the understanding of the pathogenesis of amyloidosis. Even so, the exact mechanism by which amyloid deposition occurs in the tissues starting from the precursor is still unknown. These mechanisms may vary depending on the different biochemical types of amyloidosis, with three main types of amyloid and other less common forms being defined: *AL*, previously known as primary systemic amyloidosis, the precursor protein of which comprises immunoglobulin light chains, present in β-cell clonal disorders; *AA*, formerly known as secondary amyloidosis, whose precursor protein is the serum A amyloid, present in manifestations of chronic inflammation or infectious diseases; *AF*, where the most common precursor protein is TTR (transthyretin); *Aβ2* microglobulin, in subjects with long-term end-stage renal disease (ESRD); and *Aβ*, related to Alzheimer's disease [5,6].

Amyloidosis AL

The systemic variant is the most common form of amyloidosis in developed countries. The incidence of systemic AL amyloidosis in the United States is 4.5 per 100,000 persons/year, mainly affecting those aged between 50 and 70 [7]. The precursor proteins of AL amyloidosis are intact monoclonal Ig light chains and/or their fragments (mainly the variable domain) as a result of clonal proliferation of plasma cells in bone marrow. Amino acid substitution at specific positions of the variable region destabilizes the light chains and increases the risk of these becoming amyloid. λ light chains predominate in amyloid fibrils, the λ/κ ratio being 3:1 and the most common subtype, λVI [8]. AL amyloidosis may appear in multiple myeloma and other B lymphoproliferative diseases including non-Hodgkin lymphoma and Waldestrom's macroglobulinemia. Of all patients with multiple myeloma, only 15% present amyloid deposition and less than 20% of subjects with AL amyloidosis have multiple myeloma. It is characterized by extensively attacking multiple organ systems, except the central nervous system, evolving rapidly and fatally if untreated, with an average survival from the date of diagnosis of twelve months.

In localized forms, amyloid deposits are composed of Ig light chains produced locally by the plasma cells of the infiltrate and not in bone marrow. It is usually located in the upper airway, bladder, skin and orbit. Its evolution is usually benign and presents a better prognosis than systemic forms, benefitting from surgical treatment and/or radiotherapy [9].

Amyloidosis AA

Its incidence has decreased in developed countries, constituting less than 1% of all patients with amyloidosis due to advances in medical treatment. It is the only type of systemic amyloidosis that affects children. It is associated with infectious diseases (tuberculosis or subacute bacterial endocarditis), chronic inflammatory diseases (rheumatoid arthritis, lupus, Crohn's disease) and some neoplasms (stomach cancer or renal cell carcinoma) [10].

AA amyloidosis is formed by the fibrillar protein AA (amyloid protein A), consisting of a non-immunoglobulin structure of 76 amino acids with a molecular weight of 8 kDa. The precursor of the AA protein is serum amyloid A (SAA), which is an acute-phase apoprotein synthesized in hepatocytes and transported in plasma by high-density HDL3 lipoproteins. Its production increases in inflammatory processes in response to cytokines (IL-6, IL-1). Whether due to the action of defective enzymes that give rise to incomplete proteolysis of SAA or due to a genetic defect of its molecular structure that makes it resistant to degradation by macrophages, the result is the formation of insoluble amyloid A [11].

The amyloid is generally deposited in the kidney, cardiac symptoms being rare. Its prognosis will depend on the severity of the underlying inflammatory or infectious disease [12].

Familial Amyloidosis

There are many forms of familial amyloidosis which are often associated with ethnic groups. They have an incidence of less than 1 case per 100,000 people, starting in the middle stage of life. AF amyloidoses are autosomal dominant disorders with high penetrance, in which the precursor proteins correspond to mutations of normal plasma proteins [13].

The most common form is caused by mutation of TTR or prealbumin, which is the transport protein of thyroxine and of retinol-binding. There are more than 100 different amino acid substitutions in transthyretin that can cause familial amyloidosis. Normal TTR, without mutation, can also cause amyloid deposition in what is known as senile amyloidosis, present in the elderly and usually asymptomatic, being diagnosed in 25% of autopsies. Other less common family amyloidoses are caused by mutations in apolipoprotein AI and A-II, gelsolin, lysozyme or the fibrinogen A α chain [14].

Amyloidosis Aß2 M

The precursor protein of the amyloid fibril protein is ß2-microglobulin, which has a molecular mass of 11.8 kDa and corresponds to the constant domain of the β chain of the class I major histocompatibility complex molecule (MHC-1). It is excreted by the kidneys and its concentration is found to increase in patients with ESRD. It appears as a complication of long-term haemodialysis (over 5 years), in which the proteins are poorly filtered by the dialysis membranes, with the subsequent increase in their serum levels resulting in deposition in bones and joints. New dialysis techniques have managed to reduce its incidence [15, 16].

Amyloidosis Aβ

This is the most common form of localized amyloidosis. Its precursor is the fibrillar Aβ protein arranged in the cerebral vascular walls and in the nuclei of neuritic plaques in Alzheimer patients. The Aβ protein is derived from a transmembrane glycoprotein called the amyloid β-precursor protein (AβPP), whose mutations are related to the disease [17].

DIAGNOSIS

Primary amyloidosis is a condition that must be distinguished from other forms of amyloidosis such as secondary, familial or senile amyloidosis, as its clinical course and therapy are markedly different. It must also be distinguished from localized amyloidosis, as these deposits are derived from monoclonal light chains though in the absence of clonal plasma cell disease.

In the diagnosis of amyloidosis, the characteristic clinical symptomology (nephrotic range proteinuria [18], restrictive cardiomyopathy, macroglossia in relation to the presence of monoclonal protein in serum or urine) is the first thing that leads the practitioner to suspect this condition. Confirmation is based on these two findings:

- The presence of amyloid deposits in the biopsy of abdominal fat, bone marrow, the rectum or, if applicable, the affected organ (kidney, liver, sural nerve).
- In AL amyloidosis, it is also necessary to prove the existence of a condition of clonal proliferation of plasma cells. This includes the presence of a monoclonal protein in serum and urine, monoclonal plasma cells in bone marrow, an abnormal ratio of light chains in serum and/or the presence of amyloid deposits composed of fragments of the light chain as determined by immunofluorescence or immunohistochemistry studies for kappa and lambda light chains.

Once the amyloid deposition has been shown to exist, it must be typified by immunohistochemistry using antibodies specific to the fibrillar protein precursor, a reliable technique in AA amyloidosis and less useful in AL and familial amyloidosis. DNA analysis or amino acid sequencing is used to distinguish the latter two in the absence of family history. Scintigraphy is useful in defining the extent and location of the disorder and the subsequent assessment of its severity should be made by means of non-invasive methods (blood and urine analyses, CT, MRI, etc.).

Biopsy

The presence of amyloid is necessary in histological samples in order to define the diagnosis. The biopsy is generally performed on the organ. Biopsy is positive in more than 90 per cent of cases in the kidneys or the liver. However, a high success rate can also be achieved via less invasive procedures such as abdominal fat aspirate (60 to 80 per cent), or rectal (50 to 70 per cent), bone marrow (50 to 55 per cent) or skin biopsy (50 per cent) [19,20,21,22]. Due to its ease of performance, abdominal fat aspirate and bone marrow biopsy are the preferred methods in patients suspected of amyloidosis. If fat or bone marrow results are negative, biopsy should be performed on the affected organ.

Amyloid fibrils can be identified by their characteristic appearance under electron microscopy and their ability to fix Congo red (which produces a characteristic apple-green birefringence under polarized light) or thioflavine-T (producing an intense yellow-green fluorescence) [23]. In fact, Congo red staining is considered the gold standard test in the diagnosis of amyloidosis. Under electron microscopy, the fibres are straight and unbranched with a diameter of 7.5 to 10 nm. Once the diagnosis of amyloidosis has been made using Congo red staining, there are a number of laboratory and histological tests that can help to differentiate AL amyloidosis from other forms of amyloidosis. Immunohistochemical staining can help determine the type of amyloidosis. Immunofluorescence in the kidney or other affected tissues can show the deposit of a monoclonal antibody (lambda or kappa) light chain in AL amyloidosis, although some autors have shown a low sensitivity to renal biopsy (65 per cent in a series) [24]. Immunofluorescence in bone marrow biopsy may show a clonal excess of plasma cells. This clonal excess is usually lambda in AL amyloidosis and kappa in light chain deposition disease.

Immunologic Laboratory Tests

The proliferation of plasma cells in AL amyloidosis leads to the presence of a paraprotein in serum (as an M protein in protein immunoelectrophoresis or immunofixation) or in urine (monoclonal light chains) in approximately 90 per cent of cases. Serum and urine immunofixation is required to identify the monoclonal protein (paraprotein). A positive test for monoclonal proteins must be followed up by further evaluation to determine the presence of amyloidosis or light chain deposition disease. Using techniques of immunofixation in urine or serum, monoclonal protein is found to be present in 85 to 90 per cent of cases of primary amyloidosis. In the 10 to 15 per cent of patients in whom it is not present, the following tests may be used:

- Demonstration of a clonal excess of plasma cells (lambda or kappa) in bone marrow biopsy.

- Demonstration of a clonal excess of plasma cells via serum free light chain immunoassay.

Given the finding of monoclonal gammopathy of uncertain significance (MGUS) in the general population and especially in the elderly, the existence of a monoclonal protein in the absence of the demonstration of amyloid deposits in tissue samples does not confirm diagnosis of primary amyloidosis (AL).

There exist a number of chromosomal mutations frequently present in patients with AL amyloid, such as translocation t (11; 14) (q13; q32), an abnormality that frequently occurs in multiple myeloma [25], and the deletion of (13q14), an abnormality associated with a poor prognosis in multiple myeloma [26].

Serum Amyloid P Component Scintigraphy

Serum amyloid P component (SAP) is a protein of uncertain physiological significance that has a special affinity to bind with amyloid fibres [27,28]. Regardless of the mechanism of amyloidosis, amyloid deposits in tissues can be identified with high sensitivity and specificity by means of scintigraphy after intravenous injection of technetium-labelled SAP [29,30]. This test is more accurate in AA amyloidosis and can be positive even when the pathological analysis of tissue biopsy was negative [27]. SAP scintigraphy has a sensitivity close to 100 per cent with almost universal splenic uptake, renal uptake of more than 80 per cent, adrenal uptake of 50 per cent and hepatic uptake of 25 to 50 percent [29,31]. It is less sensitive in identifying cardiac amyloidosis, due to the large amount of blood in the heart. Scintigraphy has the additional advantage of reproducibility and can be used to document the possible resolution of amyloid deposits after therapy targeted at the primary disease in patients with secondary amyloidosis [32].

The value of SAP scintigraphy is limited because it is costly, not routinely available and is less useful in the detection of cardiac amyloidosis. Another limitation is that SAP is currently obtained from blood donors, which carries a potential risk of infection.

Differential Diagnosis

Differential diagnosis in a patient with histologically confirmed amyloidosis must be made between the different clinical conditions caused by these deposits: primary amyloidosis, secondary amyloidosis, hereditary amyloidosis and senile amyloidosis. This is important from a therapeutic standpoint. The distinction between different forms of amyloidosis is suggested by history and clinical findings.

The presence of a monoclonal proliferation of plasma cells in a patient with a histology suggestive of amyloidosis is highly indicative of primary amyloidosis. However, given the prevalence of monoclonal gammopathy of uncertain significance in the general population, immunohistochemistry or immunofluorescence evidence is needed to determine whether these amyloid deposits are caused by light chains.

Secondary amyloidosis may occur in chronic inflammatory processes (rheumatoid arthritis, Crohn's disease, osteomyelitis, etc.) that increase the hepatic production of the acute-phase reactant serum amyloid A, which is degraded by macrophages favouring the deposition of fragments of amyloid in the tissues.

In senile amyloidosis, the amyloid deposit is composed of transthyretin, which accumulates in the ventricles of the heart in elderly patients, causing heart failure or arrhythmias.

Conclusion

Amyloidosis is a heterogeneous group of diseases characterized by extracellular deposition of a series of fibrillar proteins in organs and tissues, biochemically unrelated to one another, but which share common characteristics that include apple-green birefringence under polarized light after staining with Congo red and a cross-β-pleated sheet configuration under X-ray diffraction.

The amyloid deposit produces functional and structural alterations in the affected organ which vary according to location and intensity. The main mechanisms responsible for amyloid formation and deposition are not known precisely, although several factors may be important in their pathogenesis. Proteins can have a tendency to form aggregates of misfolded forms when their concentration reaches high levels, as occurs with aging (senile amyloidosis), with increased production (chronic inflammatory states, AA amyloidosis) or if protein excretion is impaired (amyloidosis associated with long-term dialysis). A mutation may give rise to a form of protein that tends to misfold and form deposits (hereditary amyloidosis). Limited proteolysis can produce a protein that forms amyloid fibrils (amyloidosis associated with Alzheimer's disease).

Clinical suspicion is essential in the diagnosis of amyloidosis. Amyloid disease should be suspected when faced with unexplained cases of nephropathy, cardiomyopathy, neuropathy or enteropathy, as well as hallmark features of periorbital ecchymoses and macroglossia. Biopsy by abdominal fat aspiration is the initial method of diagnosis, confirming the apple-green birefringence with Congo red staining under polarized light microscopy. Typifying and evaluating the extent of amyloid deposits via immunological, biochemical and genetic studies is needed to determine how to treat the disease.

References

[1] Sipe JD, Cohen AS. Review: history of the amyloid fibril. *J. Struct. Biol.* 2000;130(2-3):88-98.

[2] Merlini G, Bellotti V. Molecular mechanisms of amyloidosis. *N. Engl. J. Med.* 2003;349(6):583-96.

[3] Álvarez-Ruiz S, García-Río I, Daudén E. Systemic amyloidoses. *Actas Dermosifiliogr.* 2005;96(2):69-82.

[4] Pepys MB. Amylodosis. *Annu. Rev. Med.* 2006;57:223-41.

[5] Westermark P, Araki S, Benson MD. Nomenclature of amyloid fibril proteins. Amyloid. *Int. J. Exp. Clin. Invest.* 1999; 6:63-6.
[6] Cohen AS, Connors LH. The pathogenesis and biochemistry of amyloidosis. *J. Pathol.* 1987;151(1):1-10.
[7] Guidelines Working Group of UKMF, British Committee for Standards in Haematology BSfH. Guidelines on the diagnosis and management of AL amyloidosis. *Br. J. Haematol*. 2004;125(6):681-700.
[8] Richar A, Lebowitz MD, Luc Morris BS. *Plasma cell dyscrasia and amyloidosis Otolaryngol Clin. N. Am*. 2003;36(4):747-764.
[9] Seldin D, Skinner M. Amyloidosis. Harrison`s principles of internal medicine 17ht ed. 2008;324:2145-2149.
[10] Pettersson T, Konttinen YT, Maury CP. Treatement strategies for amyloid A amyloidosis. *Expert. Opin. Pharmacother*. 2008;9(12):2117-28.
[11] Van der Hilst J. Recent insights into the pathogenesis of type AA amyloidosis. *Scientific World Journal*. 2011;11:641-50.
[12] Lachmann HJ, Goodman HJ, Gilbertson JA, Gallimore JR, Sacabin CA, Gillmore JD, et al. Natural history and outcome in systemic AA amyloidosis. *N. Engl. J. Med.* 2007;356(23):2361-71.
[13] Saraiva MJ. Sporadic cases of hereditary systemic amyloidosis. N. *Engl. J. Med.* 2002;346:1818-9.
[14] Irigoyen MV, López M, Ureña I, Fernández- Nebro A. *Amiloidosis. Medicine.*2009; 10(33) :2185-91.
[15] Gejyo F, Yamada T, Odani S, et al. A new form of amyloid protein associated with chronic hemodialysis was identified as B2-microglobulin. *Biochem. Biophys. Res. Commun*. 1985;129:701-6.
[16] Nishi S, Alchi B, Imai N, Gejyo F. New advances in renal amyloidosis. *Clin. Exp. Nephrol.* 2008;12(2):93-101.
[17] Hardy J, Selkoe DJ. The amyloid hypothesis of Alzheimer's disease: progress and problems on the road to therapeutics. *Science* 2002;297:353-6.
[18] Ogg, CS, Cameron, JS, Willians, DG, Turner, DR. Presentation and course of primary amyloidosis of the kidney. *Clin. Nephrol*. 1981; 15:9.
[19] Kyle, RA, Greipp, PR. Amyloidosis (AL). Clinical and laboratory features in 229 cases. *Mayo Clin. Proc*. 1983; 58:665.
[20] Duston, MA, Skinner, M, Shirahama, T, Cohen, AS. Diagnosis of amyolidosis by abdominal fat pad aspiration. Analysis of four years experience. *Am. J. Med.* 1987; 82:412.
[21] Duston, MA, Skinner, M, Meenan, RF, Cohen, AS. Sensibility, specificity, and predictive value of abdominal fat aspiration for the diagnosis of amyloidosis. *Arthritis. Rheum*. 1989; 32:82.
[22] Sungur, C, Sungur, A, Ruacan, S, et al. Diagnostic value of bone marrow biopsy in patients with renal disease secondary to familian Mediterranean fever. *Kidney Int.* 1993; 44:834.
[23] Glenner, GG. Amyloid deposits and amyloidosis. The β-fibrilloses. *N. Engl. J. Med.* 1980; 302:1283.
[24] Novak, L, Cook WJ, Herrera, GA, Sanders, PW. AL-amyloidosis is underdiagnosed in renal biopsies. *Nephrol. Dial. Transplant*. 2004; 19:3050.

[25] Hayman, SR, Bailey, RJ, Jalal, SM et al. Translocations involving the immunoglobulin heavy-chain locus are possible early genetic events in patients with primary systemic amyloidosis. *Blood* 2001: 98:2266.

[26] Harrison, CJ, Mazullo, H, Ross, FM, et al. Translocations of 14q32 and deletions of 13q14 are common chromosomal abnormalities in systemic amyloidosis. *Br. J. Haematol.* 2002; 117:427.

[27] Hawkins, PN, Lavender, JP, Pepys, MB. Evaluation of systemic amyloidosis by scintigraphy with 123I-labeled serum amyloid P component. *N. Engl. J. Med.* 1990; 323:508.

[28] Gallo, G, Wisniewski, T, Choi-Miura, NH, et al. Potential role of apolipoprotein-E in fibrillogenesis. *Am. J. Pathol.* 1994; 145:526.

[29] Hawkins PN, Pepys, MB. Imaging amyloidosis with radiolabelled SAP. *Eur. J. Nucl. Med.* 1995; 22:595.

[30] Jager, PL, Hazemberg, BP, Franssen, EJ, et al. Kinetic studies with iodine-123-labeled serum amyloid P component in patients with systemic AA and AL amyloidosis and assessment of clinical value. *J. Nucl. Med.* 1998; 39:699.

[31] Lovat, LB, Persey, MR, MAdhoo, S, et al. The liver in systemic amyloidosis: Insinghts from 123I serum amyloid P component scintigraphy in 484 patients. Ut 1998; 42727.

[32] Tan, SY, Pepys, MB, Hawkins, PN. Treatment of amyloidosis. *Am. J. Kidney Dis.* 1995; 26:267.

In: Amyloids
Editors: I. P. Halcheck et al.

ISBN: 978-1-62100-538-4

Chapter 12

SERUM AND URINE FREE LIGHT CHAIN ANALYSIS IN EVALUATION OF AMYLOIDOSIS AL AND OTHER MONOCLONAL GAMMOPATHIES

Stanley S. Levinson*

Laboratory Service, Department of Veterans Affairs Medical Center, Louisville, KY, U. S.
The Department of Pathology and Laboratory Medicine, School of Medicine, University of Louisville, KY, U. S.

ABSTRACT

This article discusses approaches for testing monoclonal proteins in clinical laboratories with special emphasis on amylodosis AL and free light chain disease. Where reflex testing is possible good communication between the laboratory and the clinician leads to the most effective work-up. The article briefly reviews the major monoclonal gammopathies, including multiple myeloma, lymphocytoplasmic lymphoma, heavy chain disease, plasmacytoma, plasma cell dyscrasia associated with neuropathy, amyloidosis AL, free light chain disease, and monoclonal gammopathy of unknown significance. There is emphasis on the pathophysiology of amyloidosis AL and free light chain disease and a testing algorithm for identifying amyloidosis AL. Standard clinical laboratory techniques for identifying monoclonal proteins are discussed and examples of serum and urine protein electrophoresis and immunofixation electrophoresis are shown. A new serum free light chain assay that adds additional armament to traditional testing is discussed. The article discusses the strengths and weaknesses of this assay as compared to traditional urine testing for monoclonal free light chains. It is concluded with an algorithm that addresses approaches for monoclonal protein screening.

* Correspondence: Stanley S. Levinson, at Laboratory Service. Voice:(502) 895-3401 X5565/5572, Fax: (502) 894-6265, E mail:levinson@louisville.edu.

Introduction

Monoclonal Gammopathies

Monoclonal gammopathies may be associated with multiple myeloma (MM), lymphocytoplasmic lymphoma (WS) (previously called Waldenstrom macroglobulinemia), chronic lymphocytic leukemia, various polyneuropathies, amyloidosis AL, free light chain (FLC) deposition disease (FLCD), heavy chain disease and other B-cell malignancies, such as chronic lymphocytic leukemia, various chronic diseases or be benign - monoclonal gammopathy of undetermined significance (MGUS)[1-4].

Of these conditions the monoclonal (M) -protein directly causes the disease in amyloidosis AL, FLCD, and heavy chain deposition disease that is very rare. In most other cases, although measurement of the M-protein is helpful in diagnosis and monitoring of the disease, the plasma cells or B-cells cause the disease as a result of malignant production of cells and secretion of other substances.

Multiple Myeloma (MM)

MM is characterized as monoclonal plasma cells in the bone marrow in excess of 30% or between 10 and 30% with an appropriate monoclonal spike and/or clinical evidence of MM [5]. Multiple implies more than one site of plasma cell proliferation in contrast to solitary plasmacytoma of bone or other tissue that often has a more favorable outcome. In classical MM, the serum protein electrophoresis (SPEP) pattern shows a monoclonal (M)-protein spike > 30 g/L with a decrease in uninvolved polyclonal Ig. Such a pattern is shown in Figure 1.

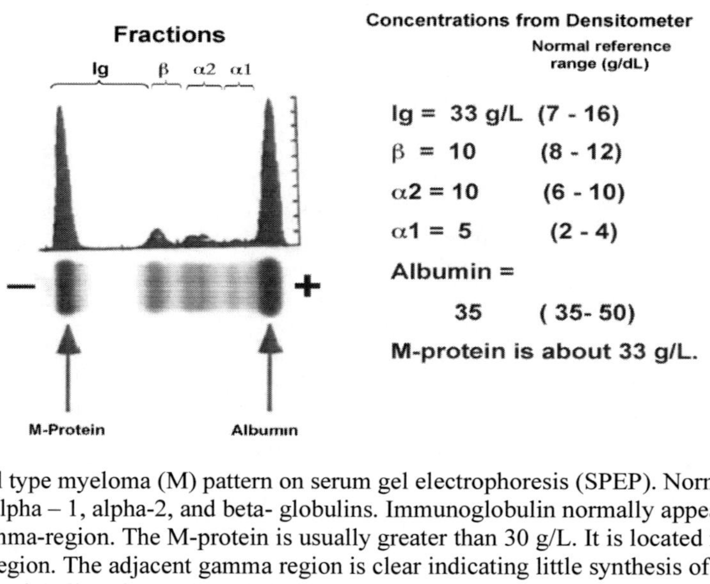

Figure 1. Classical type myeloma (M) pattern on serum gel electrophoresis (SPEP). Normal banding is seen as albumin, alpha – 1, alpha-2, and beta- globulins. Immunoglobulin normally appears as diffuse staining in the gamma-region. The M-protein is usually greater than 30 g/L. It is located in as an extra band the gamma region. The adjacent gamma region is clear indicating little synthesis of normal polyclonal immunoglobulin. Electrophoresis is towards the positive electrode. Densitometry estimates the amount of protein in each region of the gel. In this case, the M-protein is estimated as 33 g/L. The arrow indicates the location of the M-protein.

The densitometer profile is used to quantify the amount of protein in the M-spike. In some cases, MM shows few symptoms with little end organ damage and is called – asymptomatic – either indolent or smoldering MM [1]. Less frequently MM appears as oligosecretory myeloma which implies an intact M-protein less than 3 g /dL, FLC disease, where only a monoclonal FLC is secreted, and nonsecretory myeloma, where no apparent M-protein is secreted.

Lymphocytoplasmacytic Lymphoma or Waldenstrom Syndrome (WS)

WS is a low grade small-cell lymphoma that produces monoclonal IgM [6]. Clinical features are usually related to growth of the tumor. These usually include lymphadenopathy, splenomegaly, and anemia due to marrow infiltration. Many patients produce monoclonal IgM which may cause hyperviscosity syndrome, or type I or II cryoglobulinemia that may lead to Raynaud's syndrome, vasculitis, cold agglutinin hemolytic anemia, peripheral neuropathy, or immune complex disease [6, 7].

Heavy Chain Disease

Heavy chain diseases are rare variants of B-cell lymphomas that produce IgA, IgG or IgM heavy chains without light chains [4]. Alpha heavy chain disease is a form of lymphoma that develops in the intestine in persons living under conditions of poor sanitation. Campylobacter jejuni has been implicated in its pathogenesis. IgM heavy chain disease is extremely rare and is generally associated with leukemia. IgG heavy chain disease (also known as Franklin disease) is often associated with autoimmune diseases, especially rheumatoid arthritis.

Solitary Plasmacytoma

In some cases, only a single site of proliferation exists in bone. Complete remission of this disease is much more common than is complete remission in multiple myeloma [8]. Occasionally, the plasmacytoma is in tissue other than bone, often in the sinuses or nasopharynx. Here the likelihood of complete cure is high and only about two thirds of the patients exhibit a M - component protein in serum, which is usually of low concentration (< 25 g/L).

Plasma Cell Dyscrasia Associated with Neuropathies

An estimated 10% of polyneuropathies are associated with M-gammopathies [9]. These include overt multiple myeloma where IgG-κ is seen more often, WS where IgM is found, amyloidosis AL - more often lambda FLC, cryoglobulinemia, and osteolytic myeloma, more often IgG- λ or IgA-λ. POEMS syndrome is almost always associated with IgA-λ or IgG-λ. The most common monoclonal gammopathy associated with polyneuropathy is MGUS where IgG-kappa is most common. In some cases of MGUS the M- protein may harbor antibody-activity against myelin-associated glycoproteins. In other cases, the abnormal B-cells may secrete various cytokines including vascular endothelial growth factor that may cause enhanced permeability of tissues and impairment of the blood brain barrier [10-12]. Because amyloidosis AL may affect the peripheral nervous system, it may present as a polyneuropathy.

These M-protein associated neuropathies may be mistaken for Chronic inflammatory demyelinatiing poly(radiculo)neuropathy (CIDP) (13). CIDP is a chronic immune-mediated disease that is clinically similar to the acute Guillian-Barre syndrome. It is important to differentiate between these conditions because plasmapheresis and intravenous Ig that have clinical benefit in CIDP, produce little clinical benefit in many neuropathies associated with M-protein. Chemotherapy for B-cell proliferative disorders, radiation when there is isolated plasmacytoma, and stem-cell transplant are more effective [10, 11].

Amyloidosis AL and FLC Disease

Monoclonal FLC may cause amyloidosis AL or FLC disease where light chains accumulate in tissues often leading to organ failure [14]. In amyloidosis, deposition of organized proteinaceous material occurs. When viewed by X-ray diffraction, these deposits appear as fibrils in an anti-parallel conformation and β-pleated sheet structure. The exact reason that some proteins form this structure is poorly understood. Amyloid is usually recognized by biopsy where the tissue exhibits apple-green birefringence in the polarizing microscope when stained with Congo red.

Amyloid is most easily identified from a positive biopsy of the fat pads (about 85% of cases). When fat pads are not diagnostic, the effected tissue should be biopsied. The tissues affected in amyloidosis AL include: kidney, heart, liver, gastrointestinal tract, tougue leading to macroglossia and the peripheral nervous system, often manifested as carpal tunnel syndrome. If apple green birefringence is not positive, and the disease is suspected, electronmicroscopy may be necessary [15]. Once amyloidosis is diagnosed, it is important to determine the type.

In 90% of patients amyloidosis AL will exhibit a monoclonal FLC. In contrast to MM where kappa is found more often, lambda is found about twice as often as kappa. Since amyloidosis AL is the most common type of amyloidosis, the first step should be identification of a monoclonal FLC in serum or urine. Amyloidosis AL may be associated with myeloma or WS but It it is important to recognize that usually one disease predominates so that amyloidosis AL is found only about 10 -15% of the time in association with these well defined plasma cell dyscrasias [14, 16]. In the absence of MM or WS, amyloidosis AL is difficult to detect [17].

If a monoclonal FLC is not detected in serum or urine, the diagnosis of amyoidosis AL becomes more difficult because the bone marrow may show a normal or only slightly elevated number of plasma cells. A clonal predominance of either κ or λ type plasma cells can often be identified by immunohistochemical staining of a biopsy [18].

Together these two techniques of FLC detection and immunohistochemistry can allow for presumptive identification of amyloidosis AL in most cases. Figure 1 2 shows a diagnostic algorithm for identifying the type of amyloidosis. If amyloidosis AL cannot be identified, extensive additional testing is needed to identify the type. The simplest approach for the presumptive identification of amyloidosis AL is identification of a monoclonal FLC. Identification of an intact M-protein without identifying a monoclonal FLC is insufficient evidence for amyloidosis AL, since, as will be described below, benign monoclonal gammopathies are found in about 5% of older people, as a result, this conjecture has led to misidentification [17].

Monoclonal Gammopathy of Unknown Significance (MGUS)

In the general population, the most common SPEP pattern is that associated with MGUS [16] where a level of M-protein less than 30 g/L and no or little clinical or laboratory evidence for monoclonal disease is found. Very low level M-proteins, often <5 g/L, are associated with normal or even slightly elevated levels of uninvolved polyclonal Ig. These may evolved into MM over a period of years but, when the M-protein is <5 g/L, it usually does not and may even disappear upon later analysis [2, 19].

Generally, the higher the concentration of M-protein and the younger the individual, the more likely disease will develop. When the concentration of M-protein is near 30 g/L, MGUS may be difficult to distinguish from asymptomatic myeloma [1, 2]. As will be discussed below, monoclonal FLC MUGS is unusual because unlike intact monoclonal proteins, FLC are harmful to the kidney and other tissues. These are usually considered a sign of disease or impending disease [20, 21].

Monoclonal FLC and Kidney Disease (Reviewed in Reference [21])

Characteristics of Normal Polyclonal FLC

Polyclonal FLCs are synthesized in excess of heavy chains and normally very small amounts of polyclonal FLCs are found in serum and spill over into the urine. The light chains are polypeptides of about 22 kDa. They may exist as monomers or as dimers with a molecular weight of about 44 kDa. The dimers may be covalently disulfide-linked or non-covalently linked. There are about equal amounts of covalent and non-covalent κ chain dimers in normal serum, while the λ-chain dimers are largely covalent. Normally, more κ- than λ-chains are synthesized and the κ/λ ratio of intact Ig in blood is about 2 to 1 [9].

Characteristics and Excretion of Monoclonal FLC

Most commonly, monoclonal λ–FLC appear as disulfide linked covalent dimers, and monoclonal κ-FLC as noncovalent dimers, stable monomers, or mixtures of the two.

Monoclonal FLC may also be found as smaller fragments with molecular weights < 22 kDa and as anomalously larger forms > 45 kDa. Those with molecular weights of approximately 55 kDa may contain covalently linked carbohydrates such as sialic acid that are not normally associated with light chains. Although less common, monoclonal FLC have been identified as trimolecular complexes, tetramers, and even higher polymer forms.

Normally, most of the FLC is reabsorbed by the kidney tubules. Monoclonal FLC usually appear in the urine because of overflow conditions in which the concentration of FLCs in the renal tubules is greater than the amount that can be reabsorbed, ether because of high concentrations in plasma or because of damage to the kidney. Monoclonal FLC may cause renal tubular damage directly or they may cause amyloidosis AL, usually, leading to glomerular failure. It is not unusual that the disease is first thought to be due to long standing hypertension or diabetes. This error may be prolonged when only tubular failure occurs since the kidney function tests serum urea nitrogen and creatinine may not increase much above the normal range and total proteinuria may be mild.

Nephropathy Due to Monoclonal FLC

There is some evidence that cellular processing plays a differential role in the production of FLC organized deposits [22]. Nevertheless, it remains unclear whether or not this is true in human disease and how the structure of monoclonal FLC leads to different types of deposits. FLC invariably causes disease in patients with monoclonal FLC concentrations in the urine in excess of about 1g/24 hours. However, tubular disease and amyloidosis is commonly associated with lower levels of BJP as well.

Tubular damage may or may not appear as cast nephropathy (myeloma kidney). Cast nephropathy is recognized by eosinophilic, dense, and homogeneous casts that are at least partially surrounded by multinucleated foreign-body type giant cells. In the absence of definitive cast nephropathy, the disease may present as Fanconi's syndrome, with polyuria and renal loss of potassium, amino acids, phosphorous, protein and glucose. Intracytoplasmic crystals composed of monoclonal light chains have been found in kidney and in other tissue with both Fanconi's syndrome and cast nephropathy. Ultimately, organ damage may occur, especially kidney failure.

Serum and Urine Laboratory Analysis for Monoclonal Gammopathies

Analysis by Serum Electrophoresis

The great majority of monoclonal gammopathies are identified by SPEP. The SPEP profile for most patients with MM or WS is similar to that in Figure 1 where a restricted band is seen in the gamma region. In MGUS, the M-protein is more difficult to detect and appears more as shown in Figure 3. A minority of patients with MM or WS show profiles varying between those seen in Figure 1 and Figure 3.

After a band of restricted mobility is noticed, it must be identified. This is accomplished by immunofixation electrophoresis (IFE) as shown in Figure 4. The sensitivity of SPEP is about 50mg/L and that of IFE about 3 to 5X greater than SPEP. Nevertheless, monoclonal FLC are rarely identified by serum IFE because their concentration is too low.

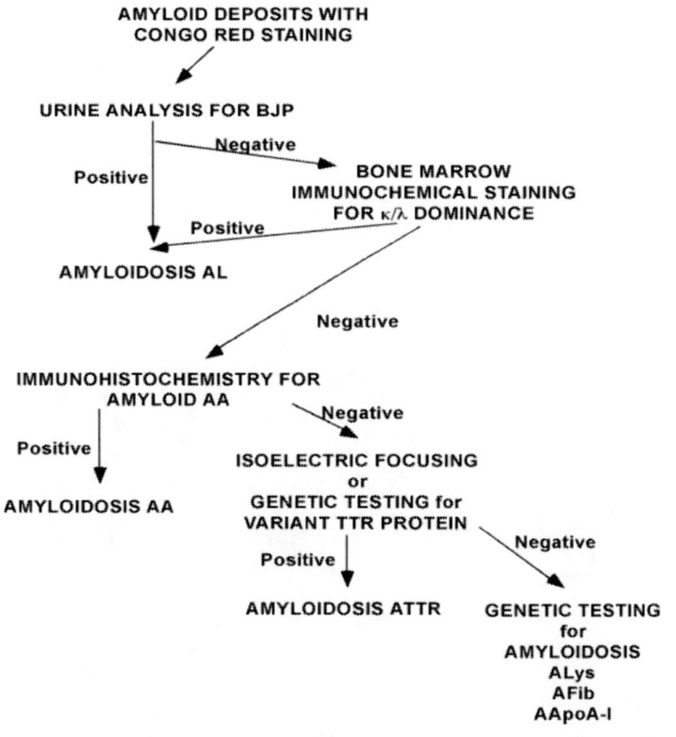

Figure 2. Laboratory testing for the types of amyloid.

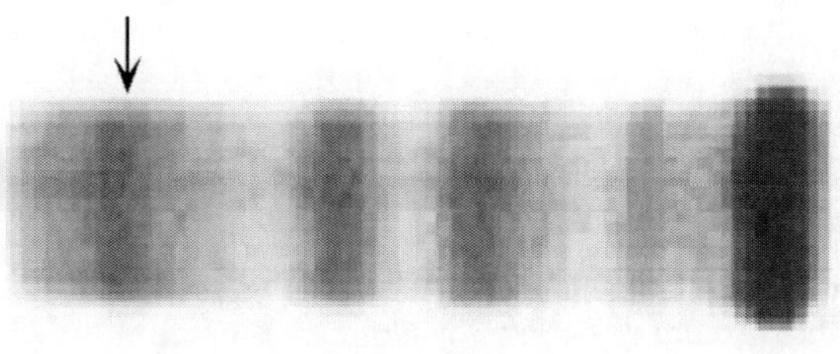

Figure 3. Nonclassical electrophoretic pattern. Compare this pattern with that in Figure 1. Notice the much smaller band as well as the diffuse staining in the gamma region that indicates normal levels of polyclonal Ig. This type of profile is commonly found in monoclonal gammopathy of undetermined significance and may be found with oligosecretory myeloma, although in the latter the polyclonal staining is usually decreased. In these cases, the concentration of M-protein in the band cannot be estimated by densitometry because of the dense staining adjacent to the band and a best estimate is made visually. The legend is similar to that in Figure 1.

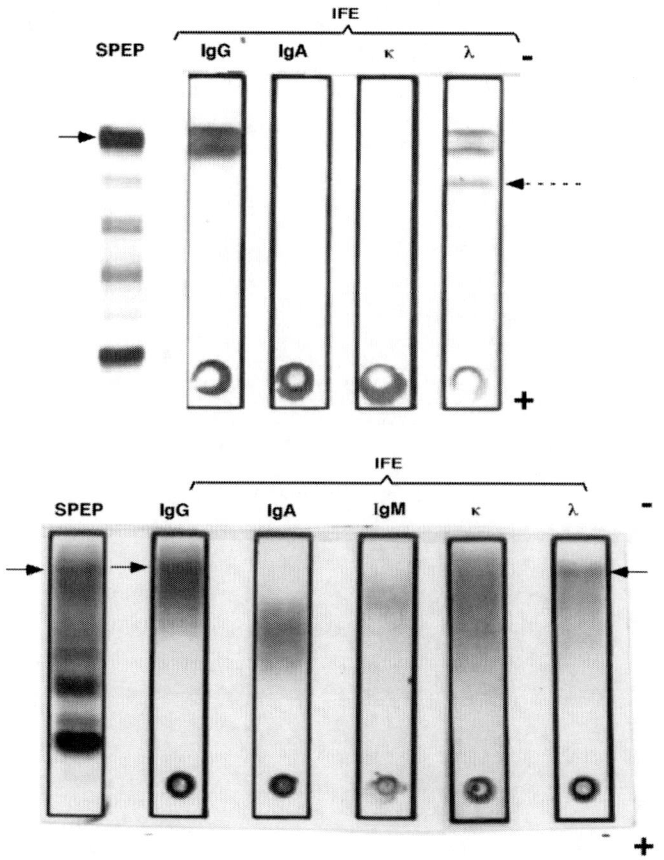

Figure 4. Serum protein electrophoresis (SPEP) with immuofixation electrophoresis (IFE). Serum protein electrophoresis is performed on several adjacent lanes. The lanes are then overlayed with antibody solutions, each has specificity for different immunoglobulins and a solution that precipitates all of the proteins (SPEP lane). In the Figure, antibodies were used against IgG, IgA, IgM, free and bound kappa and lambda light chains (these antisera react with light chain attached to heavy chain (intact Ig) and to free unattached light chain (FLC)). When a M-protein is present, it precipitates as a band in the lane in accordance with the specific antibody reaction. The band lines up at the same spot as on the SPEP lane. Above, fixation of a classical myeloma type SPEP pattern where the SPEP pattern is similar to that seen in Figure 1. In this case, the M-protein is identified as an IgG-lambda. There is also a lambda free light chain migrating just in front of the intact IgG-lambda (dashed arrow). Below, fixation of a specimen that showed a nonclassical SPEP as in Figure 3. The fixation shows the M-protein is an IgG-lambda that is more difficult to distinguish from the polyclonal staining than in the above pattern but nevertheless visually apparent. The wells at the bottom of each IFE lane are control wells that indicate a reaction has occurred.

Analysis by Urine Electrophoresis

Traditionally, monoclonal FLC of Ig have been best identified by urine IFE on a mechanically concentrated urine and measured by densitometric scan of the UPE profile as illustrated in Figure 5.

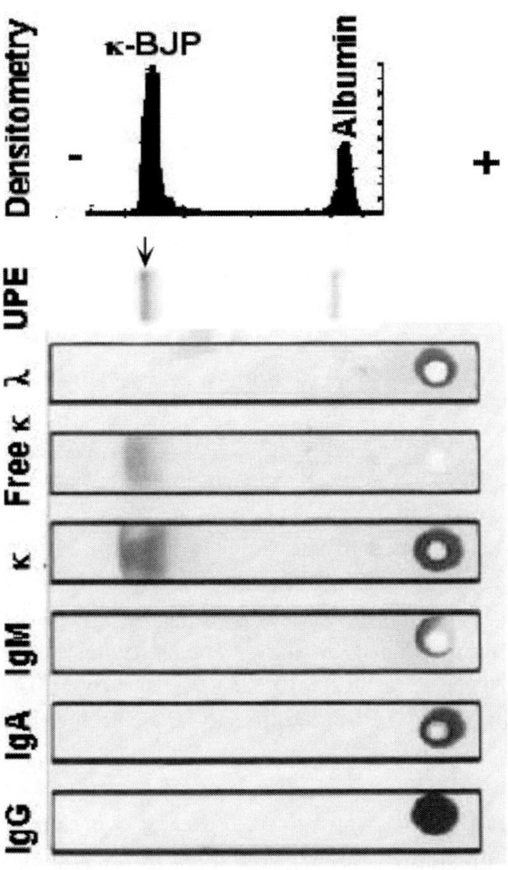

Figure 5. Urine protein electrophoresis (UPE) and immunofixation (IFE) on concentrated urine. The procedure is the same as for Figure 4, except one lane each is fixed with antibody against kappa and lambda free light chain only. The lanes labeled kappa and lambda are fixed with free and bound antibody while that labeled free is only fixed with antibody against free light chain (FLC). Otherwise details are similar to Figure 4. The densitometer tracing provides an estimate of the amount of M-protein from the UPE tracing. In this case, a light chain without a heavy chain is fixed, indicating a lambda FLC. This is confirmed by fixation in the free only lane.

Monoclonal FLC in urine are often called Bence Jones proteins (BJP) in honor of Dr. Bence-Jones who first detected them by the insensitive heat precipitation test [23]. Urine IFE is much more sensitive for identifying monoclonal FLC than serum IFE because urine can be concentrated mechanically before analysis. It is recommended that the urine be mechanically concentrated 100-fold [24]. Some protein electrophoretic methods claim greater sensitivity, nevertheless, at least a 10-fold concentration has been recommended for electrophoretic methods with greater analytic sensitivity claims [25].

Although IFE is a very sensitive technique for identifying BJP, measurement of BJP in urine by densitometry is inaccurate because it is unclear how well BJP stain with various protein dye indicators, tedious since urine may require concentration or dilution that often requires repeat analysis, and is often impossible because the BJP is migrating close to an intact monoclonal protein or accompanying beta-globulins [21, 26]. Moreover, because of the need to mechanically concentrate the urine and because IFE is not automated, these procedures are tedious and time consuming.

Analysis of FLC in Serum

A highly sensitive enhanced nephelometric commercially available automated serum FLC assay (FREELITE^{TM}) with sensitivity less than 1 mg/L is now available [27, 28]. This assay shows changes in concentrations of FLC with treatments and may be useful for following disease progression in nonsecretary and oligosecretory myeloma, and in amyloidosis AL [29-31]. A consensus committee recommended it for monitoring patients with these conditions [32]. The assay measures polyclonal FLC as well as monoclonal FLC. Since there is an elevation of kappa and lambda FLC in many conditions, the consensus committee suggested that an abnormal κ/λ ratio be used to identify a monoclonal gammopathy.

Although initial studies seemed to show that the serum FLC assay could replace urine IFE for most purposes [32-36], these conclusions may have been based on publication bias, since several later publications found that monoclonal FLC would be misidentified in many cases if the serum FLC assay was used alone. These studies suggested that urine IFE and serum FLC assay are complementary [20, 30, 37-40]. Moreover, it appears that when there is intact Ig that can be monitored by SPEP, serum FLC measurements do not add value and may be misleading [41, 42].

Among the reasons for difficulty with this sensitive assay are that the assay is not linear when monoclonal FLC is measured and that it grossly overestimates the monoclonal FLC concentration in many samples [37, 43-46]. Moreover, because the assay measures polyclonal FLC as well as monoclonal, when the concentration of monoclonal is low as compared to the concentration of polyclonal FLC, there is a masking effect on the κ/λ ratio so that it may misrepresent the contribution by monoclonal FLC [27, 47].

Also, at present, the exact reference ranges for the serum concentrations by the FLC assay for many patients remains ambiguous. Renal disease is one of the conditions that cause an elevation in polyclonal FLC. Reference ranges for persons without renal disease by the serum FLC assay are 3.3 – 19.4 mg/L for kappa FLC and 5.7 – 23.6 mg/L for lambda [48]. Reference ranges for persons with renal disease have been estimated to be as high as 251 mg/L for both serum κ-FLC and λ-FLC [49]. This is a very important problem since most patients with significant monoclonal gammopathies and especially amyloidosis AL develop renal disease at some point in the course of the monoclonal disease.

Complementarity and Comparison between the Serum FLC Assay and Urine IFE

Urine analysis by IFE is generally more specific for monoclonal FLC than analysis by the serum FLC assay. While the serum FLC assay reacts with polyclonal FLC as well as monoclonal, monoclonal proteins in urine usually migrate as a distinct entity, as shown in Figure 5, while the polyclonal FLC is dispersed and does not usually interfere with interpretation. Also, it was demonstrated by Palladini, et al. that urinary IFE was more diagnostically sensitive for identifying monoclonal FLC than the serum FLC κ/λ ratio in patients with amyloidosis in that urine IFE detected 109 of 115 patients with amyloidosis AL while FLC ratio detected only 88. This means that the serum FLC ratio missed 21 or 19% that urine IFE detected [39]. Although a number of other studies have come to the opposite conclusion, close examination of these shows that the urine samples were not appropriately concentrated or analyzed in these studies [50], while the study showing superiority of urine IFE concentrated all samples 100-fold and identified amyloidosis by the electron microscope [39].

UPE shows a sensitivity of about 100 mg/L but after a 100X mechanical concentration that would bring the sensitivity down to about 1 mg/L. This means that after the mechanical concentration, UPE shows about 1 mg/L sensitivity and since IFE produces about 3 – 5X more sensitivity than UPE [51, 52], the techniques used by Palladini were well within sensitivity claimed for the serum FLC assay. UPE is notorious for missing low concentration BJP, even with mechanical concentration [21, 52], thus, it is important to use IFE on all samples likely to have a monoclonal FLC.

Nevertheless, urine electrophoretic methods are very tedious and during the mechanical concentration step protein can be absorbed on the walls of the concentrating devise so that the final protein concentration is inexact. Also, densitometry is inaccurate because it depends on measurement of total protein in the urine that is usually obtained by inaccurate dye binding methods [21]. Besides, It is important to note that, in some cases, correct identification of BJP on IFE not only requires expert interpretation, but may require dilution and repeated analysis or even repeated analysis by other methods such as isoelectric focusing. This may occur due to interference by the polyclonal multiple (ladder) banding pattern, other artifacts that may occur due to high mechanical concentration and in some cases BJP polymers or fragments migrating as more than one band [21, 53-55].

The automated serum FLC assay is clearly less tedious than urine electrophoresis. Nevertheless, it should be pointed out that many samples need to be manually diluted and repeated as a result of a wide spread prozoning effect which leads to more tedium than originally thought [37, 40, 43, 56, 57]. Neverthelesss, growing evidence indicates that the serum FLC assay is not quantitative and may be misleading since it is expressed in units that might impart false confidence. This is because, when monoclonal FLC are measured, the serum FLC assay shows nonparallelism with calibrators and among other samples containing monoclonal FLC, nonlinearity and appears to substantially overestimate monoclonal FLC in many samples [43, 45, 46, 58, 59]. Thus, although there may be good quantification for polyclonal FLC, measurement of monoclonal FLC are fraught with problems. Hence, in spite of inaccurate measurement of FLC levels by urine electrophoresis, it remains unclear that serum FLC assay is any more accurate.

At present, the exact reference ranges for the serum concentrations by the FLC assay for many conditions remains ambiguous. Reference ranges for persons without renal disease are 3.3 – 19.4 mg/L for kappa FLC and 5.7 – 23.6 mg/L for lambda [48]. Reference ranges for persons with renal disease have been estimated to be as high as 251 mg/L for both serum κ-FLC and λ-FLC [49].

Therapy for B-cell proliferative diseases has improved greatly in the recent times. As a result, noninvasive biomarkers that suggest remission are important. Biomarkers are especially important in amyloidosis AL because the bone marrow often shows normal levels of plasma cells so that highly specialized techniques may be necessary to elucidate whether there remains a clonal dominance of plasma cells after therapy [18]. Along with other criteria, the International Myeloma Working Group suggests complete remissions be defined as a normal serum FLC ratio [60, 61]. But the Myeloma Working Group points out that, in assessing remission, either the involved FLC or the difference between the involved and uninvolved should be used because the ratios generated after therapy may reflect the degree of therapy suppression more than the tumor burden [32]. As mentioned above, the reference ranges for the concentrations of the kappa and lambda FLC vary with renal disease and there is no information regarding reference ranges for the difference between involved and uninvolved FLC. In part, this difficulty can be traced to an imperfect methodology whereby the assay is directed against polyclonal FLC, not monoclonal. Thus, when the levels of serum FLC approach the reference range, the serum assay result shows ambiguity. To assess potential complete remissions, urine IFE for BJP remains an important complementary assay.

The Consensus Opinion from the 10th Symposium on amyloid and amyloidosis regarding the hematological response to treatment stated that disappearance of the BJP detected by IFE from a concentrated urine specimen should supplement evaluation by the serum FLC ratio, even if the serum FLC ratio becomes normal [60]. Moreover, the Concensus Opinion recommended that serum FLC concentrations below 100 mg/L, although abnormal, not be considered a criterion for evaluation of the hematological response due to imperfections in the serum FLC method and that supplementary assay by IFE on a concentrated urine specimen is recommended [60].

Nevertheless, in conditions of FLC myeloma, amyloidosis AL, and nonsecretory myeloma, it is not possible to monitor progress by SPEP. Here, because of its greater simplicity, the serum FLC assay is useful for providing trend information, but it is important that both laboratory practitioners and clinicians be aware of the pitfalls. Also, when the concentrations of serum FLC fall below 100 mg/L or the serum FLC concentrations are not supported by clinical information, urine IFE should be considered.

Screening for Plasma Cell Disease

Screening for plasma cell disease takes several forms. The general screening from the non-hematologist is rule-out. Vague symptoms that could be related to plasma cell disease are ruled-out by ordering a SPEP. This would seem to be a cost effective approach since the annual incidence of MM in Western countries is only 5.6 age adjusted cases per 100,000 [5]. This accounts for the vast majority of SPEP ordered by clinicians. The nephrologist orders UPE mainly to identify whether or not proteinuria is of glomerular or tubular origin but also as a rule-out for BJP [52, 62]. Occasionally a M-protein is identified during general screening and rarely a BJP during urine screening that may suggest referral to a hematologist. The hematologist approaches screening as rule-in since referred patients have a higher likelihood of plasma cell disease. In this case, SPEP, serum IFE and urine IFE are recommended [63]. There are now recommendations that the serum FLC assay can replace urine IFE for all diagnosis except amyloidosis AL but there are no data that fully address the issue [32, 33, 35, 64].

Table 1. Algorithm for the Investigation of Monoclonal Gammopathies

(Service) Type of Screening	Initial Testing SPEP	Reflex Testing SIFE	SFLC	Urine BJP
Hematology Screening				
Classical myeloma SPEP profile (paraprotein > 30 g/L with decrease in noninvolved Ig)	Paraprotein >30	Y	Y	N
Serum FLC within reference range				Y
Paraprotein < 30 g/L Oligosecretory myloma Nonsecretary myeloma (there is often a decrease in noninvolved Ig)	Paraprotein <30	Y	Y	Y
(Service) Type of Screening	Initial Testing SPEP	Reflex Testing SIFE	SFLC	
MGUS (there is usually little or no decrease in noninvolved polyclonal Ig)	Paraprotein < 30	Y	N	N
No monoclonal banding (Normal Ig)	No monoclonal banding	Y	Y	N
Serum FLC within reference range				Y
Suspected or evidence for Amyloidosis or FLC disease	No monoclonal banding or paraprotein < 15 g/L	Y	Y	Y
Nephrology Screening Urine Profile. Description of the type of proteinuria as: Primarily tubular, glomerular or mixed.				Y
If an intact Ig or a BJP is found recommend serum analysis to better define the monoclonal gammopathy and serum FLC to establish a serum baseline value.				

(Service) Type of Screeing	***Urine BJP***			
Routine Sceening Classical myeloma SPE profile (paraprotein > 30 g/L with decrease in noninvolved Ig)	Paraprotein >30 g/L	Y	Y	N
Serum FLC within reference range				Y
Paraprotein < 3 g/dL Possible Oligosecretory meloma	< 30 g/L	Y	Y	

Table 1. (Continued)

(Service) Type of Screening	Initial Testing SPEP	Reflex Testing SIFE	 SFLC	
(there is a decrease in noninvolved Ig) or abnormal hemoglobin, calcium, kidney function, urinalysis, or bone pain or neurological complications.				Y
Serum FLC negative profile				
Possible MGUS (there is usually no little or no decrease in noninvolved Ig) Serum FLC negative profile hemoglobin, calcium, kidney function, urinalysis unremarkable and no neurological complications.	< 15 g/L	Y	N	N
No monoclonal banding No decrease in polyclonal Ig	No monoclonal Banding	N	N	N
No monoclonal banding Decreased polyclonal Ig hemoglobin, calcium, kidney function, urinalysis unremarkable and no neurological complications. Based on clinical input	No monoclonal banding	Y/N	Y/N	Y/N

Table 1 illustrates a proposed algorthm for initial screening. Although diagnosis is largely dependent on bone marrow and clinical findings, the electrophoretic findings are helpful in pointing the direction for additional testing.

The type of screening and clinical information should be interweaved. This is usually not possible in an outside reference laboratory but in a hospital laboratory where reflex testing can be performed the pathology staff should be aware of the reason for the screening and the characteristic of the patient. For example, patients with a monoclonal spike < 1.5 g/ dL and no other abnormal myeloma related end organ damage have an extremely low risk of progression so that, except for annual serum electrophoretic testing, further work-up is not needed since the risk of 10 year progression is < 11%. On the other hand, the risk rises to 34% at 3 g/dL [2, 65]. The risk of progression was <6% for monoclonal proteins < 0.5 g/dL so that older persons have little chance of developing active disease during their life time and over-investigation of a very low level M-protein in asymptomatic elderly persons has been discouraged [66]. IgM and IgA isotypes have increased risk of progression to significant disease with a relative risk (RR) of 46 for developing WS when an IgM is found as compared to a RR of 25 of developing MM, so that reflex testing at > 1g/dL may be appropriate.

In my view, all UPE ordered by nephrologists should be reflexed for an IFE since, although detection of a BJP is rare, these patients are at a higher risk for light chain disease than those without proteinuria or kidney disease. Usually, hypertension or diabetes is considered responsible for the proteinuria, but when a BJP is identified, if FLC disease or amyloidosis is found to be responsible different treatments are required. Also, the cost is not great because the number ordered is small compared to other types of screening. Fixation for free and bound kappa and lambda is all that is needed as the initial screen. If one of these shows a monoclonal band, the analysis can be repeated with fixation for a more complete battery of monoclonal proteins as shown in Figure 5.

For requests from the hematology service it is often know that there is a monoclonal gammopathy or the likelihood is much greater, so the process is not one of screening but one of investigation as to exactly what is present. It is especially important for the laboratory to have a good idea of what the hematologist is looking for. The International Myeloma Working Group guidelines recommends performing the serum FLC assay rather than urine IFE, except when screening for amyloidosis. [32]. Nevertheless, since the guidelines were issued, it has become apparent that there are a variety of conditions when the serum FLC κ/λ ratio is within the normal reference range but Bence Jones protein is identified in urine [47, 50]. Thus, Table 1 recommends urine IFE whenever the κ/λ ratio for the serum FLC assay is within the reference range. In cases of oligosecretory disease, FLC disease or amyloidosis, it is desirable to perform urine IFE, to obtain a baseline, even if the serum FLC assay is positive, since absence of a urinary BJP after successful treatment, when the serum FLC has dropped near normal levels, is helpful information.

For routine screening, if a monoclonal gammopathy is identified, it is likely the patient will be referred to the hematologist. This is true in all cases where the M-protein is greater than 30 g/L and most cases where an IgG M-protein is greater than 15 g/L. Whenever there is a significantly elevated M-protein, it is desirable to identify whether there is a low hemoglobin, renal disease, neurological dysfunction, or bone pain to help determine whether or not additional reflex testing is needed. It is likely that as more information regarding the serum FLC assay accumulates various official consensus committees will issue new guidelines. It is important to remember that a guideline is only a guideline. No algorithm can cover every case so that the laboratory should always be open to clinical input for initiating additional testing.

References

[1] Kyle RA, Remstein ED, Therneau TM, Dispenzieri A, Kurtin PJ, Hodnefield JM, et al. Clinical course and prognosis of smoldering (asymptomatic) multiple myeloma. *N. Engl. J. Med.* 2007;356:2582-90.

[2] Kyle RA, Therneau TM, Rajkumar SV, Offord JR, Larson DR, Plevak MF, Melton LJ, 3rd. A long-term study of prognosis in monoclonal gammopathy of undetermined significance. *N. Engl. J. Med.* 2002;346:564-9.

[3] Shaheen SP, Talwalkar SS, Medeiros LJ. Multiple myeloma and immunosecretory disorders: an update. *Adv. Anat. Pathol.* 2008;15:196-210.
[4] Munshi NC, Digumarthy S, Rahemtullah A. Case records of the Massachusetts General Hospital. Case 13-2008. A 46-year-old man with rheumatoid arthritis and lymphadenopathy. *N. Engl. J. Med.* 2008;358:1838-48.
[5] Palumbo A, Anderson K. Multiple myeloma. *N. Engl. J. Med.* 2011;364:1046-60.
[6] Gertz MA. Waldenstrom macroglobulinemia: a review of therapy. *Am. J. Hematol.* 2005;79:147-57.
[7] Attaelmannan M, Levinson SS. Understanding and identifying monoclonal gammopathies. *Clin. Chem.* 2000;46:1230-8.
[8] Alexanian R, Weber D, Liu F. Differential diagnosis of monoclonal gammopathies. *Arch. Pathol. Lab. Med.* 1999;123:108-13.
[9] Ropper AH, Gorson KC. Neuropathies associated with paraproteinemia. *N. Engl. J. Med.* 1998;338:1601-7.
[10] Dispenzieri A, Kyle RA. Neurological aspects of multiple myeloma and related disorders. *Best Pract. Res. Clin. Haematol.* 2005;18:673-88.
[11] Ropper AH, Raje NS, Lawrimore TM, Camelo-Piragua S, Sohani AR. Case records of the Massachusetts General Hospital. Case 7-2010. A 49-year-old man with peripheral neuropathy and ascites. *N. Engl. J. Med.* 2010;362:929-40.
[12] Watanabe O, Arimura K, Kitajima I, Osame M, Maruyama I. Greatly raised vascular endothelial growth factor (VEGF) in POEMS syndrome. *Lancet* 1996;347:702.
[13] Sghirlanzoni A, Solari A, Ciano C, Mariotti C, Fallica E, Pareyson D. Chronic inflammatory demyelinating polyradiculoneuropathy: long-term course and treatment of 60 patients. *Neurol. Sci.* 2000;21:31-7.
[14] Guidelines on the diagnosis and management of AL amyloidosis. *Br. J. Haematol.* 2004;125:681-700.
[15] Herrera GA, Turbat-Herrera EA. Renal diseases with organized deposits: an algorithmic approach to classification and clinicopathologic diagnosis. *Arch. Pathol. Lab. Med.* 2010;134:512-31.
[16] Rajkumar SV, Dispenzieri A, Kyle RA. Monoclonal gammopathy of undetermined significance, Waldenstrom macroglobulinemia, AL amyloidosis, and related plasma cell disorders: diagnosis and treatment. *Mayo. Clin. Proc.* 2006;81:693-703.
[17] Lachmann HJ, Booth DR, Booth SE, Bybee A, Gilbertson JA, Gillmore JD, et al. Misdiagnosis of hereditary amyloidosis as AL (primary) amyloidosis. *N. Engl. J. Med.* 2002;346:1786-91.
[18] Falk RH, Comenzo RL, Skinner M. The systemic amyloidoses. *N. Engl. J. Med.* 1997;337:898-909.
[19] Strobel SL. Transient paraproteinemia: an intriguing immunological anomaly. *Ann. Clin. Lab. Sci.* 2003;33:265-70.
[20] Beetham R, Wassell J, Wallage MJ, Whiteway AJ, James JA. Can serum free light chains replace urine electrophoresis in the detection of monoclonal gammopathies? *Ann. Clin. Biochem.* 2007;44:516-22.
[21] Levinson SS, Keren DF. Free light chains of immunoglobulins: clinical laboratory analysis. *Clin. Chem.* 1994;40:1869-78.
[22] Herrera GA, Teng J, Turbat-Herrera EA. Renal amyloidosis: current views on pathogenesis and impact on diagnosis. *Contrib. Nephrol.* 2011;169:232-46.

[23] Bence Jones H. Papers on chemical pathology: lecture III. *Lancet* 1847;ii:88-92.

[24] Keren DF, Alexanian R, Goeken JA, Gorevic PD, Kyle RA, Tomar RH. Guidelines for clinical and laboratory evaluation of patients with monoclonal gammopathies. *Arch. Pathol. Lab. Med.* 1999;123:106-7.

[25] Roden AC, Lockington KS, Tostrud LJ, Katzmann JA. Urine protein electrophoresis and immunoelectrophoresis using unconcentrated or minimally concentrated urine samples. *Am. J. Clin. Pathol.* 2008;130:141-5.

[26] Levinson SS. Urine protein electrophoresis and immunofixation electrophoresis supplement one another in characterizing proteinuria. *Ann. Clin. Lab. Sci.* 2000;30:79-84.

[27] Bradwell AR, Carr-Smith HD, Mead GP, Harvey TC, Drayson MT. Serum test for assessment of patients with Bence Jones myeloma. *Lancet* 2003;361:489-91.

[28] Katzmann JA. Quantitative free light chain assays for the diagnosis and monitoring of monoclonal gammopathies. *J. Clin. Ligand. Assay* 2004;27:246-55.

[29] Matsuda M, Yamada T, Gono T, Shimojima Y, Ishii W, Fushimi T, et al. Serum levels of free light chain before and after chemotherapy in primary systemic AL amyloidosis. *Intern. Med.* 2005;44:428-33.

[30] Sanchorawala V, Seldin DC, Magnani B, Skinner M, Wright DG. Serum free light-chain responses after high-dose intravenous melphalan and autologous stem cell transplantation for AL (primary) amyloidosis. *Bone Marrow Transplant.* 2005;36:597-600.

[31] Drayson M, Tang LX, Drew R, Mead GP, Carr-Smith H, Bradwell AR. Serum free light-chain measurements for identifying and monitoring patients with nonsecretory multiple myeloma. *Blood* 2001;97:2900-2.

[32] Dispenzieri A, Kyle R, Merlini G, Miguel JS, Ludwig H, Hajek R, et al. International Myeloma Working Group guidelines for serum-free light chain analysis in multiple myeloma and related disorders. *Leukemia* 2009;23:215-24.

[33] Hill PG, Forsyth JM, Rai B, Mayne S. Serum free light chains: an alternative to the urine Bence Jones proteins screening test for monoclonal gammopathies. *Clin. Chem.* 2006;52:1743-8.

[34] Katzmann JA. Screening panels for monoclonal gammopathies: time to change. *Clin. Biochem. Rev.* 2009;30:105-11.

[35] Katzmann JA, Dispenzieri A, Kyle RA, Snyder MR, Plevak MF, Larson DR, et al. Elimination of the need for urine studies in the screening algorithm for monoclonal gammopathies by using serum immunofixation and free light chain assays. *Mayo Clin. Proc.* 2006;81:1575-8.

[36] Piehler AP, Gulbrandsen N, Kierulf P, Urdal P. Quantitation of serum free light chains in combination with protein electrophoresis and clinical information for diagnosing multiple myeloma in a general hospital population. *Clin. Chem.* 2008;54:1823-30.

[37] Levinson SS. Hook Effect with Lambda Free Light Chain in Serum Free Light Chain Assay. *Clin. Chim. Acta* 2010;411:1834-6.

[38] Shaheen SP, Levinson SS. Serum free light chain analysis may miss monoclonal light chains that urine immunofixation electrophoreses would detect. *Clin. Chim. Acta* 2009;406:162-6.

[39] Palladini G, Russo P, Bosoni T, Verga L, Sarais G, Lavatelli F, et al. Identification of amyloidogenic light chains requires the combination of serum-free light chain assay with immunofixation of serum and urine. *Clin. Chem.* 2009;55:499-504.
[40] Tate JR, Gill D, Cobcroft R, Hickman PE. Practical considerations for the measurement of free light chains in serum. *Clin. Chem.* 2003;49:1252-7.
[41] Singhal S, Vickrey E, Krishnamurthy J, Singh V, Allen S, Mehta J. The relationship between the serum free light chain assay and serum immunofixation electrophoresis, and the definition of concordant and discordant free light chain ratios. *Blood* 2009;114:38-9.
[42] Uljon SN, Richardson PG, Schur PH, Anderson KC, Tanasijevic MJ, Lindeman NI. Serial serum free light chain measurements do not detect changes in disease status earlier than electrophoretic M-spike measurements in patients with intact immunoglobulin myeloma. *Clin. Chim. Acta* 2011;412:562-8.
[43] Tate JR, Mollee P, Dimeski G, Carter AC, Gill D. Analytical performance of serum free light-chain assay during monitoring of patients with monoclonal light-chain diseases. *Clin. Chim. Acta* 2007;376:30-6.
[44] Levinson SS. Studies of Bence Jones proteins by immunonephelometry. *Annals Clin. Lab. Sci.* 1992;22:100-9.
[45] Briand PY, Decaux O, Caillon H, Grosbois B, Le Treut A, Guenet L. Analytical performance of the serum free light chain assay. *Clin. Chem. Lab. Med.* 2010;48:73-9.
[46] Levinson SS. Regarding the Overestimation of Serum {kappa} Free Light Chains. *Clin. Chem.* 2011;57:775-7.
[47] Levinson SS. Polyclonal free light chain of Ig may interfere with interpretation of monoclonal free light chain kappa/lambda ratio. *Ann. Clin. Lab. Sci.* 2010;40:348-53.
[48] Katzmann JA, Clark RJ, Abraham RS, Bryant S, Lymp JF, Bradwell AR, Kyle RA. Serum reference intervals and diagnostic ranges for free kappa and free lambda immunoglobulin light chains: relative sensitivity for detection of monoclonal light chains. *Clin. Chem.* 2002;48:1437-44.
[49] Hutchison CA, Harding S, Hewins P, Mead GP, Townsend J, Bradwell AR, Cockwell P. Quantitative assessment of serum and urinary polyclonal free light chains in patients with chronic kidney disease. *Clin. J. Am. Soc. Nephrol.* 2008;3:1684-90.
[50] Levinson S. Urine immunofixation electrophoresis remains important and is complementary to serum free light chain. *Clin. Chem. Lab. Med.* 2011; 49 (10) in press.
[51] Levinson SS, Elin RJ, Yam L. Light chain proteinuria and lysozymuria in a patient with acute monocytic leukemia. *Clin. Chem.* 2002;48:1131-2.
[52] Kaplan IV, Levinson SS. Misleading urinary protein pattern in a patient with hypogammaglobulinemia: effects of mechanical concentration of urine. *Clin. Chem.* 1999;45:417-9.
[53] Harrison HH. The "ladder light chain" or "pseudo-oligoclonal" pattern in urinary immunofixation electrophoresis (IFE) studies: a distinctive IFE pattern and an explanatory hypothesis relating it to free polyclonal light chains. *Clin. Chem.* 1991;37:1559-64.
[54] MacNamara EM, Aguzzi F, Petrini C, Higginson J, Gasparro C, Bergami MR, et al. Restricted electrophoretic heterogeneity of immunoglobulin light chains in urine: a cause for confusion with Bence Jones protein. *Clin. Chem.* 1991;37:1570-4.

[55] Hess PP, Mastropaolo W, Thompson GD, Levinson SS. Interference of polyclonal free light chains with identification of Bence Jones proteins. Clin Chem 1993;39:1734-8.

[56] Murata K, Clark RJ, Lockington KS, Tostrud LJ, Greipp PR, Katzmann JA. Sharply increased serum free light-chain concentrations after treatment for multiple myeloma. *Clin. Chem*. 2010;56:16-8.

[57] Tate J, Bazeley S, Sykes S, Mollee P. Quantitative serum free light chain assay--analytical issues. *Clin. Biochem. Rev*. 2009;30:131-40.

[58] de Kat Angelino CM, Raymakers R, Teunesen MA, Jacobs JF, Klasen IS. Overestimation of serum kappa free light chain concentration by immunonephelometry. *Clin. Chem*. 2010;56:1188-90.

[59] Tate J, Mollee P, Johnson R. Monoclonal gammopathies--clinical and laboratory issues. *Clin. Biochem. Rev*. 2009;30:89-91.

[60] Durie BG, Harousseau JL, Miguel JS, Blade J, Barlogie B, Anderson K, et al. International uniform response criteria for multiple myeloma. *Leukemia* 2006;20:1467-73.

[61] Mollee P. Current trends in the diagnosis, therapy and monitoring of the monoclonal gammopathies. *Clin. Biochem. Rev*. 2009;30:93-103.

[62] Singhal S, Stein R, Vickrey E, Mehta J. The serum-free light chain assay cannot replace 24-hour urine protein estimation in patients with plasma cell dyscrasias. *Blood* 2007;109:3611-2.

[63] The international myeloma working group. Criteria for the classification of monoclonal gammopathies, multiple myeloma and related disorders: a report of the myeloma working group. *Br. J. Haematol*., Vol. 121, 2003:749-57.

[64] Abadie JM, van Hoeven KH, Wells JM. Are renal reference intervals required when screening for plasma cell disorders with serum free light chains and serum protein electrophoresis? *Am. J. Clin. Pathol.* 2009;131:166-71.

[65] Blade J. Clinical practice. Monoclonal gammopathy of undetermined significance. *N. Engl. J. Med*. 2006;355:2765-70.

[66] Smith A, Wisloff F, Samson D. Guidelines on the diagnosis and management of multiple myeloma 2005. *Br. J. Haematol*. 2006;132:410-51.

In: Amyloids
Editors: I. P. Halcheck et al.

ISBN: 978-1-62100-538-4

Chapter 13

TRANSTHYRETIN VAL30MET FAMILIAL AMYLOID POLYNEUROPATHY IN JAPAN

Haruki Koike*, Fumiaki Tanaka and Gen Sobue
Department of Neurology,
Nagoya University Graduate School of Medicine,
Nagoya, Japan

ABSTRACT

Recent biochemical and molecular analyses suggest that familial amyloid polyneuropathy (FAP) is not as rare as previously thought. Transthyretin Val30Met-associated FAP (FAP ATTR Val30Met) is the most common form of FAP. Although patients with FAP ATTR Val30Met are geographically concentrated to Japan, Portugal, and Sweden, a late-onset form of this type of FAP was discovered in non-endemic areas and is widely distributed throughout the world.

In this article, we describe the clinical, electrophysiologic, and histopathologic features of FAP ATTR Val30Met in Japanese patients by comparing conventional early-onset cases from endemic foci and late-onset cases from non-endemic areas. When compared with the classical phenotype of FAP, the clinicopathologic features of FAP patients from non-endemic areas tend to be nonspecific.

As a result, a diagnosis of FAP may be missed unless amyloid deposition is seen on sural nerve biopsy. Therefore, an in depth knowledge of the characteristics of FAP ATTR Val30Met as well as a high degree of clinical suspicion is required to achieve early diagnosis.

Keywords: amyloid, amyloidosis, familial amyloid polyneuropathy, transthyretin.

* Address correspondence: Haruki Koike, MD, PhD or Fumiaki Tanaka, MD, PhD, Department of Neurology, Nagoya University Graduate School of Medicine, Nagoya 466-8550 Japan. Fax:+81-52-744-2393, Tel:+81-52-744-2391, E-mail: koike-haruki@med.nagoya-u.ac.jp or ftanaka@med.nagoya-u.ac.jp

INTRODUCTION

Familial amyloid polyneuropathy (FAP) is a hereditary form of amyloidosis arising from mutated amyloidogenic proteins, such as transthyretin (ATTR), gelsolin (Agel), and apolipoprotein A-I (AApoA-I) (Benson et al., 2007; Koike and Sobue, 2010a). Of these mutated proteins, ATTR is the most common cause of FAP. Since the first report by Andrade in 1952, endemic foci of FAP have been discovered in Portugal, Japan, and Sweden (Andrade, 1952; Araki et al., 1968; Andersson et al., 1976). Subsequent development of biochemical and molecular analyses revealed that FAP cases in these endemic foci were caused by a substitution of guanine by adenine in exon 2, leading to a substitution mutation of valine to methionine in position 30 of TTR (Tawara et al., 1983). Initially, FAP was thought to be a disease restricted to the endemic foci in these countries. However, approximately 100 other mutations in the TTR gene have since been identified throughout the world, and FAP is now considered to be more prevalent in Japan and elsewhere than was originally thought. Nevertheless, FAP ATTR Val30Met is still the most common type of FAP (Benson, 1977; Ikeda et al., 1987, 2002; Nakazato et al., 1992a; Holmgren et al., 1994; Sousa et al., 1995; Reilly et al., 1995; Plante-Bordeneuve et al., 2007).

In Japan, patients with FAP ATTR Val30Met are concentrated to two endemic foci, Arao and Ogawa (Ikeda et al., 1987, 2002; Nakazato et al., 1992a). Cases in these areas show characteristic features, including age of onset from the late 20s to the early 40s, a high penetrance rate, marked autonomic dysfunction, loss of superficial sensation, including nociception and thermal sensation (i.e., sensory dissociation), atrioventricular conduction block requiring pacemaker implantation, and anticipation in the age of onset (Ikeda et al., 1987; Nakazato et al., 1992a; Ando et al., 1998; Misu et al., 2000; Koike et al., 2002).

By contrast, a late-onset form of FAP ATTR Val30Met unrelated to the endemic foci has also been described (Misu et al., 1999; Koike et al., 2002, 2004, 2011; Plante-Bordeneuve et al., 2007). The typical features of these cases are distinct from those related to the endemic foci and include an age of onset over 50 years, a low penetrance rate, relatively mild autonomic dysfunction, loss of all sensory modalities rather than sensory dissociation, frequent presence of cardiomegaly, extreme male preponderance, and the absence of anticipation in the age of onset (Misu et al., 1999, 2000; Koike et al., 2002, 2004; Sobue and Koike, 2003; Koike and Sobue 2010b). Clinical differences between early- and late-onset FAP ATTR Val30Met have also been reported in Portugal (Conceição et al., 2007). Through the development of genetic diagnostic techniques, the late-onset type of FAP ATTR Val30Met has been determined to be more common than previously thought.

Because the features of classical FAP are more widely known than those of late-onset FAP, a diagnosis of sporadic late-onset FAP is often missed or delayed (Ando et al., 1993; Ikeda et al., 2007; Koike and Sobue 2010a, 2010b; Koike et al., 2011). Liver transplantation is an established treatment for FAP (Holmgren et al., 1993), but patients diagnosed at older ages may not be candidates for such therapy. Furthermore, recent advances in other therapeutic strategies, such as the prevention of amyloid fibril formation, the reduction of TTR deposition by immunization, or gene therapy (Sekijima et al., 2005, 2008; Terazaki et al., 2006; Cardoso et al., 2010; Macedo et al., 2008, 2010), increases the value of early diagnosis of patients with FAP more important. Thus, clinicians should be aware of the variability in the clinical, electrophysiologic, and histopathologic features of FAP in order to facilitate early diagnosis, particularly for sporadic cases of late-onset FAP ATTR Val30Met in non-endemic areas.

This article reviews the clinical, electrophysiologic, and histopathologic features of FAP ATTR Val30Met by comparing conventional early-onset cases from endemic foci to late-onset cases from non-endemic areas in Japan.

Clinical Features

In Japan, there are conventionally well-known endemic foci of FAP ATTR Val30Met in Arao City in Kumamoto Prefecture and in Ogawa Village in Nagano Prefecture (Araki et al., 1968; Ikeda et al., 1987, 2002; Nakazato et al., 1992a; Koike et al., 2002; Ando et al., 2005). The clinical features of FAP ATTR Val30Met in these two endemic foci are similar (Araki et al., 1968; Ikeda et al., 1987). Although there are some exceptions, the characteristics of cases in these endemic foci include equal distribution among males and females, an age of onset between late 20s to the early 40s, polyneuropathy with sensory dissociation, and autonomic dysfunction. The initial symptoms of these cases are primarily dysesthesias of the lower limbs or digestive symptoms, such as vomiting, diarrhea, and constipation. The sensory deficit starts in the distal portions of the lower extremities and progresses in an ascending fashion, representing a symmetric polyneuropathy pattern. Superficial sensation, such as nociception and thermal sensations, is affected in the early stage of the disease, while deep sensation, such as vibrational and positional sensations, is lost in the later phase. This characteristic mode of sensory deficit is known as sensory dissociation. A burning sensation or a shooting pain in the lower limbs is common. Muscle weakness and atrophy occur several years following the onset of the sensory deficits. As for the autonomic symptoms, alternating episodes of diarrhea and constipation are the most common manifestation, and are followed by orthostatic hypotension and bladder dysfunction.

The clinical features of FAP ATTR Val30Met cases with no relationship to the endemic foci are much different from those of the conventional early-onset cases in the endemic foci (Misu et al., 1999; Koike et al., 2002, 2011). A previous study investigated the characteristics of 35 late-onset FAP ATTR Val30Met cases with no relationship to endemic foci (Misu et al., 1999). Inclusion criteria of this study were polyneuropathy, Val30Met mutation of the *TTR* gene, onset at age 50 years or older, and no relationship to endemic foci for at least the two most prior generations of family members. A sural nerve biopsy specimen was obtained from 20 of 35 patients. Congo red staining was performed in these cases, and amyloid deposition was detected in 18 cases. Features included an extreme male preponderance (male:female, 10.7:1), polyneuropathy with loss of all sensory modalities, and mild autonomic dysfunction. The initial symptoms were paresthesia in 30 patients and weakness in five patients. In the majority of patients, the sensory involvement was similar for the superficial and deep sensory modalities. Dissociation of the sensory impairment between modalities was not prominent in most cases, and autonomic dysfunction was not seen as an initial symptom in any of the patients. At the time of the initial examination, orthostatic hypotension was present in 18 of the 35 patients. Mild to moderate alternating episodes of diarrhea and constipation were noted in 13 patients, and mild to moderate disturbance of urination (mainly hesitancy) was present in 15 patients. Decreased sweating, particularly in the legs, was observed in 19 patients. Erectile dysfunction was encountered in 17 out of 23 male patients.

These autonomic symptoms were generally mild to moderate in severity and did not substantially affect the quality or level of activity of the patients' daily life, except for in three patients who had syncopal attacks. Disease duration in these patients tended to be shorter when compared with that of early-onset cases from the endemic foci (Koike et al., 2009). The cause of death in these patients was mainly due to heart failure (Koike et al., 2009).

To confirm the presence of the late-onset type of FAP ATTR Val30Met unrelated to the endemic foci, we conducted a nationwide survey in Japan (Koike et al., 2002). In this survey, we compared the clinical and geographic features of the two groups of FAP ATTR Val30Met as stratified by age of onset (before or after 50 years of age). This study included 82 families with early-onset disease and 59 families with late-onset disease. The results of this nationwide survey confirmed earlier impressions that late-onset FAP ATTR Val30Met with atypical features was more prevalent than previously believed. Most patients included in the early-onset group of this survey had the conventional form of FAP ATTR Val30Met in the endemic foci (Arao and Ogawa). The distinctive features of early- and late-onset FAP ATTR Val30Met are summarized in Table 1.

Table 1. Charateristics of FAP ATTR Val30Met in Japan

Features	Early-onset cases from endemic foci	Late-onset cases from non-endemic areas
Age of onset	34.4 ± 6.4*	64.0 ± 6.4*
Men / women	0.9 / 1*	4.5 / 1*
Presence of family history	Common	Rare
Penetrance rate	High	Low
Initial symptoms	Autonomic dysfunction Sensory deficit	Sensory deficit
Sensory dissociation	Common	Rare
Autonomic dysfunction in early stages	Severe with ADL disturbance	Mild without ADL disturbance
Cardiac involvement	Atrioventricular conduction block	Cardiomegaly
Anticipation of age at onset	Present	Absent

Values are expressed as the mean ± SD. ADL = activities of daily living.
*Based on a previous report (Koike et al., 2008).

Genetic Features

The penetrance rate of early-onset FAP ATTR Val30Met in the Japanese endemic foci is high (Koike et al., 2002; Koike and Sobue 2010a); in fact, a nearly completely penetrant autosomal dominant trait with almost equal sex distribution is observed in patients from endemic foci (Ikeda et al., 1987; Tashima et al., 1995; Yamamoto et al., 1998).

By contrast, the penetrance rate is low in the late-onset FAP ATTR Val30Met that occurs in the individuals outside of the endemic foci (Koike et al., 2002; Koike and Sobue 2010a).

According to a previous report of 35 families with late-onset FAP ATTR Val30Met unrelated to the endemic foci (Misu et al., 1999), only 11 probands had an apparent family history of FAP. The clinical features of the symptomatic siblings were similar to those of the probands. The age of onset was over 50 years except in two of the 25 symptomatic siblings. Thirty-six cases among 21 families were asymptomatic carriers as demonstrated by DNA analysis, and most of these carriers were identified late in life.

A nationwide survey for FAP ATTR Val30Met in Japan supported these findings (Koike et al., 2002). A family history of the disease was present in 94% of patients with early-onset disease and in 48% of patients with late-onset disease, indicating that the penetrance rate was low in the late-onset group. A relationship to the two Japanese endemic foci was noted in 91% of the early-onset cases and in only 10% of the late-onset cases. The geographical distribution of the early- and late-onset groups of patients differs markedly; the early-onset patients are largely restricted to the two major foci, while the late-onset patients are widely distributed throughout Japan.

Another characteristic difference is that anticipation in the age of onset is present in patients with early-onset FAP ATTR Val30Met but not in patients with late-onset FAP ATTR Val30Met (Nakazato et al., 1992a; Tashima et al., 1995; Yamamoto et al., 1998; Misu et al., 2000). Specifically, onset occurs at progressively earlier ages in successive generations in the families from the endemic foci in Japan, while the age of onset in patients in families with late-onset disease occurs at a uniformly late age among siblings when a family history of the disease is present. Although the mechanisms underlying anticipation are unknown, this observation suggests that the late-onset form of FAP ATTR Val30Met without anticipation that is present throughout Japan may be the primary form of FAP ATTR Val30Met. Furthermore, endemic foci of the disease may have developed due to anticipation resulting from the influence of unknown environmental or genetic factors.

Outside Japan, endemic foci of FAP ATTR Val30Met are present in Portugal and Sweden (Andrade et al., 1952; Andersson et al., 1976). Although patients with FAP ATTR Val30Met have been considered to be concentrated in geographically restricted areas of Japan, Portugal, and Sweden, a late-onset form of this type of FAP was discovered in non-endemic areas throughout the world (Kincaid et al., 1989; Reilly et al., 1995; Koike et al., 2002; Koike and Sobue, 2010a). Recent studies revealed differences in the penetrance rate of FAP ATTR Val30Met among various regions throughout the world (Holmgren et al., 1994; Sousa et al., 1995; Hellman et al., 2008; Dardiotis et al., 2009; Saporta et al., 2009). In addition, recent epidemiological studies identified the presence of new endemic foci in Ishikawa Prefecture of Japan and in Cyprus (Kato-Motozaki et al., 2008; Dardiotis et al., 2009). In this context, an undiagnosed or underdiagnosed population of FAP ATTR Val30Met may be present throughout the world.

Studies of FAP ATTR Val30Met patients in an endemic area of Portugal showed an age of onset and phenotype similar to that of the early-onset FAP ATTR Val30Met cases seen in Japan (Andrade et al., 1952; Coutinho et al., 1980; Sousa et al., 1995). However, the features of FAP ATTR Val30Met in a Swedish endemic focus were not similar to those in the Japanese and Portuguese endemic foci, but rather resembled those of the late-onset cases seen in the non-endemic areas of Japan. These include a mean age of onset in the fifth decade of life, a low penetrance rate, a tendency to manifest sensorimotor neuropathy rather than autonomic symptoms as an initial symptom, and extensive amyloid deposition in the heart (Benson et al., 1977; Holmgren et al., 1994). In some Swedish FAP ATTR Val30Met patients, symptoms of cardiac failure are more prominent than those of neuropathy (Ihse et al., 2008). The age of onset in these patients tends to be late, and amyloid deposits in such patients show characteristics similar to those observed in senile systemic amyloidosis (Bergström et al., 2005; Ihse et al., 2008). In addition, a recent study of a new endemic focus (Ishikawa Prefecture) outside of the conventional endemic foci (Arao and Ogawa) in Japan showed similar findings (Kato-Motozaki et al., 2008). According to a previous study assessing 25 FAP ATTR Val30Met cases from 10 families, the features of patients in this new focus are different from those in the conventional endemic foci, but resemble those seen in the Swedish cases (Kato-Motozaki et al., 2008).

According to these observations, FAP ATTR Val30Met patients in Japan could be classified into: (1) conventional early-onset cases in the endemic foci (Arao and Ogawa); (2) late-onset cases in an endemic focus (Ishikawa); and (3) late-onset cases in the non-endemic areas. Furthermore, the cases in the Swedish endemic focus and Ishikawa Prefecture could be classified as an intermediate form, with characteristics between those seen in cases in the non-endemic areas and those in the other endemic foci. As the late-onset form of FAP ATTR Val30Met in non-endemic areas may be more prevalent in Japan and elsewhere than was generally appreciated (Koike et al., 2002, 2008), this entity may represent the prototypical form of FAP ATTR Val30Met, with some indigenous environmental or genetic factors resulting in the endemic focus of the early-onset form of FAP ATTR Val30Met with anticipation and higher penetrance rate.

ELECTROPHYSIOLOGIC FEATURES

Nerve conduction studies in patients with FAP ATTR Val30Met indicate mainly axonal neuropathy with some slowing of conduction velocities and prolongation of distal latency (Koike et al., 2008, 2009). We previously compared the electrophysiologic features of late-onset FAP ATTR Val30Met cases unrelated to the endemic foci and compared them to conventional early-onset cases in the endemic foci (Koike et al., 2008). The extent of reduction of the compound muscle action potentials and sensory nerve action potentials was more profound in the late-onset group, even when the decline of these indices with aging in normal control subjects was taken into consideration, and the involvement of the lower-limbs tended to be more conspicuous in the late-onset group. In the early-onset group, electrophysiologic indices tended to be preserved in the early phase of the neuropathic symptoms, but progressed as the duration of neuropathic symptoms became longer. By contrast, electrophysiologic indices tended to be severely affected in the late-onset group even in cases with a short duration of neuropathic symptoms. These findings may reflect the different underlying mechanisms that cause neuropathy in these two forms of FAP ATTR Val30Met.

Another issue that physicians should take into consideration is the presence of carpal tunnel syndrome in amyloidosis. In general, amyloidosis is considered to increase the risk of carpal tunnel syndrome (Benson et al., 1975; Kyle et al., 1975, 1989; Kelly et al., 1983). Carpal tunnel syndrome is the first manifestation in gelsolin- and some transthyretin-associated forms of FAP (Kiuru et al., 1994; Ikeda et al., 2002), and a recent report described patients with FAP ATTR Val30Met starting with upper limb neuropathy resembling carpal tunnel syndrome (Tojo et al., 2010). In a large-scale study of patients with FAP ATTR Val30Met, however, the degree of abnormality in the median nerve was similar to that in the ulnar nerve (Koike et al., 2009). This result indicates that nerve damage at the carpal tunnel is not the sole distinctive feature in the majority of FAP ATTR Val30Met patients, although variability was present to some extent in individual patients. Therefore, the electrophysiologic abnormality at the distal portion of the median nerve may be a consequence of polyneuropathy rather than an entrapment injury in the majority of FAP ATTR Val30Met cases. However, we cannot exclude the possibility of isolated or predominant carpal tunnel syndrome features in patients with FAP ATTR Val30Met. Indeed, TTR-positive amyloid deposits have been reported in some patients with carpal tunnel syndrome (Kyle et al., 1992; Nakamichi et al., 1996; Kodaira et al., 2008). Further study is needed to resolve this issue.

Pathological Features

Pathological findings in patients with early-onset FAP ATTR Val30Met from the two conventional endemic foci in Japan are similar (Sobue et al, 1990; Koike et al., 2004). We previously reported that the pathological differences between the early-onset FAP ATTR Val30Met from the endemic foci and late-onset FAP ATTR Val30Met from the non-endemic areas corresponded with the clinical differences (Koike et al., 2004). Both groups showed amyloid deposition throughout the length of the nerves and in sympathetic and sensory ganglia, but the amounts were greater in the early-onset cases.

Amyloid deposition and neuronal cell loss were greater in sympathetic than in dorsal root ganglia in the early-onset cases; the opposite was true in the late-onset cases. Size assessment of the remaining neurons in these ganglia suggested a predominant loss of small neurons in the early-onset cases, while there was a loss of neurons of all sizes in the late-onset cases. Non-fibrillar deposits of TTR were abundantly present in the subperineurial space of the nerve trunk of the late-onset case and were present to a lesser extent in the early-onset cases.

Sural nerve biopsy specimens from patients with FAP ATTR Val30Met show a tendency of predominant small-fiber loss in early-onset cases, especially in the early phase of the disease, while variable fiber-size distribution with the presence of axonal sprouting and relatively preserved unmyelinated fibers are observed in late-onset cases (Figure 1) (Koike et al., 2004; Koike and Sobue, 2010b).

The relative preservation of unmyelinated fibers in comparison to myelinated fibers correlates with the low incidence of sensory dissociation and the relatively mild autonomic symptoms that are characteristic of late-onset cases. Teased fiber preparations show prominent axonal degeneration and only mild segmental demyelination and remyelination, and amyloid deposits are detected to a minimal or mild degree in the endoneurium. Amyloid deposition is sometimes negative by gross examination in these cases (Koike et al., 2004, 2011); thus, careful histopathologic examination is needed.

In terms of other organ system, amyloid deposition in the thyroid gland, gastrointestinal tract, pancreas, adrenal gland, and kidney is more prominent in early-onset cases than in late-onset cases. Amyloid deposition in the thyroid gland and kidney is especially prominent in early-onset cases and is relatively scant in late-onset cases. Conversely, amyloid deposition in the heart and the anterior pituitary lobe is greater in late-onset cases than in early-onset cases, making heart weight significantly greater in the late-onset cases (Koike et al., 2004, 2009). Cardiac amyloid deposition is more prominent in the atrium and the subendocardial region in early-onset cases, while it is prominent throughout the myocardium in the late-onset cases.

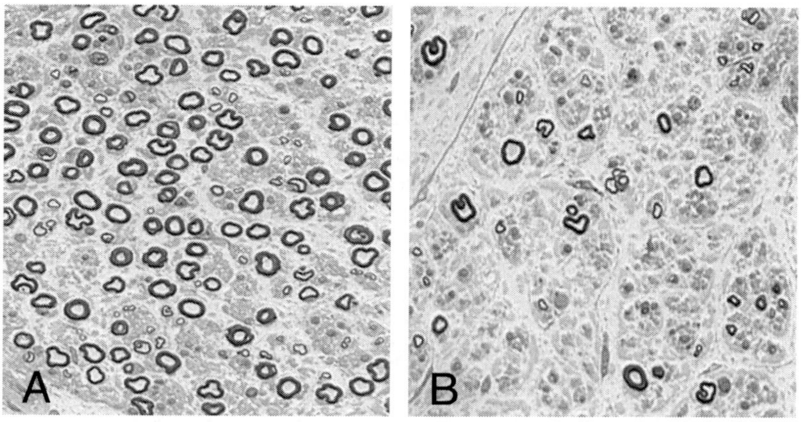

Figure 1. Representative photographs of sural nerve biopsy specimens from a patient with early-onset FAP ATTR Val30Met from an endemic focus (A) and from a patient with late-onset FAP ATTR Val30Met from a non-endemic area (B, C, and D) of Japan. Toluidine blue stain. (A) There is a greater loss of small myelinated fibers than large myelinated fibers. (B) Predominant small-fiber loss as in (A) is not observed.

Characteristics of the amyloid deposits in these two types of FAP ATTR Val30Met are also different (Koike et al., 2009). Amyloid deposits in the early-onset cases tend to be highly congophilic, show strong apple-green birefringence with Congo red staining, and are arranged in long, parallel fibrils in most organs. Conversely, those in the late-onset cases are generally weakly congophilic, show faint apple-green birefringence with Congo red staining, and have short, haphazard fibrils. Surface-enhanced desorption/ionization time-of-flight mass spectrometry (SELDI TOF-MS) also shows distinct characteristics for the amyloid deposits; TTR deposits in the heart are composed mostly variant TTR (more than 80%) in the early-onset cases, while abundant wild-type TTR (approximately 60%) is present in the late-onset cases (Koike et al., 2009).

This abundance of wild-type TTR in the late-onset cases suggests that the mechanism of amyloid deposition is similar to that observed in senile systemic amyloidosis, in which wild-type TTR is an amyloid precursor protein (Gustavsson et al., 1995). The extreme male predominance in late-onset FAP ATTR Val30Met in the non-endemic areas, as opposed to a nearly 1:1 male-female ratio in the early-onset cases in the endemic foci, is also similar to that observed in senile systemic amyloidosis (Kyle et al., 1996; Misu et al., 1999; Koike et al., 2002; Ng et al., 2005). In addition, amyloid deposition in the heart and anterior pituitary gland that is characteristic of the late-onset FAP ATTR Val30Met cases in the non-endemic areas is sometimes found in the normal elderly population (Westermark et al., 1997; Koike et al., 2004). These observations suggest that age-dependent changes in the microenvironment of interstitial tissues in various organs determine the severity and distribution pattern of amyloid deposition in each group.

DIAGNOSIS AND ITS IMPEDIMENTS

Since FAP is a disease caused by the deposition of amyloid, diagnosis should be made by tissue biopsy in combination with biochemical and molecular testing. The phenotype of TTR-related FAP is diverse and can involve manifestations arising from the peripheral nervous system, the heart, and the central nervous system (Ikeda et al., 2002). Heterogeneity of the clinical manifestations has been recognized even among individuals with the same mutation of the *TTR* gene (Koike et al., 2002). Tissue biopsy should be obtained from a clinically affected organ, such as a peripheral nerve or the heart. Biopsies from nonspecific sites, including the gastrointestinal tract, abdominal fat, or minor salivary gland, may also be considered.

Congo red staining is an essential procedure to detect amyloid deposits. Immunohistochemical studies, including immunogold electron microscopy, are useful to identify the causative proteins (Bergström et al., 2005; Misumi et al., 2009). Pathological deposition of wild-type TTR may occur in various sites, including the heart, lung, kidney, gastrointestinal tract, and tenosynovial tissue in the carpal tunnel of subjects without mutation of the *TTR* gene (Gustavsson et al., 1995; Kyle et al., 1996; Takei et al., 2002; Westermark et al., 2003; Kodaira et al., 2008). Thus, in addition to the detection of amyloid deposition, the demonstration of a TTR abnormality at the protein or DNA level is necessary to make a definitive diagnosis of FAP. For FAP ATTR Val30Met, a specific radioimmunoassay was developed in 1984 to detect this variant form of TTR in the serum (Nakazato et al., 1984). Southern blot DNA analysis also became available at same year (Sasaki et al., 1984). Since the introduction of polymerase chain reaction (PCR) for the allele-specific enzymatic amplification of genomic DNA, restriction fragment length analysis and DNA sequencing can be carried out more rapidly and easily, which has made it possible to identify the various mutations of the TTR gene related to the development of FAP (Nakazato et al., 1992b). Although DNA sequencing is currently the most common technique, the application of mass spectrometry or isoelectric focusing is also useful to identify the variant form of TTR (Kishikawa et al., 1996; Benson et al., 2002; Altland et al., 2007; Klein et al., 2011). SELDI TOF-MS without an immunoprecipitation step enables the quantitative evaluation of TTR variants in tissue amyloid deposits and body fluids (Ueda et al., 2009).

Plante-Bordeneuve et al. (2007) described the diagnostic pitfalls related to sporadic cases of FAP mainly French origin. According to this study, decreased nerve conduction velocity, increased cerebrospinal fluid protein, negative biopsy findings, and false immunolabeling of amyloid deposits were the main causes of diagnostic errors. Some cases were misdiagnosed as chronic inflammatory demyelinating polyneuropathy (CIDP) (Plante-Bordeneuve et al., 2007). That study concluded that DNA testing should be performed in patients with a progressive length-dependent small fiber polyneuropathy of unknown origin, especially when associated with autonomic dysfunction.

We previously described the clinical, electrophysiologic, and histopathologic characteristics of FAP ATTR Val30Met in 15 Japanese patients without apparent family history of the disease who were referred to our institution for diagnostic nerve biopsy (Koike et al., 2011). Typical features of conventional FAP are the presence of sensory dissociation and marked autonomic dysfunction reflected by the predominant loss of small fibers, including unmyelinated fibers (Dyck et al., 1969; Thomas et al., 1974; Said et al., 1984;

Sobue et al., 1990; Takahashi et al., 1991; Koike et al., 2004). However, patients in this series had nonspecific manifestations that did not match these typical symptoms; as such, referring physicians had not considered the possibility of FAP prior to sural nerve biopsy. In this series, neuropathy initiated as those of somatic neuropathy in the extremities in all cases. It was notable that 3 of 15 cases in this study manifested upper limb neuropathy as initial manifestation as sensory symptoms in patients with FAP ATTR Val30Met from endemic foci in Japan invariably starts from the lower extremities. Prior to referral to our institution, CIDP was considered as the putative diagnosis in eight of 15 patients. In one patient, pain in the lower limbs and regular intake of large amount of alcohol led to an initial diagnosis of alcoholic neuropathy (Koike et al., 2001a, 2003; Koike and Sobue, 2006). Although symptoms of somatic neuropathy were more predominant than those of autonomic dysfunction in these patients, autonomic symptoms gradually became more apparent. Therefore, careful attention to elicit a history of autonomic symptoms is critical at the time of the initial evaluation. Increased cerebrospinal fluid protein levels and abnormalities of nerve conduction indices (e.g., conduction velocity and distal latency) may lead to an erroneous diagnosis of CIDP in the absence of biopsy. On the other hand, such features are not so much conspicuous in typical axonal neuropathies, including alcoholic and beriberi neuropathies (Koike et al., 2001b, 2003), although the presence of secondary segmental demyelination due to axonal atrophy is reported in association with these neuropathies (Koike et al., 2003; Koike and Sobue, 2006).

Another obstacle to diagnosis is the interpretation of the pathological findings of sural nerve biopsy specimens (Koike et al., 2011). In a study of 15 cases with FAP ATTR Val30Met without a family history of the disease, only four cases showed more predominant loss of small myelinated fibers than large myelinated fibers (Koike et al., 2011). The density of unmyelinated fibers also tended to be preserved considering the extent of the reduction of myelinated fibers. Therefore, clinicians should consider the possibility of FAP when axonal neuropathy of undetermined etiology occurs in elderly patients, even if small-fiber predominant loss is not observed. Congo red staining should be performed in such cases. Because amyloid deposition is sometimes negative on gross examination (Koike et al., 2004, 2011), careful histopathologic examination is needed.

Biopsy of other organs to detect amyloid deposition may be considered, but the positive rate may be lower than that for nerve biopsy; according to a previous study of patients with FAP without a family history of the disease, amyloid deposition was found by nerve biopsy in 51 of 69 patients, by salivary gland biopsy in 17 of 27 patients, by rectal biopsy in four of 14 patients, and by abdominal fat aspirate in one of five patients (Plante-Bordeneuve et al., 2007). Therefore, genetic analysis for FAP should be considered if amyloid is not detected on sural nerve biopsy.

One practical clue that may lead to the diagnosis of FAP ATTR Val30Met in patients without a family history in non-endemic areas may be subclinical cardiac involvement. This may include findings suggestive of cardiomyopathy, such as an increased cardiothoracic ratio on chest X-ray or thickened interventricular septum on echocardiography, which can be observed in most of these patients even in the absence of symptoms of heart failure (Koike et al., 2011). Although most patients with late-onset FAP ATTR Val30Met from non-endemic areas show somatic neuropathy as the initial manifestation of the disease, some patients experience symptoms of heart failure in the initial phase of the disease (Misu et al., 1999); in a nationwide survey of FAP ATTR Val30Met in Japan, three of 59 late-onset cases had heart failure as the initial manifestation of the disease (Koike et al., 2002). In addition, the main cause of death in patients with FAP ATTR Val30Met from non-endemic areas was heart failure due to severe amyloid deposition (Koike et al., 2004, 2009). Therefore, detection of subclinical cardiac involvement may help to diagnose late-onset FAP ATTR Val30Met in patients without a family history of the disease.

CONCLUSION

FAP ATTR Val30Met is the most common form of FAP. Although patients with FAP ATTR Val30Met were previously thought to be geographically restricted to areas of Japan, Portugal, and Sweden, recent advances in molecular diagnostic techniques have enabled the recognition of this type of FAP throughout the world. Although FAP ATTR Val30Met is now considered to be more prevalent in Japan and elsewhere than was originally thought, awareness of the sporadic form of the disease is relatively low, which can lead to delayed diagnosis. Thus, an appreciation of the variability in the clinical, electrophysiologic, and histopathologic features of FAP, especially sporadic late-onset FAP ATTR Met30 cases in the non-endemic areas, is critical.

Previous studies suggested that the characteristic finding in sural nerve biopsy specimens from patients with FAP ATTR Val30Met in endemic foci is predominant loss of small fibers, including unmyelinated fibers, in association with the presence of sensory dissociation and marked autonomic dysfunction (Takahashi et al., 1991; Said et al., 1994; Koike et al., 2004). Conversely, such features are not common in the late-onset cases from the non-endemic areas (Koike et al, 2004, 2011). Therefore, the clinicopathologic features in these cases tend to be nonspecific. Indeed, physicians may not take the possibility of FAP into consideration until amyloid deposition is shown on sural nerve biopsy.

Of note, amyloid deposition is sometimes not seen on gross examination in these cases (Koike et al., 2004, 2011), so careful histopathologic examination is needed. Because of these difficulties in diagnosis, a high degree of clinical suspicion is required at the time of the initial clinical and electrophysiologic evaluation of neuropathy with an undetermined etiology for successful early diagnosis of sporadic late-onset FAP ATTR Val30Met.

ACKNOWLEDGMENTS

This work was supported by grants from the Ministry of Health, Labor and Welfare and the Ministry of Education, Culture, Sports, Science and Technology of Japan.

REFERENCES

Altland K, Benson MD, Costello CE, Ferlini A, Hazenberg BP, Hund E, Kristen AV, Linke RP, Merlini G, Salvi F, Saraiva MJ, Singer R, Skinner M, Winter P. Genetic microheterogeneity of human transthyretin detected by IEF. *Electrophoresis* 2007; 28: 2053-64.

Andersson R. Familial amyloidosis with polyneuropathy. A clinical study based on patients living in northern Sweden. *Acta Med. Scand. Suppl.* 1976; 590: 1-64.

Ando Y, Ueyama H, Watanabe S, Miyaji M, Tanaka Y, Ando M, Araki S. Early involvement of motor nerve dysfunction in atypical cases with familial amyloidotic polyneuropathy (FAP) type I. *Muscle Nerve* 1993; 16: 978-9.

Ando Y, Suhr OB. Autonomic dysfunction in familial amyloid polyneuropathy (FAP). *Amyloid: Int. J. Exp. Clin. Invest* 1998; 5: 288-300.

Andrade C. A peculiar form of peripheral neuropathy: Familial generalized amyloidosis with special involvement of the peripheral nerves. *Brain* 1952; 75: 408-27.

Araki S, Mawatari S, Ohta M, Nakajima A, Kuroiwa Y. Polyneuritic amyloidosis in a Japanese family. *Arch. Neurol.* 1968; 18: 593-602.

Benson MD, Cohen AS, Brandt KD, Cathcart ES. Neuropathy, M components, and amyloid. *Lancet* 1975; 1: 10-2.

Benson MD, Cohen AS. Generalized amyloid in a family of Swedish origin. A study of 426 family members in seven generations of a new kindship with neuropathy, nephropathy, and central nervous system involvement. *Ann. Intern. Med.* 1977; 86: 419-24.

Benson MD, Yazaki M, Magy N. Laboratory assessment of transthyretin amyloidosis. *Clin. Chem. Lab Med.* 2002; 40: 1262-5.

Benson MD, Kincaid JC. The molecular biology and clinical features of amyloid neuropathy. *Muscle Nerve* 2007; 36: 411-23.

Berghoff M, Kathpal M, Khan F, Skinner M, Falk R, Freeman R. Endothelial dysfunction precedes C-fiber abnormalities in primary (AL) amyloidosis. *Ann. Neurol.* 2003; 53: 725-30.

Bergström J, Gustavsson A, Hellman U, Sletten K, Murphy CL, Weiss DT, et al. Amyloid deposits in transthyretin-derived amyloidosis: cleaved transthyretin is associated with distinct amyloid morphology. *J. Pathol.* 2005: 206: 224-32.

Coelho T, Sousa A, Lourenco E, Ramalheira J. A study of 159 Portuguese patients with familial amyloidotic polyneuropathy (FAP) whose parents were both unaffected. *J. Med. Genet.* 1994; 31: 293-9.

Conceição I, De Carvalho M. Clinical variability in type I familial amyloid polyneuropathy (Val30Met): comparison between late- and early-onset cases in Portugal. *Muscle Nerve* 2007; 35: 116-8.

Coutinho P, DeSilva AM, Lima JL, Barbosa AR. Forty years of experience with type I amyloid neuropathy: review of 483 cases. In: Glenner GG, Costa PP, Freitass F, eds. *Amyloid and amyloidosis*. Amsterdam, Pa: Exerpta Medica; 1980: 88-98.

Dardiotis E, Koutsou P, Papanicolaou EZ, Vonta I, Kladi A, Vassilopoulos D, Hadjigeorgiou G, Christodoulou K, Kyriakides T. Epidemiological, clinical and genetic study of familial amyloidotic polyneuropathy in Cyprus. *Amyloid: J. Protein Folding Disord.* 2009; 16: 32-7.

Dyck PJ, Lambert EH. Dissociated sensation in amyloidosis. Compound action potential, quantitative histologic and teased-fiber, and electron microscopic studies of sural nerve biopsies. *Arch. Neurol.* 1969; 20: 490-507.

Gustavsson A, Jahr H, Tobiassen R, Jacobson DR, Sletten K, Westermark P. Amyloid fibril composition and transthyretin gene structure in senile systemic amyloidosis. *Lab Invest* 1995; 73: 703-8.

Hellman U, Alarcon F, Lundgren HE, Suhr OB, Bonaiti-Pellié C, Planté-Bordeneuve V. Heterogeneity of penetrance in familial amyloid polyneuropathy, ATTR Val30Met, in the Swedish population. *Amyloid: J. Protein Folding Disord* 2008; 15: 181-6.

Holmgren G, Costa PM, Andersson C, Asplund K, Steen L, Beckman L, Nylander PO, Teixeira A, Saraiva MJ, Costa PP. Geographical distribution of TTR met30 carriers in northern Sweden: discrepancy between carrier frequency and prevalence rate. *J. Med. Genet.* 1994; 31: 351-4.

Ihse E, Ybo A, Suhr O, Lindqvist P, Backman C, Westermark P. Amyloid fibril composition is related to the phenotype of hereditary transthyretin V30M amyloidosis. *J. Pathol.* 2008; 216: 253-61.

Ikeda S, Hanyu N, Hongo M, et al. Hereditary generalized amyloidosis with polyneuropathy: clinicopathological study of 65 Japanese patients. *Brain* 1987; 110: 315-37.

Ikeda S, Nakazato M, Ando Y, Sobue G. Familial transthyretin-type amyloid polyneuropathy in Japan: clinical and genetic heterogeneity. *Neurology.* 2002; 58: 1001-7.

Ikeda S. Is familial amyloid polyneuropathy rare? DNA testing is changing the concept of this disease. *Neurology.* 2007; 69; 627-8.

Kato-Motozaki Y, Ono K, Shima K, Morinaga A, Machiya T, Nozaki I, et al. Epidemiology of familial amyloid polyneuropathy in Japan: Identification of a novel endemic focus. *J. Neurol. Sci.* 2008; 270: 133-40.

Kishikawa M, Nakanishi T, Miyazaki A, Shimizu A, Nakazato M, Kangawa K, Matsuo H. Simple detection of abnormal serum transthyretin from patients with familial amyloidotic polyneuropathy by high-performance liquid chromatography/electrospray ionization mass spectrometry using material precipitated with specific antiserum. *J. Mass Spectrom.* 1996; 31: 112-4.

Kelly JJ Jr. The electrodiagnostic findings in peripheral neuropathy associated with monoclonal gammopathy. *Muscle Nerve* 1983; 6: 504-9.

Kishikawa M, Nakanishi T, Miyazaki A, Shimizu A, Nakazato M, Kangawa K, Matsuo H. Simple detection of abnormal serum transthyretin from patients with familial amyloidotic polyneuropathy by high-performance liquid chromatography/electrospray ionization mass spectrometry using material precipitated with specific antiserum. *J. Mass Spectrom.* 1996; 31: 112-4.

Kiuru S, Seppäläinen AM. Neuropathy in familial amyloidosis, Finnish type (FAF): electrophysiological studies. *Muscle Nerve* 1994; 17: 299-304.

Klein CJ, Vrana JA, Theis JD, Dyck PJ, Dyck PJ, Spinner RJ, Mauermann ML, Bergen HR 3rd, Zeldenrust SR, Dogan A. Mass spectrometric-based proteomic analysis of amyloid neuropathy type in nerve tissue. *Arch. Neuro*l. 2011; 68: 195-9.

Kodaira M, Sekijima Y, Tojo K, Tsuchiya A, Yazaki M, Ikeda S, Sekijima Y, Hoshii Y, Tachibana S. Non-senile wild-type transthyretin systemic amyloidosis presenting as bilateral carpal tunnel syndrome. *J. Peripher. Nerv. Syst.* 2008; 13: 148-50.

Koike H, Mori K, Misu K, Hattori N, Ito H, Hirayama M, Sobue G. Painful alcoholic polyneuropathy with predominant small-fiber loss and normal thiamine status. *Neurology* 2001a; 56(: 1727-32.

Koike H, Misu K, Hattori N, Ito S, Ichimura M, Ito H, Hirayama M, Nagamatsu M, Sasaki I, Sobue G. Postgastrectomy polyneuropathy with thiamine deficiency. *J Neurol Neurosurg Psychiatry* 2001b; 71: 357-62.

Koike H, Misu K, Ikeda S, Ando Y, Nakazato M, Ando E, Yamamoto M, Hattori N, Sobue G. Type I (Transthyretin Met30) familial amyloid polyneuropathy in Japan: early- versus late-onset form. *Arch. Neurol.* 2002; 59: 1771-6.

Koike H, Iijima M, Sugiura M, Mori K, Hattori N, Ito H, Hirayama M, Sobue G. Alcoholic neuropathy is clinicopathologically distinct from thiamine-deficiency neuropathy. *Ann Neurol* 2003; 54: 19-29.

Koike H, Misu K, Sugiura M, Iijima M, Mori K, Yamamoto M, Hattori N, Mukai E, Ando Y, Ikeda S, Sobue G. Pathology of early- vs late-onset TTR Met30 familial amyloid polyneuropathy. *Neurology* 2004; 63: 129-38.

Koike H, Sobue G. Alcoholic neuropathy. *Curr Opin Neurol* 2006; 19: 481-6.

Koike H, Kawagashira Y, Iijima M, Yamamoto M, Hattori N, Tanaka F, Hirayama M, Ando Y, Ikeda S, Sobue G. Electrophysiological features of late-onset transthyretin Met30 familial amyloid polyneuropathy unrelated to endemic foci. *J. Neurol.* 2008; 255: 1526-33.

Koike H, Ando Y, Ueda M, Kawagashira Y, Iijima M, Fujitake J, Hayashi M, Yamamoto M, Mukai E, Nakamura T, Katsuno M, Hattori N, Sobue G. Distinct characteristics of amyloid deposits in early- and late-onset transthyretin Val30Met familial amyloid polyneuropathy. *J. Neurol. Sci.* 2009; 287: 178-84.

Koike H, Morozumi S, Kawagashira Y, Iijima M, Yamamoto M, Hattori N, Tanaka F, Nakamura T, Hirayama M, Ando Y, Ikeda SI, Sobue G. The significance of carpal tunnel syndrome in transthyretin Val30Met familial amyloid polyneuropathy. *Amyloid: J. Protein Folding Disord.* 2009; 15: 1-7.

Koike H, Sobue G. Diagnosis of familial amyloid polyneuropathy: wide-ranged clinicopathological features. *Expert Opin Med Diagn* 2010a; 4: 323-31

Koike H, Sobue G. Late-onset transthyretin Val30Met familial amyloid polyneuropathy unrelated to endemic foci. In: Dyck PJ, Engelstad JK, Low PA, Amrami KK, Spinner RJ, Klein CJ, editors. *Companion to peripheral neuropathy: Illustrated cases and new developments.* Philadelphia: Elsevier Saunders 139-43, 2010b.

Koike H, Hashimoto R, Tomita M, et al. Diagnosis of sporadic transthyretin Val30Met familial amyloid polyneuropathy: a practical analysis. *Amyloid* 2011; 18: 53-62.

Kyle RA, Bayrd ED. Amyloidosis: review of 236 cases. *Medicine* (Baltimore) 1975; 54: 271-99.

Kyle RA, Eilers SG, Linscheid RL, Gaffey TA. Amyloid localized to tenosynovium at carpal tunnel release. Natural history of 124 cases. *Am. J. Clin. Pathol.* 1989; 91: 393-7.

Kyle RA, Gertz MA, Linke RP. Amyloid localized to tenosynovium at carpal tunnel release. Immunohistochemical identification of amyloid type. *Am. J. Clin. Pathol.* 1992; 97: 250-3.

Kyle RA, Spittell PC, Gertz MA, Li CY, Edwards WD, Olson LJ, et al. The premortem recognition of systemic senile amyloidosis with cardiac involvement. *Am. J. Med.* 1996; 101: 395-400.

Maruyama K, Ikeda S, Yanagisawa N, Nakazato M. Diagnostic value of abdominal fat tissue aspirate in familial amyloid polyneuropathy. *J. Neurol. Sci.* 1987; 81: 11-8.

Misu K, Hattori N, Ando Y, Ikeda S, Sobue G. Anticipation in early- but not late-onset familial amyloid polyenuropathy (TTR Met 30) in Japan. *Neurology.* 2000; 55: 451-2.

Misu K, Hattori N, Nagamatsu M, et al. Late-onset familial amyloid polyneuropathy type I (transthyretin Met 30-associated familial amyloid polyneuropathy) unrelated to endemic focus in Japan; clinicopathological and genetic features. *Brain*. 1999; 122: 1951-62.

Nakamichi KI, Tachibana S. Amyloid deposition in the synovium and ligament in idiopathic carpal tunnel syndrome. *Muscle Nerve* 1996; 19: 1349-51.

Nakazato M, Kangawa K, Minamino N, Tawara S, Matsuo H, Araki S. Radioimmunoassay for detecting abnormal prealbumin in the serum for diagnosis of familial amyloidotic polyneuropathy (Japanese type). *Biochem. Biophys. Res. Commun*. 1984; 122: 719-25.

Nakazato M, Shiomi K, Miyazato M, Matsukura S. Type 1 familial amyloidotic polyneuropathy in Japan. *Intern. Med.* 1992a; 31: 1335-8.

Nakazato M, Ikeda S, Shiomi K, Matsukura S, Yoshida K, Shimizu H, Atsumi T, Kangawa K, Matsuo H. Identification of a novel transthyretin variant (Val30----Leu) associated with familial amyloidotic polyneuropathy. *FEBS Lett*. 1992b; 306: 206-8.

Ng B, Connors LH, Davidoff R, Skinner M, Falk RH. Senile systemic amyloidosis presenting with heart failure: a comparison with light chain-associated amyloidosis. *Arch. Intern. Med*. 2005; 165: 1425-9.

Plante-Bordeneuve V, Ferreira A, Lalu T, Zaros C, Lacroix C, Adams D, Said G. Diagnostic pitfalls in sporadic transthyretin familial amyloid polyneuropathy (TTR-FAP). *Neurology* 2007; 69: 693-8.

Reilly MM, Adams D, Booth DR, et al. Transthyretin gene analysis in European patients with suspected familial amyloid polyneuropathy. *Brain* 1995; 118: 849-56.

Said G, Ropert A, Faux N. Length-dependent degeneration of fibers in Portuguese amyloid polyneuropathy: a clinicopathologic study. *Neurology* 1984; 34: 1025-32.

Saporta MA, Zaros C, Cruz MW, André C, Misrahi M, Bonaïti-Pellié C, Planté-Bordeneuve V. Penetrance estimation of TTR familial amyloid polyneuropathy (type I) in Brazilian families. *Eur. J. Neurol*. 2009; 16: 337-41.

Saravia MJ, Costa PP, Goodman DS. Genetic expression of a tansthyretin mutation in typical and late-onset Portuguese families with familial amyloidotic polyneuropathy. *Neurology* 1986; 36: 1413-7.

Sasaki H, Sakaki Y, Matsuo H, Goto I, Kuroiwa Y, Sahashi I, Takahashi A, Shinoda T, Isobe T, Takagi Y. Diagnosis of familial amyloidotic polyneuropathy by recombinant DNA techniques. *Biochem. Biophys. Res. Commun.* 1984; 125: 636-42.

Sobue G, Nakao N, Murakami K, et al. Type I familial amyloid polyneuropathy. a pathological study of the peripheral nervous system. *Brain* 1990; 113: 903-19.

Sobue G, Koike H, Misu K, Hattori N, Yamamoto M, Ikeda S, Ando Y, Nakazato M, Inukai A. Clinicopathologic and genetic features of early- and late-onset FAP type I (FAP ATTR Val30Met) in Japan. *Amyloid: J. Protein Folding Disord.* 2003; 10 Suppl 1: 32-8.

Sousa A, Coelho T, Barros J, Sequeiros J. Genetic epidemiology of familial amyloidotic polyneuropathy (FAP)-type I in Povoa do Varzim and Vila do Conde (north of Portugal). *Am. J. Med. Genet.* 1995; 60: 515-21.

Sousa MM, Cardoso I, Fernandes R, Guimaraes A, Saraiva MJ. Deposition of transthyretin in early stages of familial amyloidotic polyneuropahty: evidence for toxicity of nonfibrillar aggregates. *Am. J. Pathol.* 2001; 159: 1993-2000.

Tawara S, Nakazato M, Kangawa K, Matsuo H, Araki S. Identification of amyloid prealbumin variant in familial amyloidotic polyneuropathy (Japanese type). *Biochem. Biophys. Res. Commun.* 1983; 116: 880-8.

Takahashi K, Yi S, Kimura Y, Araki S. Familial amyloidotic polyneuropathy type 1 in Kumamoto Japan: a clinicopathologic, histochemical, immunohistochemical, and ultrastructural study. *Hum. Pathol.* 1991; 22: 519-27.

Tashima K, Ando Y, Tanaka Y, Uchino M, Ando M. Change in age of onset in patients with familial amyloidotic polyneuropathy type I. *Intern. Med.* 1995; 34: 748-50.

Terazaki H, Ando Y, Fernandes R, Yamamura K, Maeda S, Saraiva MJ. Immunization in familial amyloidotic polyneuropathy: counteracting deposition by immunization with a Y78F TTR mutant. *Lab. Invest* 2006; 86: 23-31.

Tojo K, Sekijima Y, Kelly JW, Ikeda S. Diflunisal stabilizes familial amyloid polyneuropathy-associated transthyretin variant tetramers in serum against dissociation required for amyloidogenesis. *Neurosci. Res.* 2006; 56: 441-9.

Ueda M, Misumi Y, Mizuguchi M, Nakamura M, Yamashita T, Sekijima Y, Ota K, Shinriki S, Jono H, Ikeda S, Suhr OB, Ando Y. SELDI-TOF mass spectrometry evaluation of variant transthyretins for diagnosis and pathogenesis of familial amyloidotic polyneuropathy. *Clin. Chem.* 2009. 55: 1223-7.

Yamamoto K, Ikeda S, Hanyu N, Takeda S, Yanagisawa N (1998). A pedigree analysis with minimized ascertainment bias shows anticipation in Met30-tansthyretin related familial amyloid polyneuropathy. *J. Med. Genet.* 35, 23-30.

Westermark P, Eriksson L, Engström U, Eneström S, Sletten K. Prolactin-derived amyloid in the aging pituitary gland. *Am. J. Pathol.* 1997; 150: 67-73.

In: Amyloids
Editors: I. P. Halcheck et al.
ISBN: 978-1-62100-538-4

Chapter 14

AMYLOID IMAGING AND BIOMARKERS IN MCI AND ALZHEIMER'S DISEASE

***Katsutoshi Furukawa*[1], *Nobuyuki Okamura*[2], *Manabu Tashiro*[3], *Shozo Furumoto*[2], *Ren Iwata*[4], *Kazuhiko Yanai*[2], *Yukitsuka Kudo*[5], *and Hiroyuki Arai*[1]**

[1] Department of Geriatrics and Gerontology, Division of Brain Sciences, Institute of Development, Aging and Cancer, Tohoku University, Japan

[2] Department of Pharmacology, Tohoku University Graduate School of Medicine, Japan

[3] Division of Cyclotron Nuclear Medicine, Cyclotron and Radioisotope Center, Japan

[4] Division of Radiopharmaceutical Chemistry, Cyclotron and Radioisotope Center, Japan

[5] Department of NeuroImaging Research, Innovation New Biomedical Engineering Center, Japan

ABSTRACT

Biomarkers for early and accurate diagnosis of Alzheimer's disease (AD) have been paid great attention over the past several years. Alzheimer's Disease NeuroImaging Initiative (ADNI), which follows up precise and longitudinal changes in biomarkers, has begun in the USA, Europe, Australia and Japan, and there have been several important findings [1]. Quite a few studies and projects including ADNI revealed that amyloid β (Aβ) should be one of the most important biomarkers of AD. Accumulation and distribution of Aβ in the brain and the measurement of its concentration in the cerebrospinal fluid can be analyzed by imaging techniques using positron emission tomography (PET) and ELISA, respectively. An amyloid imaging modality using a novel PET tracer was developed by Klunk et al. in 2004 [2]. Since then some of the investigators including us have released new amyloid tracers. Our group recently developed a novel PET tracer, ^{11}C-labeled 2-(2-[2-dimethylaminothiazol- 5-yl] ethenyl) - 6- (2- [fluoro]ethoxy) benzoxazole ([^{11}C]BF-227), and succeeded in *in vivo* detection of amyloid plaques in the AD brains [3,4]. Needless to say, biomarkers in biofluid now also play important roles in the diagnosis of AD. Our group for the first time in the world reported that tau protein in the cerebrospinal fluid (CSF) is elevated in AD patients compared to normal controls and other neurological diseases [5]. In addition to tau, Aβ (especially Aβ1-42) in CSF is a sensitive and specific biomarker of AD. Both the elevated total-tau (t-tau) and phosphorylated-tau (p-tau) and the decreased Aβ1-42 in CSF can distinguish not only AD from normal controls but also can discriminate between the subjects with mild cognitive impairment who convert to AD later (MCI-converters) and the ones with MCI who do not convert to AD (MCI-non-converters) while they are in the MCI stage [6]. It has been reported that Aβ1-42 in CSF and amyloid PET show a change in the pathological process of AD earliest among the biomarkers, therefore these are considered to be the most sensitive ones for the early diagnosis of AD [6].

INTRODUCTION

Alzheimer's disease (AD) is a one of the diseases researchers are eager to pacify in the 21st century. The number of AD patients is now reported to be more than 24 million in the world, and societies all around the world would collapse because of the flood of patients with dementia without any effective solutions. A very unique novel project, Alzheimer's Disease NeuroImaging Initiative (ADNI), which is a longitudinal study employing neuro-imaging, biofluid biomarkers, and neuropsychological examinations, has started in the USA in 2005 [1]. The project has now spread to Europe, Australia and Japan, and is releasing strong series of evidence. From previous studies and projects on AD including ADNI, amyloid β (Aβ) is believed to play a central role of the pathogenesis of AD. Senile plaques are a pathological hallmark of the AD brain as well as neurofibrillary tangles, and aggregated Aβ is a major component of the senile plaques. It is considered that a single monomer of Aβ itself is not strongly toxic or pathogenic while Aβ oligomers consisting of 2-12 Aβ molecules yield robust neurotoxicity and deteriorate neurons in the cerebrum. Although any reliable methods to quantify Aβ oligomers have not been established yet, amyloid imaging using PET and the concentration of Aβ1-42 in cerebrospinal fluid (CSF) are now considered to be a sensitive and specific biomarker for AD diagnosis. In recent years several laboratories, including ours, have succeeded in visualizing Aβ deposition in living patients' brains with AD using PET probes [2, 3, 4]. In addition to amyloid PET, regional cerebral glucose metabolism (rCMRglu) has been investigated by several investigators [4] using [^{18}F] 2-fluoro-deoxy-D-glucose (FDG) and PET in AD. Ronald Petersen addressed the concept of mild cognitive impairment (MCI), which is an intermediate state between normal aging and AD [7]. We performed FDG-PET as well as BF-227 on the same subjects, which are aged normal (AN), MCI, and AD, and carefully analyzed, and compared the results with these two kinds of PET.

Biomarkers in biofluid (CSF, blood, and urine) for diagnosis of AD have been investigated for a long time. To diagnose AD and other diseases that cause dementia easily and accurately by a single marker must be a physicians' dream. Although many molecules in AD have been examined, no markers in blood (serum, plasma, and blood cells) have reached strong evidence. So far only total-tau (t-tau), phosphorylated tau (p-tau) and Aβ1-42 in CSF have reached a consensus. We also would like to describe these biomarkers in biofluid including Aβ to help AD diagnose in this article.

METHODS

The diagnosis of MCI and probable AD followed the MCI clinical criteria presented by "Petersen et al." [7] and "the National Institute of Neurological and Communicative Disorders and Stroke - Alzheimer's Disease and Related Disorders Association" [8], respectively.

The PET procedure for BF-227 was described precisely before [3]. BF-227 and its *N*-desmethylated derivative (a precursor of [^{11}C]BF-227) were custom-synthesized by Tanabe R&D Service Co. After intravenous injection with 211-366 mBq of [^{11}C]BF-227, dynamic PET images were obtained for 60 min with each subject's eyes closed. Standardized uptake value (SUV) images of [^{11}C]BF-227 were obtained by normalizing tissue radioactivity concentration by injected dose and body weight. The FDG-PET procedure was described previously [9]. Subjects were scanned in a quiet and dimly-lit room with their eyes closed after at least four hours of food restriction. The emission data were corrected for tissue attenuation using the transmission data. Regions of interest (ROIs) were placed on individual axial magnetic resonance (MR) images. Because there was neither senile plaques nor glucose hypometabolism in the cerebellum of AD, ratios of regional SUV to cerebellar SUV (SUVR) were calculated as an index of [^{11}C]BF-227 retention and CMRglu.

RESULTS

BF-227 Retention in MCI

First, we analyzed PET images with [^{11}C]BF-227 in 12 aged normal (AN) subjects, 15 MCIs and 15 ADs, and representative brain PET images are shown in Figure 1. As indicated in the figure, some MCI subjects showed strong retention of [^{11}C]BF-227, but other MCI subjects did not. Most AD cases, however, indicated strong accumulation of [^{11}C]BF-227 especially in frontal, temporal and parietal cortices. If the retention pattern of [^{11}C]BF-227 is compared to that of PIB, the accumulation of [^{11}C]BF-227 in the frontal lobe looks much weaker than that of PIB.

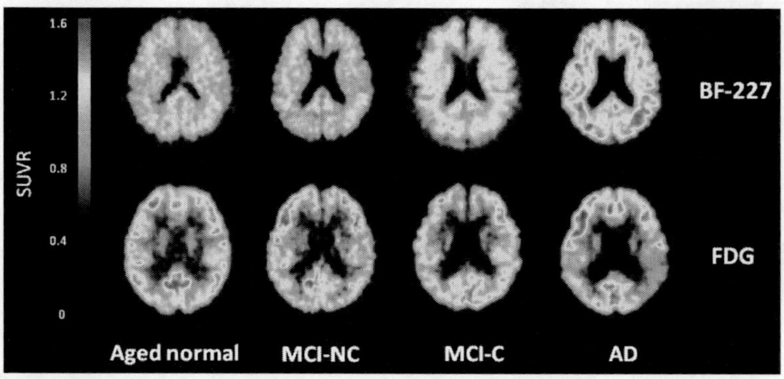

Figure 1. Representative axial brain PET images with BF-227. AD: an aged normal, MCI-NC: an MCI subject who did not convert to AD, MCI-C: an MCI subject who did convert to AD. AD: Alzheimer's disease.

Figure 2 shows the mean neocortical and regional SUVRs of [^{11}C]BF-227 for the three groups. Both the mean neocortical SUVRs for MCI and AD are significantly higher than that for AN. As we previously reported [3], significantly higher SUVRs were observed in most cerebral regions in AD compared to AN except for the medial temporal lobe. A significantly lower SUVR in MCI was observed in the parietal cortex compared to AD. In the other neocortical regions, MCI subjects showed a tendency towards milder retention of BF-227 than that in AD.

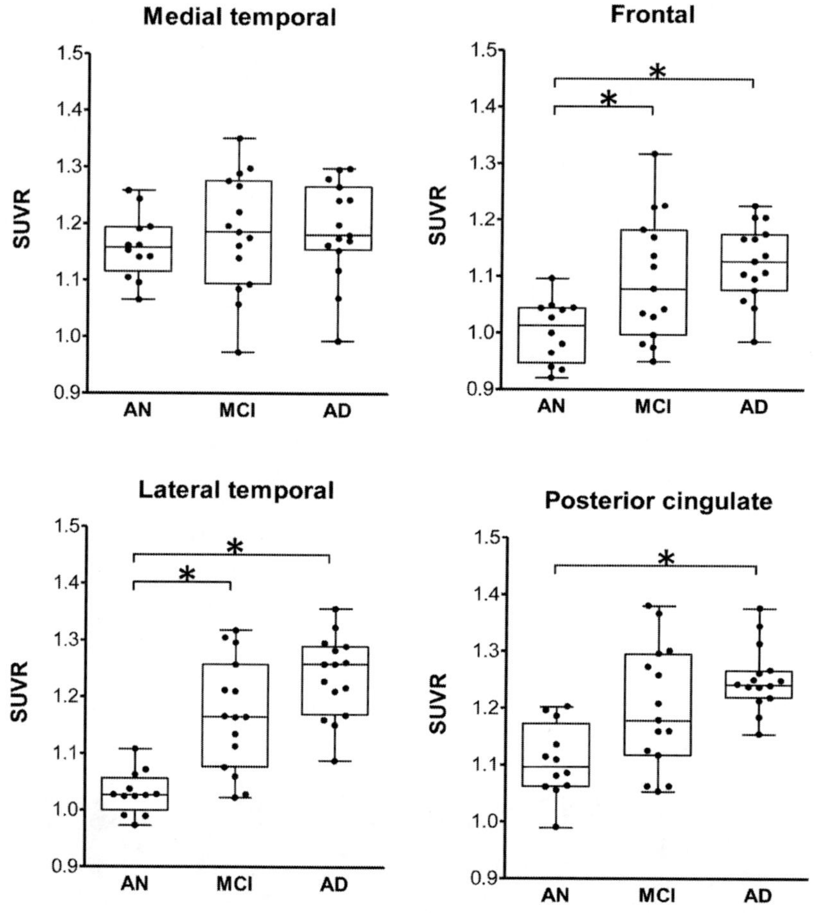

Figure 2 continued on next page.

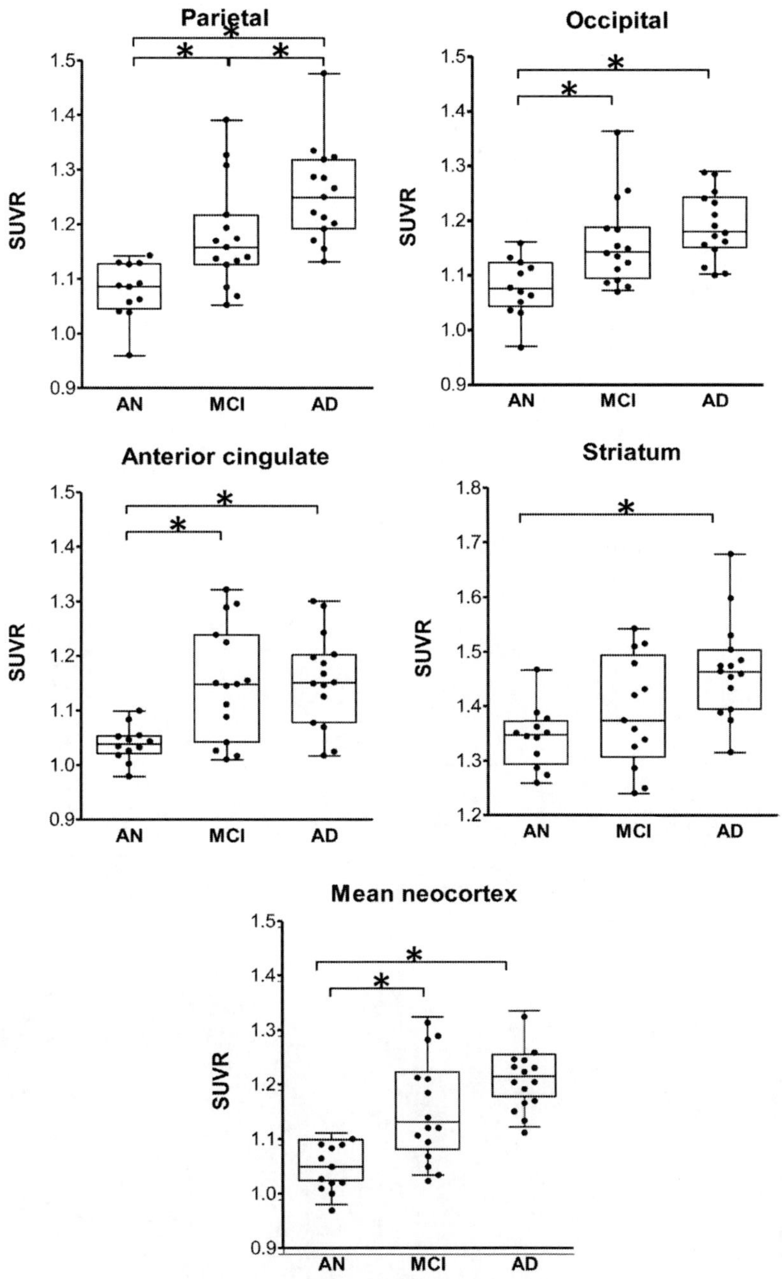

Figure 2. Box plots of SUVR values with BF-227 PET for AN, MCI and AD.

Cerebral Glucose Metabolism in AN, MCI and AD

We also applied FDG to three groups (AN, MCI, and AD) in order to elucidate the status of amyloid plaque deposition and synaptic functions in MCI and compared the diagnostic performance of BF-227-PET with that of FDG-PET in AD cases.

We studied with PET using both BF-227 and FDG. PET images were obtained after administration of BF-227 or FDG and the regional standardized uptake value (SUV) and the ratio of regional to cerebellar SUV were calculated as an index of BF-227 and FDG binding. AD patients showed increased uptake of [^{11}C]BF-227 in the neocortical areas and striatum as well as decreased glucose metabolism in temporo-parietal, posterior cingulate and medial temporal areas. MCI subjects showed a significant increase in BF-227 uptake in the neocortical areas similar to AD, and the most significant difference of BF-227 retention was observed in the parietal lobe if its retentions for MCI were compared to those for AD and AN.

On the other hand hypometabolism of glucose in MCI was confined to cingulate and medial temporal cortices. The uptake of BF-227 in the cerebral cortex negatively correlated well with glucose metabolism. Receiver operating characteristic (ROC) analysis indicated higher specificity and sensitivity with BF-227-PET than those with FDG-PET for a differential diagnosis between AD and normal control. A significant reduction of neocortical SUVR was observed in both MCI and AD patients compared to AN in FDG-PET (Figure 3).

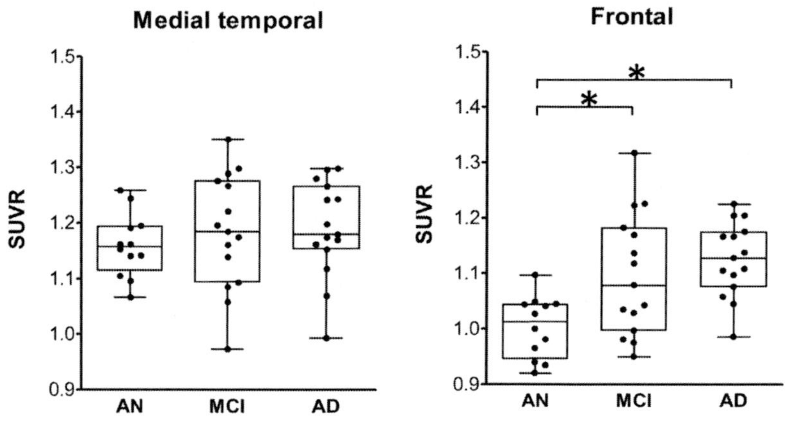

Figure 3 continued on next page.

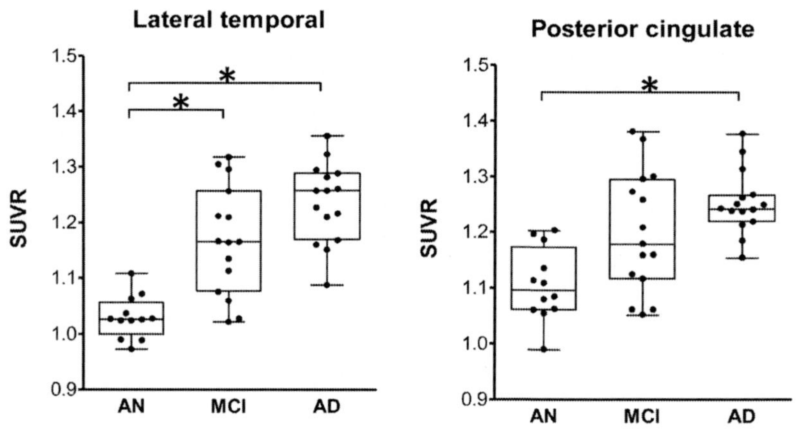

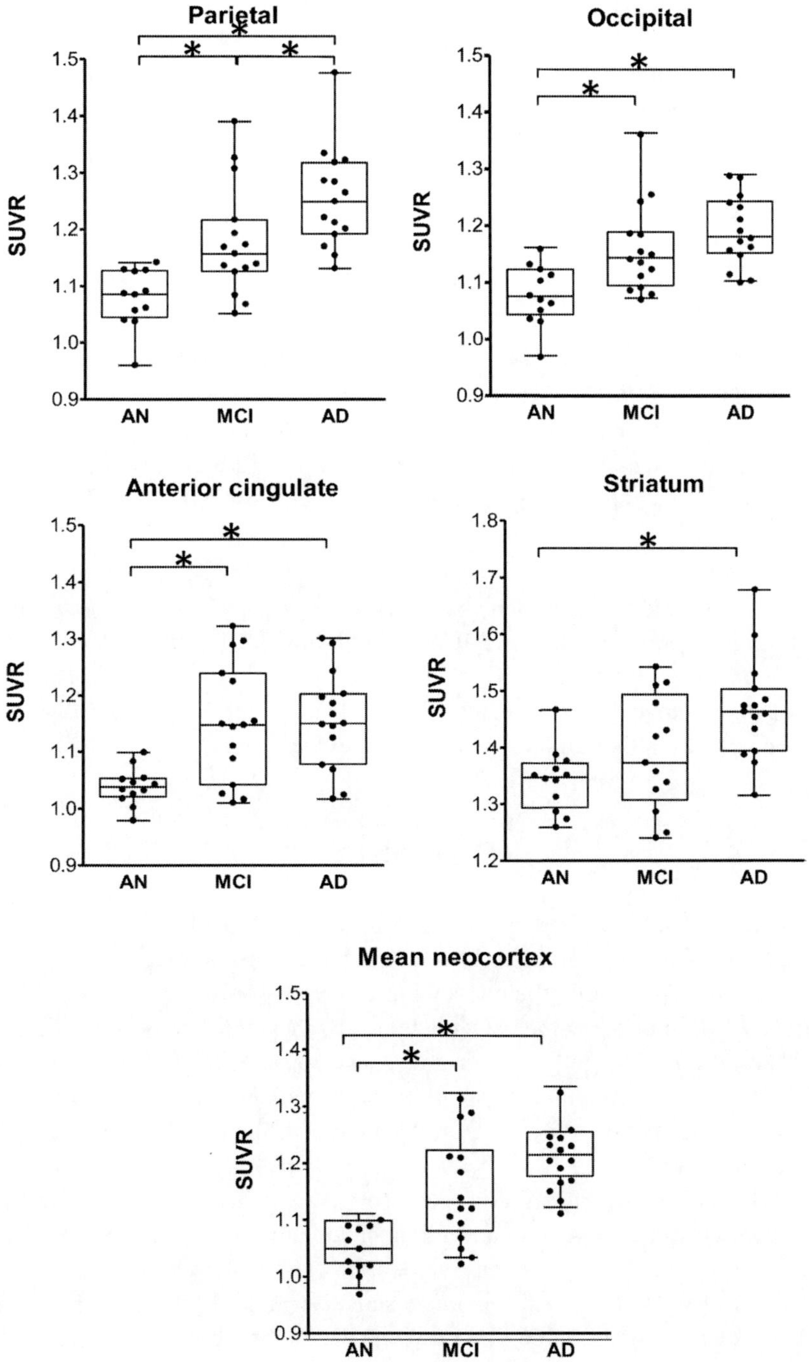

Figure 3. Box plots of SUVR values with FDG-PET for AN, MCI and AD.

Neocortical SUVR of FDG-PET for each subject was plotted against neocortical SUVR of BF-227-PET. ROC analysis was performed for the lateral temporal SUVR of BF-227 and posterior cingulate SUVR of FDG. The AUC for BF-227 (0.994) is much higher than that for FDG (0.839). It is concluded that [^{11}C]BF-227-PET is a strong tool for early detection of AD pathology and also even in the MCI stage.

CSF Biomarkers

Previous reports including the ones from ADNI indicate that Aβ1-42, t-tau, and p-tau are strong biomarkers for diagnosis of AD. Monomers of Aβ1-42 dynamically form Aβ oligomers leading to their aggregation. Although it was reported that 75% of the total soluble Aβ1-42 is oligomeric [10, 11], quantification of Aβ is based on measurement of Aβ monomers. Therefore an increase in Aβ oligomers could cause decreased measurable levels of Aβ. Also senile plaques in AD brains are considered as a reservoir of Aβ, that is, if the more senile plaques are generated, the lower CSF Aβ will be. On the other hand t-tau in CSF is surmised to be released from damaged neurons and CSF t-tau is increased in several neurological disorders. Some studies indicate p-tau, which reflects pathological change in AD, is more sensitive and specific than t-tau. Decreased Aβ1-42 in addition to increased t-tau and p-tau can predict conversion from MCI to AD with 83% sensitivity and 72% specificity [12]

Numeral studies have been done with biomarkers in blood (plasma, serum and blood cells) for AD diagnosis, although there is no strong evidence yet.

DISCUSSION

It was presented that FDDNP can detect a high signal in MCI by binding not only for amyloid plaques but also tau neurofibrillary tangles [13], and the retention level for MCI is between AN and AD. On the other hand, several groups reported that about a half of the MCI subjects showed PIB uptake in the AD range, and other MCI subjects indicated retention levels lower than the AD range [14]. Our study also revealed higher retention of BF-227 in 60-70% of MCI subjects and in almost all the AD patients. Therefore, the amyloid PET technique is considered to be a highly useful and strong method for early detection of AD patients in the MCI stage.

FDG-PET has been used in investigations for MCI, and low rCMRglu in the temporo-parietal and medial frontal cortices and hippocampus was reported as one of the most prominent predictors of subsequent cognitive decline [15]. Our results indicate, however, that amyloid retention detected by BF-227 is more sensitive and specific than FDG-PET for AD diagnosis. Therefore it is reasonable that amyloid PET is more sensitive than FDG-PET for detecting MCI, which is regarded as a prodromal state of dementia or early AD, and similar results were also reported from the ADNI [1].

Concerning the biomarkers in biological fluid, Aβ1-42, t-tau and p-tau in CSF reached strong evidence in AD diagnosis. One problem with these must be invasiveness of lumbar puncture. It is not easy to take lumbar punctures many times during the disease progression because of its pain and some other risks. Some people even have difficulties taking lumbar puncture if they take anti-thrombotic drugs or have deformation of vertebrates. ADNI is still going on around the world, and examine biomarkers in blood very carefully, therefore they expect some novel good biomarkers will come out in AD.

REFERENCES

[1] Weiner MW, Aisen PS, Jack CR Jr, et al. (2010) Alzheimer's Disease Neuroimaging Initiative. The Alzheimer's disease neuroimaging initiative: progress report and future plans. *Alzheimers Dement*. 6: 202-211.

[2] Klunk WE, Engler H, Nordberg A, et al. (2004) Imaging brain amyloid in Alzheimer's disease with Pittsburgh Compound-B. *Ann Neurol* 55: 306-319.

[3] Kudo Y, Okamura N, Furumoto S, et al. (2007) 2-(2-[2-Dimethylaminothiazol- 5-yl]ethenyl)-6-(2-[fluoro]ethoxy)benzoxazole: a novel PET agent for in vivo detection of dense amyloid plaques in Alzheimer's disease patients. *J Nucl Med* 8: 553-561.

[4] Furukawa K, Okamura N, Tashiro M, et al. (2010) Tau in cerebrospinal fluid: a potential diagnostic marker in Alzheimer's disease. Amyloid PET in mild cognitive impairment and Alzheimer's disease with BF-227: comparison to FDG-PET. *J Neurol*. 257: 721-727

[5] Arai H, Terajima M, Miura M, et al. (1995) Tau in cerebrospinal fluid: a potential diagnostic marker in Alzheimer's disease. *Ann Neurol*. 38: 649-652.

[6] Shaw LM, Vanderstichele H, Knapik-Czajka M, et al. (2011) Alzheimer's Disease Neuroimaging Initiative. Qualification of the analytical and clinical performance of CSF biomarker analyses in ADNI. *Acta Neuropathol*. 121: 597-609.

[7] Petersen RC, Smith GE, Waring SC, et al. (1999) Mild cognitive impairment: clinical characterization *Arch Neurol* 56: 303-308.

[8] McKhann G, Drachman D, Folstein M, et al. (1984) Clinical diagnosis of Alzheimer's disease: report of the NINCDS-ADRDA Work Group under the auspices of Department of Health and Human Services Task Force on Alzheimer's Disease. *Neurology* 34: 939-944

[9] Okamura N, Arai H, Higuchi M, et al. (2001) [18F]FDG-PET study in dementia with Lewy bodies and Alzheimer's disease. *Prog Neuropsychopharmacol Biol Psychiatry*. 25: 447-456.

[10] Mathis CA, Lopresti BJ, Klunk WE. (2007) Impact of amyloid imaging on drug development in Alzheimer's disease. *Nucl Med Biol*. 34: 809-822.

[11] Cummings JL. (2011) Biomarkers in Alzheimer's disease drug development. *Alzheimers Dement*. 7: e13-44.

[12] Mattsson N, Zetterberg H, Hansson O, et al. (2009) CSF biomarkers and incipient Alzheimer disease in patients with mild cognitive impairment. *JAMA*. 302: 385-393.

[13] Small GW, Kepe V, Ercoli LM et al. (2006) PET of brain amyloid and tau in mild cognitive impairment. *N Engl J Med* 355: 2652-2663.

[14] Kemppainen NM, Aalto S, Wilson IA, et al. (2007) PET amyloid ligand [11C]PIB uptake is increased in mild cognitive impairment. *Neurology*. 68: 1603-1606

[15] Drzezga A, Lautenschlager N, Siebner H, et al. (2003) Cerebral metabolic changes accompanying conversion of mild cognitive impairment into Alzheimer's disease: a PET follow-up study. *Eur J Nucl Med Mol Imaging*. 30: 1104-1113.

INDEX

C

F

G

H

I

K

L

M

Q

R

S

T

U

V

W

Y

Z